Activity-Based Intelligence

Principles and Applications

For a complete listing of titles in the
Artech House Electronic Warfare Library,
turn to the back of this book.

Activity-Based Intelligence

Principles and Applications

Patrick Biltgen

Stephen Ryan

ARTECH HOUSE

BOSTON | LONDON
artechhouse.com

Library of Congress Cataloging-in-Publication Data
A catalog record for this book is available from the U.S. Library of Congress.

British Library Cataloguing in Publication Data
A catalogue record for this book is available from the British Library.

Cover design by John Gomes

ISBN 13: 978-1-60807-876-9

685 Canton Street
Norwood, MA 02062

10 9 8 7 6 5 4 3 2 1

Contents

Foreword

ABI is the most important development in intelligence analysis to come out of the wars in Iraq and Afghanistan. I first encountered it a few years ago when I was engaged in a RAND project looking at how the intelligence community was exploiting social media, like Twitter and Facebook. A team at the National Geospatial-Intelligence Agency (NGA) heard about the project and asked me to meet with them. Their use of social media was relatively incidental—they were scraping openly available sources, like Wikimapia and Google Earth, for geolocated data. But their work was stunning—all the more so for someone like me whose experience with intelligence had focused on strategic issues for which data usually was in short supply.

The team was providing support to warfighters by assembling data about particular locations. Then, if some event of potential interest occurred—say, a truck pulling up to a farmhouse—they could search the database for information about that location to see if this current event was of interest. If it was, then analysts could dig deeper into the data. If it wasn't, then the activity would be recorded in the database in case later events made it of interest. In either case, the video was unimportant; it was the activity that mattered.

ABI represents a fundamentally different way of doing intelligence analysis, one that is important in its own terms but that also offers the promise of creatively disrupting what is by now a pretty tired paradigm for thinking about the intelligence process. Cold war targets of intelligence often were large (e.g., missiles or army formations). Perhaps more important, they had a signature: We knew what a T-72 Soviet tank looked like. And typically, those signatures were embedded in doctrine: If imagery captured one T-72, it was likely to be accompanied by a number of its brethren.

The terrorist targets that drove the creation of ABI are utterly different. They are small—networks or even individuals—not large, and they have neither signature nor doctrine. If a suspected terrorist has a signature, a cell phone number for instance, it is fleeting, for he or she can change it at will. The fight against terror marked the transition from reporting on known targets to discovering the unknown. ABI enables discovery as a core principle. Discovery—how to do it and what it means—is an exciting challenge, one that the intelligence community is only beginning to confront, and so this book is especially timely.

The prevailing intelligence paradigm is still very linear when the world is not: Set requirements, collect against those requirements, then analyze. Or as one wag put it: "Record, write, print, repeat." That paradigm may have made sense when trying to solve the puzzles surrounding the Soviet Union, when the target was secretive but ponderous. It makes much less sense now, and ABI disrupts that linear collection, exploitation, dissemination cycle of intelligence. It is focused on combining data—any data—where it is found. It does not prize data from secret sources but combines unstructured text, geospatial data, and sensor-collected intelligence. It marked an important passage in intelligence fusion and was the first manual evolution of "big data" analysis by real practitioners. ABI's initial focus on counterterrorism impelled it to develop patterns of life on individuals by correlating their activities, or events and transactions in time and space.

ABI is based on four fundamental pillars that are distinctly different from other intelligence methods. The first is georeference to discover. Sometimes the only thing data has in common is time and location, but that can be enough to enable discovery of important correlations, not just reporting what happened. The second is sequence neutrality: We may find a critical puzzle piece before we know there is a puzzle. Think how often that occurs in daily life, when you don't really realize you were puzzled by something until you see the answer.

The third principle is data neutrality. Data is data, and there is no bias toward classified secrets. ABI does not prize exquisite data from intelligence sources over other sources the way that the traditional paradigm does. The fourth principle comes full circle to the first: integrate before exploitation. The data is integrated in time and location so it can be discovered, but that integration happens before any analyst turns to the data.

ABI necessarily has pushed advances in dealing with "big data," enabling technologies that automate manual workflows, thus letting analysts do what they do best. In particular, to be discoverable, the metadata, like time and location, have to be normalized. That requires techniques for filtering metadata and drawing correlations. It also requires new techniques for visualization, especially geospatial visualization, as well as tools for geotemporal pattern analysis. Automated activity extraction increases the volume of georeferenced data available for analysis. ABI is also enabled by new algorithms for correlation and fusion, including rapidly evolving advanced modeling and machine learning techniques.

This book is an introduction to the core principles of ABI as described by ABI practitioners. It introduces students to ABI-enhancing technologies and provides a survey overview of the advanced analytics associated with ABI, as well as rich, unclassified examples of data exploration techniques with a focus on analytic discovery. The examples apply the principles of ABI to open data sets.

To be sure, ABI still suffers growing pains. As a catchphrase currently hot in intelligence circles, there is a temptation to make everything ABI, which risks making it nothing. Thus, this book's purpose is to carefully lay out ABI's core methods, enhancing technologies, and related applications. One of ABI's watchwords is "mass is more." With cellphones in every pocket and cameras on every street corner, data is more and more ubiquitous. Collecting it en masse, though, runs first into technical challenges. The data sets are larger. The metadata is not standardized. There are more varieties of stovepiped and isolated data sets. The ultimate challenge, though, is enabling access while protecting privacy.

ABI came of age in the fight against terror, but it is an intelligence method that can be extended to other problems—especially those that require identifying the bad guys among the good in areas like counternarcotics or maritime domain awareness. Beyond that, ABI's emphasis on correlation instead of causation can disrupt all-too-comfortable assumptions. Sure, analysts will find lots of spurious correlations, but they will also find intriguing connections in interesting places, not full-blown warnings but, rather, hints about where to look and new connections to explore. At least, by using the methods and techniques in this book, analysts will spend more time unraveling mysteries and less time digging for data.

Gregory F. Treverton
Washington, DC

Preface

Writing about a new field, under the best of circumstances, is a difficult endeavor. This is doubly true when writing about the field of intelligence, which by its nature must operate in the shadows, hidden from the public view. Developments in intelligence, particularly in analytic tradecraft, are veiled in secrecy in order to protect sources and methods; some of these are technical, like spy satellites; others are people, brave individuals risking their lives. This textbook describes a revolutionary intelligence analysis methodology using approved, open-source, or commercial examples to introduce the student to the basic principles and applications of activity-based intelligence (ABI).

We view this as the beginning of the conversation on ABI—a "new" term that has proliferated across the intelligence vernacular over the past five years. Although the use of the term is new, the principles of this discipline are endemic and enduring in the study of intelligence.

A recent survey, *Foundational Technologies for Activity-Based Intelligence: A Review of the Literature,* by Dr. James Llinas, research professor at the State University of New York at Buffalo and Dr. James Scrofani of the Naval Postgraduate School noted that "much has been said about ABI in the trade publications of the intelligence community...but there has been very little published in the scientific literature that describes particular and novel technical methods that are explicitly supportive of and traceable to an ABI application." Industry and academic experts lamented the lack of a seminal text on this emergent discipline despite its initial developments more than 10 years ago. The training of analysts and engineers to support intelligence requirements has lagged demand from government and industry. We have developed this text through extensive research, synthesis, discussion, and collaboration with more than two dozen experts in this shadowy and nascent field.

Activity-Based Intelligence: Principles and Applications is aimed at students of intelligence studies, entry-level analysts, technologists, and senior-level policy makers and executives who need a basic primer on this emergent series of methods. This text is authoritative in the sense that it documents, for the first time, an entire series of difficult concepts and processes used by analysts during the wars in Iraq and Afghanistan to great effect. It also summarizes basic enabling techniques, technologies, and methodologies that have become associated with ABI. We believe that this text provides the security, defense, and law enforcement communities with a common base upon which to further expand development.

The authors are in many ways products of our environment. Ryan is a career intelligence professional with a master's degree in security studies from Georgetown University's Walsh School of Foreign Service and a bachelor's degree in international affairs from The George Washington University's Elliott School of International Affairs. Ryan provides the analytic and operational perspective: He was on the ground for almost a year in Afghanistan, where he practiced ABI as an analytic methodology. Between 2013 and 2014, he worked to institutionalize the concepts for other analysts and issue types beyond counterterrorism and counterinsurgency operations. He supports numerous studies and panels as a subject-matter expert on ABI techniques and methods. Ryan brings this analytic perspective to his work in mission-oriented engineering and technology development, where he leads a diverse research portfolio for a major defense contractor and works with government and military clients to solve their hardest problems.

Biltgen holds a Ph.D. in aerospace engineering from the Georgia Institute of Technology with a focus on modeling, simulation, and analysis of complex systems. He was introduced to the ABI method while working on the conceptual design of a processing system for persistent surveillance system in 2010. As a trained engineer, Biltgen brings a process and technology perspective to ABI, focusing on sensor data and making it usable to analysts. Today, he supports government clients in the design and development of processing and analysis software for multi-intelligence data.

Biltgen and Ryan were introduced in 2012 through a mutual government colleague and mentor, Gerry. They began collaborating on the new field and its implications for intelligence that has spanned their combined service across intelligence agencies and the special operations community. Perhaps most importantly, their professional collaboration grew into a personal friendship that today stretches from Washington to Los Angeles.

The pair joined forces based on a mutual passion for the topic, but their discussions often diverged (constructively) based on their collective experiences, creating profound discussions on the subjects contained in this book, including automation, big data analytics, and intelligence analysis. The interplay between agreement and disagreement crystallized many of the topics in this book, and their differing perspectives enrich the experience for the intelligence student. In the end, the authors believe that this combined perspective offers a highly complete understanding of a difficult field, and hope that academia, government, industry, and law enforcement can identify with these perspectives on ABI.

Stephen has numerous people to thank, far more than there is room for here. Numerous intelligence officers, still serving in government capacities, were and remain instrumental in the continuing development and practical application of ABI on the battlefield. Victoria Nguyen, formerly of NGA and now of Whitespace Solutions, was Stephen's partner in helping bring ABI to the broader national security enterprise; she made critical contributions day in and day out. Without her efforts, this book would not exist. Timothy and Charles were also part of the core team that helped define much of this work for the National Geospatial-Intelligence Agency, without whom Stephen could not have written this text. Dr. Gregory Treverton, first of RAND Corporation and now chairman of the National Intelligence Council, was and continues to be a valued colleague and friend whose research on ABI contributed much to the field. A superb team of engineers from the MITRE Corporation created many of the early technology prototypes and wrote original white papers on the technology enabling ABI, which contributed much to the foundation of the second half of this book. Jackie Barbieri and Schuyler Kellogg, now of Whitespace Solutions, among others still serving in government, helped breathe life into the very first course to train intelligence analysts in these principles. Melanie Corcoran, now of Analytic Fusions, was and remains an incredible colleague and friend whose tireless efforts to create technology and programs for the nascent field of ABI are worthy of recognition. Stephen's colleagues at Northrop Grumman Corporation provided a wonderfully collaborative environment and encouraged Stephen's work on this project. Bruce Hoffman, Paul Pillar, John Gannon, Anthony Arend, Chuck Cushman, and Matt O'Gara were world-class teachers and mentors in studying terrorism, insurgency, intelligence, international law, and national security. Susan Kaplan and Helen Crowley fostered Stephen's lifelong love of social sciences, history, and politics many years ago. Finally, Stephen would like to thank his parents Adrianne and John for a lifetime of love and support, without which this undertaking would have been impossible.

Patrick would like to thank the engineers and technologists at BAE Systems, especially Don Miller, Terri Ward, and Curtis McConnell who led the development of one of the first systems built for use by ABI analysts. His work in this area would not have been possible without the friendship and mentoring of Kent Murdoch. He is eternally grateful to Jim MacLeay, another friend and mentor, who recruited him into the field of intelligence. Patrick would also like to thank George Logue and Barry Barlow of Vencore for their mentorship and support throughout the production of this work. Dr. Dimitri Mavris, regents professor at the Georgia Institute of Technology, taught Patrick how to innovate and tackle open-ended problems, which was critical in making his transition from aerospace engineering to intelligence processing and analysis. Patrick would like to thank his parents, Bill and Judy, who provided him with every opportunity, often through great personal sacrifice. Most importantly, Patrick thanks his wonderful wife, Janel, not only for her encouragement and support but also for her key insights and collaboration across the breakfast table.

Both authors would like to thank Mark Phillips, author of the USD(I) Strategic Advantage series of papers, which coined the phrase "activity-based intelligence" and brought it to the attention of the entire national security establishment, contributing much to the ideas herein through frequent discussion and collaboration. Mark served as mentor to both authors over the last five years and was instrumental in guiding both authors' careers along a positive and rewarding trajectory.

David Gauthier of NGA engaged regularly with both authors and, through his work leading the ABI roundtable, helped foster many of the discussions that lead to this book. Representing the public face of ABI since 2012, Gauthier's descriptions of complex concepts frequently form the basis of educational lesson in this book. He continues to serve as a friend and mentor to both authors.

The authors would also like to acknowledge the tireless efforts of Jason Moses from the Office of the Director of National Intelligence (ODNI), who today continues to address the difficult challenge of coordinating a common approach to ABI across different agencies and military services with far different institutional perspectives and incentives.

Ed Waltz, series editor for *Intelligence and Information Warfare* and legendary luminary in the field of intelligence, brought his immense perspective and experience to bear in the development of this text. It was at his insistence that the authors came together and put finger to keyboard, bringing this into being.

Many colleagues provided insights and examples to enhance this work. Jeff Wilson from ClearTerra helpfully provided expert opinion and software illustrations for technology enabling georeference to discover. Gregg Sypeck of Mav6 contributed material on the Blue Devil II airship. Heidi Buck of Navy SPAWAR contributed material on the RAPIER automated processing system. Paul Runkle, Levi Kennedy, Jonathan Woodworth, and Peter Shargo of the Signal Innovations Group contributed extensive material on tracking technologies and methods. David Waldrop of the Illumina Consulting Group developed an example using the LUX processor. Greg Bottomley from Northrop Grumman conducted some of the important work cited in Chapter 14 regarding mathematical approaches to data fusion. Diana Levey of the SAS Institute contributed the JMP analysis software used for many of the visual analytic examples throughout the text.

Rich LaValley reviewed and provided input on the entire manuscript, an onerous task for which both Biltgen and Ryan are immensely grateful. Marissa Koors at Artech House was responsive and positive throughout the entire writing process and ensured that the authors stayed on task and schedule. Mark Phillips, William Raetz, Alex Shernoff, Sarah Hank, and David Gauthier all contributed chapters to this text, and it is the richer for it.

The last acknowledgement—and perhaps most important—is for Gerry. He laid the intellectual foundation for what this book now brings to paper. His innovative thinking, tireless pursuit of reforming the analytical process, numerous overseas deployments, and keen technological skills set the stage for both the artists and technologists who would follow in his footsteps. He has been a personal friend, confidante, and mentor to both authors. The nation owes him a great debt for his service.

This book has been reviewed by the National Geospatial-Intelligence Agency (Case #15-198), Central Intelligence Agency, and National Reconnaissance Office (Case #2015-01706 and Case #2015-01987) to prevent the release of classified information. Any factual errors or omissions, however, are the responsibility of the authors. The opinions, assessments, and conclusions contained herein are ours alone.

Patrick Biltgen,
Clifton, VA

Stephen Ryan
Los Angeles, CA

1

Introduction and Motivation

1.1 The Fourth Age of Intelligence

The first age of intelligence began during World War II when President Franklin D. Roosevelt created the Office of Strategic Services (OSS) in 1942 [1]. The wartime intelligence agency was primarily charged with human intelligence, covert action, special operations, and sabotage. The National Security Act of 1947 formalized the creation of the Central Intelligence Agency (CIA) as a professional civilian intelligence agency. The early years of agency operations focused heavily on anticommunism and containment of the Soviet Union. Because of the difficulty in penetrating the Soviet Union with human assets, CIA fielded the first advanced technical collection capabilities like the U-2 high-altitude spy plane.

By the early 1960s, the United States had entered the second age of intelligence, characterized by increasing budgets and revolutionary technologies like imagery satellites, sonar, and digital cryptography. Large complex systems dominated the military and intelligence establishment while government agencies ballooned in size and stature. A single-minded focus on Cold War adversaries dominated intelligence operations and military planning. Although intelligence professionals wax philosophically about the "good ol' days" with a defined, nation-state actor with predictable doctrine, the world remained on the brink of global thermonuclear war for decades.

The second age came crashing down on September 11, 2001. A team of 19 foreigners trained for a military operation at commercial flight schools inside our own borders and launched a coordinated attack with military precision that took the world by surprise. A ragtag group of terrorists defeated a trillion-dollar defense enterprise with two-dollar box cutters and a handful of flight lessons. Congressional investigations faulted an inability to "connect the dots" and recognized that intelligence methodologies developed for Cold War adversaries and World War II nation-states were ill-equipped to defeat nonstate actors and asymmetric threats that followed no discernable doctrine or discoverable pattern.

The subsequent 10 years saw the buildup of new collection systems and a shift to expeditionary, dispersed operations. Hundreds of unmanned aerial vehicles filled the skies beaming thousands of hours of video all over the world. In 2009, Lt. General David Deptula, U.S. Air Force deputy chief of staff for intelligence, surveillance, and reconnaissance said, "We're going to find ourselves in the not too distant future swimming in sensors and drowning in data" [2]. Deptula was right. That year, the air force collected nearly 24 years worth of video if watched continuously [3]. The word *terabyte*—one trillion bytes or one thousand gigabytes—first entered the vernacular and then became commonplace as data-filled hard drives were exfiltrated from theater and processed by trailer-sized computer systems. Multisensor platforms with high-definition video and signal geolocation capabilities were rushed to theater as quick reaction capabilities to hunt suspected terrorists hiding in the noise [4].

Thousands of new and inexperienced analysts flooded into the intelligence community in droves. A 2007 study found that "roughly half of the analytical work force has five or fewer years experience on the job" [5]. What these analysts lacked in experience they made up for with computer skills. These "digital natives"—the children of Nintendo and Atari—were in middle school when the World Wide Web went live. As *Wall Street Journal* writer Ron Alsop points out, the "millennials" are intensely tech-savvy, social, collaborative, and multitasking [6]. Military commanders in Operation Iraqi Freedom were surprised to learn that maneuver forces had eschewed combat ratios for chat rooms. "War is fought on chat," said Col. Paul Miller, USMC assistant chief of staff for the 1st Marine Expeditionary Force [7].

The third age rose with a swell of information technology and social media, but it ended with a tweet as Sohaib Athar, the owner of a small coffee shop in Abbottabad, Pakistan posted: "Helicopter hovering over Abbottabad at 1AM (is a rare event)." On May 2, 2011, a team of U.S. special operations forces raided a three-story compound and killed Al Qaeda leader Osama Bin Laden. The still-classified intelligence gathering that located the reclusive leader was an innovative mix of diverse perspectives and deep analysis of multiple sources of intelligence or *multi-INT.*

By mid 2014, the community was once again at a crossroads: the dawn of the fourth age of intelligence. This era is dominated by diverse threats, increasing change, and increasing rates of change. This change also includes an explosion of information technology and a convergence of telecommunications, location-aware services, and the Internet with the rise of global mobile computing. Tradecraft for intelligence integration and multi-INT dominates the intelligence profession. New analytic methods for "big data" analysis have been implemented to address the tremendous increase in the volume, velocity, and variety of data sources that must be rapidly and confidently integrated to understand increasingly dynamic and complex situations. Decision makers in an era of streaming real-time information are placing increasing demands on intelligence professionals to anticipate what may happen…against an increasing range of threats amidst an era of declining resources. This textbook is an introduction to the methods and techniques for this new age of intelligence. It leverages what we learned in the previous ages and introduces integrative approaches to information exploitation to improve decision advantage against emergent and evolving threats. The dominant characteristics of the four ages are shown in Figure 1.1.

1.1.1 An Era of Dynamic Change and Diverse Threats

Increasing change—and increasing rates of change—dominates the fourth age. The current volume, velocity, and variety of threats to national security are unprecedented. Advanced, asymmetric, networked enemies are better equipped

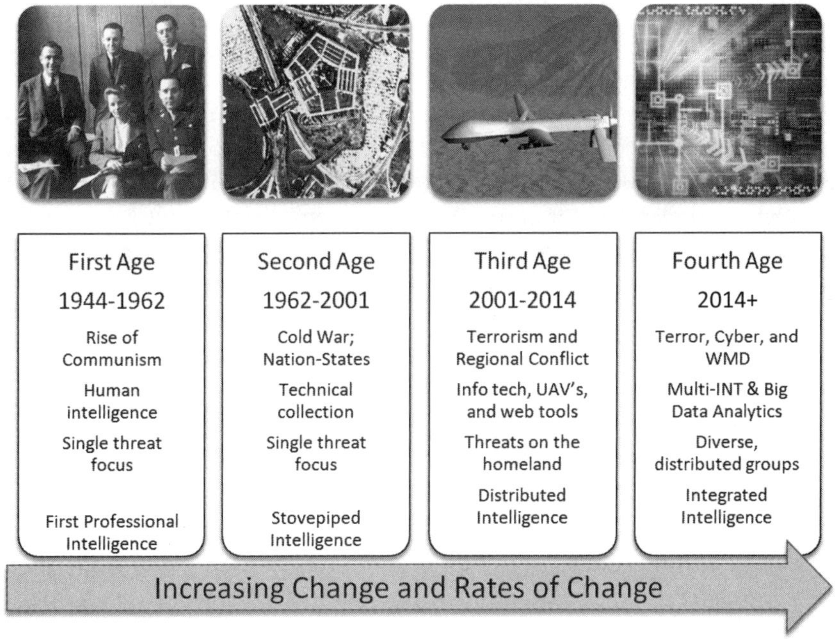

Figure 1.1 Primary drivers of the four ages of intelligence. (Imagery: CIA, NRO, USAF, and DARPA.)

and better prepared to identify and exploit vulnerabilities in our monolithic national security establishment. In the 21st century, the world has entered an era of persistent conflict.

Between December 2010 and 2013, uprisings related to the movement known as the Arab Spring forced rulers from power in Tunisia, Egypt, Libya, and Yemen. After over four years of uprisings and violent protests, Syria has erupted into full-scale civil war with over 120,000 civilians killed and over 4 million Syrians displaced as refugees. In August 2013, Syrian president Bashar al-Assad launched the first chemical weapons attack of the 21st century, killing over 3,000 civilians. In 2014, a jihadist militant organization called the Islamic State of Iraq and the Levant (ISIL) invaded Iraq from Syria and took control of much of Anbar province through violent clashes with local security forces, defying the Iraqi parliament and claiming the establishment of an Islamic caliphate in northern Iraq that by late 2014 stretched from the Syrian border to the edge of Baghdad.

Large nations with conventional and expeditionary militaries are resurgent. In 2014, amidst instability in Ukraine, Russia annexed the Crimean peninsula and strengthened military capabilities along its western border and at the strategic naval base in Sevastopol. China, along with Brunei, the Philippines, Vietnam, Indonesia, Taiwan, and Malaysia, escalated involvement in territorial disputes in the South China Sea especially around the Spratly and Paracel islands. Much of the dispute involves trade routes, fishing, and suspected oil and natural gas deposits in the region.

Cyberattacks—both those by nation-states and criminal organizations—which were considered mostly theoretical in the early 2000s, have become commonplace daily occurrences. In 2013, cybersecurity firm Mandiant outed China's 2nd Bureau of the People's Liberation Army (PLA) General Staff Department's 3rd Department as the prime protagonist in the multiyear advanced persistent threat operation that hacked over 140 high-profile defense and commercial firms [8]. Later that year, a data breach at Target department stores, in which over 70,000,000 pieces of customer data were stolen by manipulation of an insider's logon credentials, rattled investors, diminished consumer confidence, and negatively impacted the company's earnings by almost 50% [9]. According to the Privacy Rights Clearinghouse, similar hacks have stolen personal data from the Sony Playstation Network, TJ Maxx, ebay, RSA security, the Montana Health Department, AOL, the California Department of Motor Vehicles, Mt. Gox Bitcoin Exchange, Experian, and over 4,000 others totaling nearly a billion personal records [10].

Rampant emboldened piracy threatens vessels in the Indian Ocean. Increasingly violent drug gangs have taken control of large swaths of central and western Mexico. Nigerian militants in the Boko Haram terrorist organization kidnapped 219 girls from a local school to be sold into slavery, an act that some believe was intended to provoke western entities to a response that would destabilize the country. Transnational criminal organizations, terrorist groups, cyberactors, counterfeiters, and drug lords increasingly blend together; multipolar statecraft is being rapidly replaced by groupcraft.

The impact of this dynamism is dramatic. In the Cold War, intelligence focused on a single nation-state threat coming from a known location. During the Global War on Terror, the community aligned against a general class of threat coming from several known locations, albeit with ambiguous tactics and methods. The fourth age is characterized by increasingly asymmetric, unconventional, unpredictable, proliferating threats menacing and penetrating from multiple vectors, even from within. Gaining a strategic advantage against these diverse threats requires a new approach to collecting and analyzing information.

1.1.2 The Convergence of Technology and the Dawn of Big Data

Information processing and intelligence capabilities are becoming democratized. On September 17, 2011, the National Reconnaissance Office (NRO) declassified the KH-9 HEXAGON spy satellite. The 60-foot long, 30,000-pound "big bird" was a film-return imagery intelligence (IMINT) system that operated from 1971 to 1986 [11]. Designed and built by Lockheed Martin and a team of subcontractors, the super-secret $3.2-billion program delivered space-based reconnaissance capabilities unmatched by any other nation for decades [12].

Fast-forward to the fourth age of intelligence where countries like China, Israel, India, South Korea, Japan, France, and Germany control military imaging satellites [13]. Commercial imagery firms like DigitalGlobe (United States) and SPOT (France) provide high-resolution color imagery that one can buy with a credit card. Canada, Germany, and Italy have even lofted advanced commercial radar satellites capable of imaging through clouds and at night [14]. They also create high-resolution terrain maps used for dozens of civil, military, and commercial purposes.

By late 2013, Skybox Imaging, a Silicon Valley satellite imagery start-up founded by four Stanford University graduate students, launched a 220-pound imagery satellite that was built for $50 million with off-the-shelf components. In December 2013, the company released the world's first high-definition video from space, captivating the world with its revolutionary low-cost capability. In June 2014, Google acquired Skybox Imaging for $500 million with a plan to improve real-time high-resolution imagery streaming and data analytics to commercial customers. Intelligence capabilities that were unimagined 10 years ago will soon be available on an Internet-connected smart watch. In 2010, the Navy described the challenge of "information dominance," citing the increasing volume of data produced by air- and space-based sensor systems and the lack of infrastructure to store, transport, and exploit it (Figure 1.2).

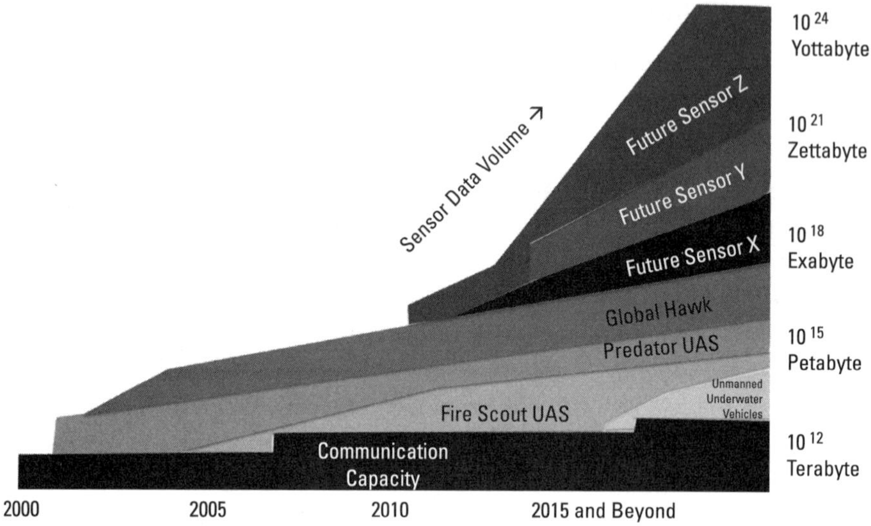

Figure 1.2 Projected explosion in intelligence, surveillance and reconnaissance (ISR) sensor data. (Adapted from [15].)

In addition to rapidly proliferating intelligence collection capabilities, the fourth age of intelligence coincided with the introduction of the term "big data." Big data refers to high-volume, high-velocity data that is difficult to process, store, and analyze with traditional information architectures. It is thought that the term was first used in an August 1999 article in *Communications of the ACM* [16]. The McKinsey Global Institute calls big data "the next frontier for innovation, competition, and productivity" [17]. New technologies like crowdsourcing, data fusion, machine learning, and natural language processing are being used in commercial, civil, and military applications to improve the value of existing data sets and to derive a competitive advantage. A major shift is under way from technologies that simply store and archive data to those that process it—including real-time processing of multiple "streams" of information.

Big data is largely created by increasingly powerful and proliferated Internet-connected computing devices. The International Telecoms Union found that almost 40% of the world's population is online and predicted that by the end of 2014, there would be more mobile phones than people on the planet [18]. Cisco predicts that there will be 21 billion Internet-connected, data-producing and -consuming devices like mobile phones, tablets, video game consoles, and even smart cars by 2018 [19]. The explosion of digital data produced by these devices created a tipping point in information management—for the first time data was being created faster than humans and machines could process, analyze, integrate, consume, and understand it—leading to new challenges for intelligence analysis and decision making.

1.1.3 Multi-INT Tradecraft: Visualization, Statistics, and Spatiotemporal Analysis

During the first, second, and third ages of intelligence, advances in analytics and processing capabilities were catalyzed by the military. Today, the most powerful computational techniques are being developed for business intelligence, high-speed stock trading, and commercial retailing. These are analytic techniques—which intelligence professionals call their "tradecraft"—developed in tandem with the "big data" information explosion. They differ from legacy analysis techniques because they are visual, statistical, and spatial.

The emerging field of visual analytics is "the science of analytical reasoning facilitated by visual interactive interfaces" [20, p. 4]. It recognizes that humans are predisposed to recognize trends and patterns when they are presented using consistent and creative cognitive and perceptual techniques. Technological advances like high-resolution digital displays, powerful graphics cards and graphics processing units, and interactive visualization and human-machine interfaces have changed the way scientists and engineers analyze data. These methods include three-dimensional visualizations, clustering algorithms, data filtering techniques, and the use of color, shape, and motion to rapidly convey large volumes of information.

Next came the fusion of visualization techniques with statistical methods. Statistical analysis software like SAS, R, SPSS, and Minitab is used by engineers and scientists to analyze data to discover statistically significant trends. These tools originally functioned as programming languages where reports and tables were produced by introducing data-processing commands into a terminal window. In the 1980s, SAS introduced a new type of statistics package called JMP, which took advantage of the graphical user interface (GUI) of the new Macintosh computer to produce colorful and interactive graphs and charts based on the same statistical processing commands. This capability evolved to focus on interactive graphs and charts that present information in both a visual and statistically sound manner. Analysts introduced methods for statistical storytelling, where mathematical functions are applied through a series of steps to describe interesting trends, eliminate infeasible alternatives, and discover anomalies so that decision makers can visualize and understand a complex decision space quickly and easily.

Geographic information systems (GISs) and the science of geoinformatics had been used since the late 1960s to display spatial information as maps and charts. Early uses were primarily focused on digital cartography, but the third major revolution in analysis methods was the integration of spatial and temporal techniques with visualization and statistical analysis. Data sets collected over long time spans could be analyzed using histograms and time series plots to identify change, tipping points, and temporal trends. Using statistical mapping tools, ethnographers could study how migration patterns change over time.

Seasonal land use, trade routes, weather patterns, real estate, tribal conflicts, and thousands of other data sets could easily be analyzed based on place.

Google's acquisition of Keyhole, Inc. in 2005 and the subsequent development of Google Maps and Google Earth made it easier for developers and scientists to manipulate maps, charts, and satellite imagery. Elections were reported using wall-sized touchscreen maps that described the minute-by-minute battle for "red states" and "blue states" [21]. Increasingly, software tools like JMP, Tableau, GeoIQ, MapLarge, and ESRI ArcGIS have included advanced spatial and temporal analysis tools that advance the science of data analysis. The ability to analyze trends and patterns over space and time is called spatiotemporal analysis.

1.1.4 The Need for a New Methodology

The fourth age of intelligence is characterized by the changing nature of threats, the convergence in information technology, and the availability of multi-INT analytic tools—three drivers that create the conditions necessary for a revolution in intelligence tradecraft. This class of methods must address nonstate actors, leverage technological advances, and shift the focus of intelligence from reporting the past to anticipating the future. We refer to this revolution as ABI, a method that former RAND analyst and National Intelligence Council Greg Treverton chairman has called the most important intelligence analytic method coming out of the wars in Iraq and Afghanistan [22].

1.2 Introducing ABI

Intelligence analysts deployed to Iraq and Afghanistan to hunt down terrorists found that traditional intelligence methods were ill-suited for the mission. The traditional intelligence cycle begins with the target in mind (Figure 1.3), but terrorists were usually indistinguishable from other people around them. The analysts—digital natives savvy in visual analytic tools—began by integrating already collected data in a geographic area. Often, the only common metadata between two data sets was time and location so they applied spatiotemporal analytic methods to develop trends and patterns from large, diverse data sets. These data sets described *activities:* events and transactions conducted by entities (people or vehicles) in an area. Sometimes, the analysts would discover a series of unusual events that correlated across data sets. When integrated, it represented the pattern of life of an entity. The entity sometimes became a target. The subsequent collection and analysis on this entity, the resolution of identity, and the anticipation of future activities based on the pattern of life produced a new range of intelligence products that improved the effectiveness of the counterterrorism mission. This is how ABI got its name.

Figure 1.3 The intelligence cycle. (Image source: Central Intelligence Agency [25].)

ABI is a new methodology—a series of analytic methods and enabling technologies—based on the following four empirically derived principles, which are distinct from traditional intelligence methods.

- Georeference to discover: Focusing on spatially and temporally correlating multi-INT data to discover key events, trends, and patterns.
- Data neutrality: Prizing all data, regardless of the source, for analysis.
- Sequence neutrality: Realizing that sometimes the answer arrives before you ask the question.
- Integration before exploitation: Correlating data as early as possible, rather than relying on vetted, finished intelligence products, because seemingly insignificant events in a single INT may be important when integrated across multi-INT.

These four fundamental pillars are described in detail in Chapter 3.

While various intelligence agencies, working groups, and government bodies have offered numerous definitions for ABI, we define it as "a set of spatiotemporal analytic methods to discover correlations, resolve unknowns, understand networks, develop knowledge, and drive collection using diverse multi-INT data sets."

ABI's most significant contribution to the fourth age of intelligence is a shift in focus of the intelligence process from reporting the known to discovery of the unknown. The Sections 1.2.1–1.2.6 in this chapter summarize the key breakthroughs and novel properties that distinguish ABI from its predecessors in an attempt to highlight what's new about this methodology.

1.2.1 The Primacy of Location

When you think about it, everything and everybody has to be somewhere.
—The Honorable James R. Clapper[1], 2004 [23]

The primacy of location is the central principle behind the new intelligence methodology ABI. Since everything happens somewhere, all activities, events, entities, and relationships have an inherent spatial and temporal component whether it is known a priori or not.

Hard problems cannot usually be solved with a single data set. The ability to reference multiple data sets across multiple intelligence domains— multi-INT—is a key enabler to resolve entities that lack a signature in any single domain of collection. In some cases, the only common metadata between two data sets is location and time—allowing for location-based correlation of the observations in each data set where the strengths of one compensate for the weaknesses in another.

Real estate agents have long told us that the three most important factors are "location, location, location," but the tipping point for the fourth age and key breakthrough for the ABI revolution was the ability and impetus to integrate the concept of location into visual and statistical analysis of large, complex data sets. This was the key breakthrough for the revolution that we call ABI.

1.2.2 From Target-Based to Activity-Based

The paradigm of intelligence and intelligence analysis has changed, driven primarily by the shift in targets from the primacy of nation-states to transnational groups or irregular forces
—Greg Treverton, RAND, [24, p. ix]

The traditional intelligence cycle, shown in Figure 1.3, is target-centric. A target can be a physical location like an airfield or a missile silo. Alternatively, it can be an electronic target, like a specific radio-frequency emission or a telephone number. Targets can be individuals, such as spies who you want to recruit. Targets might be objects like specific ships, trucks, or satellites. In the cyberdomain, a target might be an e-mail address, an Internet protocol (IP) address, or even a specific device. The target is the subject of the intelligence question. The linear cycle of planning and direction, collection, processing and exploitation, analysis and production, and dissemination begins and ends with the target in

1. At the time, Clapper was the director of the National Imagery and Mapping Agency (NIMA), which was later renamed as the National Geospatial-Intelligence Agency (NGA). The then served as the undersecretary of defense for intelligence (USD-I) and was named as the fourth director of National Intelligence in 2010.

mind. But what if you can't locate, identify, or describe your target? How do you even begin?

The term "activity-based" is the antithesis of the "target-based" intelligence model. This book describes methods and techniques for intelligence analysis when the target or the target's characteristics are not known a priori. In ABI, the target is the output of a deductive analytic process that begins with unresolved, ambiguous entities and a data landscape dominated by events and transactions.

Targets in traditional intelligence are well-defined, predictable adversaries with a known doctrine. If the enemy has a known doctrine, all you have to do is steal the manual and decode it, and you know what they will do. Analysis that applies inductive reasoning is valid for this class of problems. If you observe A, one can reason that B will come next and then C. If you observe C, you can assume that A and B have already happened. This reasoning model is not appropriate for dynamic, unpredictable adversaries that eschew these predictable patterns.

The focus on doctrine-based enemies evolved in harmony with technical collection capabilities that focused on defined signatures. If you know that the enemy must do C, you can define the signatures for event C. Then, you can build a collection system to sense the signatures associated with C. A scheduled, target-based collection deck is appropriate because you know what to look for and where to find it. The sensor knows it when it sees it. Analysts induce all the missing pieces based on the predefined model.

In the ABI approach, instead of scheduled collection, incidental collection must be used to gather many (possibly irrelevant) events, transactions, and observations across multiple domains. In contrast to the predictable, linear, inductive approach, analysts apply deductive reasoning to eliminate what the answer is *not* and narrow the problem space to feasible alternatives. When the target blends in with the surroundings, a durable, "sense-able" signature may not be discernable. Proxies for the entity, such as a communications device, a vehicle, a credit card, or a pattern of actions, are used to infer patterns of life from observations of activities and transactions.

The output of a traditional intelligence cycle is a disseminated intelligence report. Because of multiple reviews, competing analysis, disagreements across agencies, and the institutionalized fear of failure, serialized formal intelligence reporting is almost never timely [26]. While long-term analysis is needed to inform national policy, NGA director Robert Cardillo addressed these concerns in a public speech in March of 2015, saying, "no intelligence product is ever 'finished.' So add finished intelligence product to the list of extinct terms at NGA" [27]. Informal collaboration and information sharing evolved as geospatial analysis tools became more democratized and distributed. Analysts share their observations—layered as dots on a map—and tell spatial stories about entities, their activities, their transactions, and their networks.

Often, the output of ABI is a resolved entity, a defined pattern of life, or an understanding of an unknown behavior of phenomenon. While these conclusions are difficult to document concretely, the "knowledge" gained from the analytic process informs subsequent collection and, ultimately, action against the resolved target.

1.2.3 Shifting the Focus to Discovery

All truths are easy to understand once they are discovered; the point is to discover them.

—Galileo Galilei

Intelligence is the business of resolving uncertainty and understanding the unknown. ABI shifts the focus of the intelligence process from reporting on known locations and targets to discovery of the unknown. Using visual, statistical, and spatial analysis, patterns that were otherwise undetectable can be discerned. Ultimately, the focus of ABI is on resolving unknowns—especially the identity and behavior of individual entities.

Figure 1.4 summarizes four domains of analytic focus. The vertical axis represents behaviors and signatures: observable phenomena that can be collected

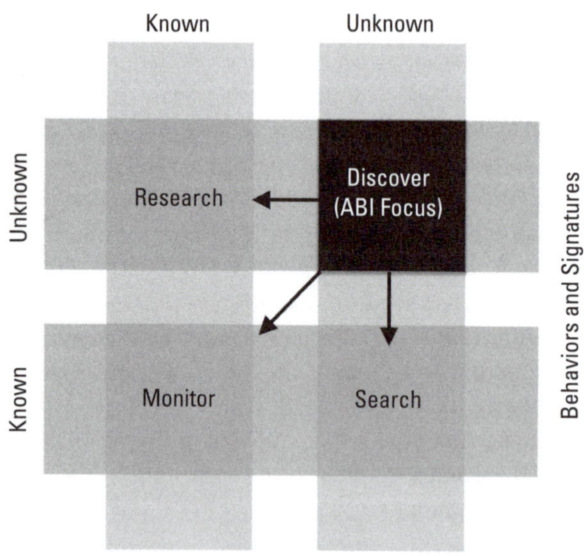

Figure 1.4 The focus of ABI is on discovering unknowns. (Source: National Geospatial-Intelligence Agency [29, p. 7].)

against. The horizontal axis is associated with locations and targets: the subject of collection activities. While traditional intelligence has long implemented techniques for researching, monitoring, and searching, the primary focus of ABI methods is on discovery of the unknown, which represents the hardest class of intelligence problems.

The lower left corner of Figure 1.4 represents known-knowns: monitoring. These are known locations or targets, and the focus of the analytic operation is to monitor them for change. For example, the fleet headquarters of the Russian Northern Fleet is located at Severomorsk near Murmansk [28]. Naval analysts know what submarines look like (their signature), and they know how they operate (their behavior). In this problem, the targets, location, behaviors, and signatures are all known. The intelligence task is monitoring the location for change and alerting when naval vessels depart.

The next quadrant of interest is in the upper left of Figure 1.4. Here, the behaviors and signatures are unknown, but the targets or locations are known. Continuing the previous example, analysts would research the Severomorsk site and attempt to identify any new vessels (new signature) or new operational activities (new behavior) at the known location. Have the Russians changed their procedures for ship maintenance? Is there new equipment on the ship? Have new facilities been constructed? Do the ships arrive or depart according to a distinctive pattern? The research task builds deep contextual analytic knowledge to enhance understanding of known locations and targets, which can then be used to identify more targets for monitoring and enhance the ability to provide warning.

The lower right quadrant of Figure 1.4, search, requires looking for a known signature/behavior in an unknown location. For example, during the Cuban Missile Crisis in 1962, analysts searched U-2 imagery for the telltale layout of Russian surface-to-air missile sites and medium-range ballistic missile (MRBM) installations [30]. Once a new type of equipment is identified, for example a new main battle tank, analysts searching imagery would attempt to locate this equipment. Looking in known exercise areas is monitoring. Searching previously undiscovered areas for the new equipment is search. For obvious reasons, this laborious task is universally loathed by analysts.

The "new" function and the focus of ABI methods is the upper right. You don't know what you're looking for, and you don't know where to find it. This has always been the hardest problem for intelligence analysts, and we characterize it as "new" only because the methods, tools, policies, and tradecraft have only recently evolved to the point where discovery is possible outside of simple serendipity.

Discovery is a data-driven process. Analysts, ideally without bias, explore data sets to detect anomalies, characterize patterns, investigate interesting threads, evaluate trends, eliminate the impossible, and formulate hypotheses.

The focus of this book is primarily on the methods and tools used to enhance the discovery process and then move discoveries into one of the other three quadrants for further analysis.

Typically, analysts who excel at discovery are detectives. They exhibit unusual curiosity, creativity, and critical thinking skills. Generally, they tend to be rule breakers. They get bored easily when tasked in the other three quadrants. New tools are easy for them to use. Spatial thinking, statistical analysis, hypothesis generation, and simulation make sense. This new generation of analysts—largely comprised of millennials hired after 9/11—catalyzed the evolution of ABI methods because they were placed in an environment that required a different approach. Frankly, their lack of experience with the traditional intelligence process created an environment where something new and different was possible.

1.2.4 Discovery Versus Search

Discovery and search are two closely related but distinctly different phenomena. To search is to find. Popular media widely reported the "discovery" of the Higgs Boson by European laboratory CERN in 2013. The existence of the sensationalized "God particle" was postulated by Peter Higgs and five others in 1964 as a missing piece in the standard model of physics. The Large Hadron Collider, a $9-billion, 10-year project was constructed almost specifically to search for the Higgs: a known signature (an elementary particle with a mass between 125 and 127 GeV/c^2) in a relatively unknown and heretofore unexplored region of subatomic particles. The scientists found exactly what they were looking for.

In contrast, Christopher Columbus discovered the New World in 1492. The Italian explorer set sail on the premise that one could reach the East Indies by sailing westward. His unconventional approach discovered the continents that form 28.5% of the Earth's landmass. This discovery literally changed a long-held worldview and forever altered the socioeconomic, political, cultural, and technological history of the planet. In 1928, Fleming discovered a new world of medicine when he noticed that a fungus called *Penicillium rubens* exuded a substance that killed bacteria. These are discoveries of unknown unknowns. They are often surprising, but they are also the most profound because of their implications and consequences.

1.2.5 Discovery: An Example

To illustrate the discovery process, consider how we shop for houses. You go to the store, buy a newspaper, and begin reading the classified ads from beginning to end, right? Perhaps you call a real estate agent and say, "Take me to see houses!" These legacy workflows illustrate how people used to accomplish the task. Today, you might go to an Internet site like Yahoo!, Trulia, or Zillow. They begin with a search box asking for an area of interest. You type "Falls Church, VA," and hit search.

These sites all display the results of the query on a map and provide all the available data. You zoom in, pan around, and get a feeling for high- and low-density areas, transit corridors, shopping centers, and schools. The geospatial context is critical to discovery.

The next step is to filter the results based on some criteria. Metadata like price, lot size, and number of bedrooms help focus the search. You think you know what you're looking for, but not exactly. You click on a few samples over here and read a short text description over there. Then, you explore a bit. What do these have in common? How are these different? You review pictures. The ones with four bedrooms are really far away. The houses in this neighborhood are too small. How much is that guestroom for Grandma really going to cost you? The process of eliminating samples that cannot be the answer is an example of deductive reasoning.

Then, you stumble upon something unusual: two houses with identical floor plans, 500 feet apart. One, however, has a list price $100,000 less than the other. Induction: It must be a dump. You review dozens of pictures. Actually, "el cheapo" appears to be nicer. A moment of discovery: The only difference between these two houses is their zip code.

The initial search, tipped off by your friend's advice, focused the area of interest to lovely and popular Falls Church, VA; however, adjacent Annandale and the zip code 22003 are strongly correlated with lower prices. (As a good ABI analyst, you don't try to postulate causality.) You dig further into this correlation and find that it holds over a wide area. You cast the net again and pull back new candidates that match your filtered criteria in the new area...

In this process, you set out to find a house, but after performing some analysis, you discovered what you really wanted. Reconsidering Figure 1.4, the initial metadata filtering and the process of trying different criteria highlights the fact that your perfect house has an unknown signature at the beginning of the search. You don't know exactly what you're looking for. Because the address of your perfect house is not known a priori, your dream house has an unknown location. You don't know where to find it. The process we just described is discovery.

So what happens next? Casting the net again with new criteria illustrates how analysts move problems from the upper right to the lower right, search. Identifying a few known targets, you might visit them, take pictures, pull crime reports from the area, check out the schools, examine tax records, and the like. The deep dive into a few focused (known) areas is research. Finally, you might just add a few discovered candidates to your favorites list and wait for the price to change. That is monitor in action.

Are we saying that hunting terrorists is the same as house shopping? Of course not, but the processes have their similarities. Location (and spatial analysis) is central to the search, discovery, research, and monitoring process. Browsing metadata helps triage information and focus the results. The problem constantly changes as new entities appear or disappear. Resources are limited and it's impossible to action every lead…

However, the two cases also differ in many significant ways. The house records are objects, not events and transactions. The metadata is clean and structured. The spatial database was constructed for you. The entities don't move. They don't intentionally try to deceive you. They don't change identities. In fact, they loudly proclaim, "I am a house that would like to be purchased!" Only a handful of terrorists have been equally boisterous. Imagine the challenge extended worldwide across dozens of intelligence problems dominated by difficult and deceitful adversaries with unknown signature and doctrine.

That is why you are reading this book.

1.2.6 Summary: The Key Attributes of ABI

Over the past several years, debate in the intelligence community centered around whether ABI was a "new" method or whether we have always exploited "activity." Though the latter statement is true, the empirically derived four pillars of ABI, the convergence of the three forces in the fourth age of intelligence (Section 1.1), and the primacy of location, targets as outputs, and the focus on discovery represent the intelligence revolution we sought in Section 1.1.4.

The key attributes that distinguish ABI from traditional intelligence are summarized in Table 1.1. ABI is a new tradecraft, focused on discovering the unknown, that is well-suited for advanced multi-INT analysis of nontraditional threats in a "big data" environment.

1.3 Organization of this Textbook

This textbook is directed at entry-level intelligence professionals, practicing engineers, and research scientists familiar with general principles of intelligence and analysis. It takes a unique perspective on the emerging methods and techniques of ABI with a specific focus on spatiotemporal analytics and the associated technology enablers. Figure 1.5 introduces an organizing model for ABI methods and techniques that places analysts at the center of a dynamic intelligence environment where they have access to data (left), analytical methods (below), and multi-INT knowledge (right).

The introductory section in Chapters 1 and 2 describes the origins, history, and evolution of ABI. Chapter 3 introduces the four pillars of ABI, a series of fundamental tradecraft breakthroughs that were developed over a

Table 1.1
Key Attributes of Traditional Intelligence and ABI

Attribute	Traditional Intel	ABI
Adversary	Nation-states; predictable; doctrine-based	Asymmetric threats; unpredictable; motivation-based
Signature	Durable; physical; definite	Non-durable; proxies
Smallest Unit	Class of equipment/object	Individual entity with unique identifier
Analytic Reasoning	Inductive; linear	Deductive; non-linear
Data focus	Single-INT; compartmented	Cross-domain multi-INT
Analysis model	Phased; linear; segregated; pattern-analysis; exploit	Sequence-neutral; forensic; pattern-of-life; discovery
Target model	Facilities & targets; coordinate; targeted	Area of interest; population; region; incidental collection
Motivation	Collection-driven	Analysis-driven
Reporting	Finished serial reporting	In-work products; layers; files
Collection frequency	Scheduled; deck-based	Persistent and pervasive; multi-INT

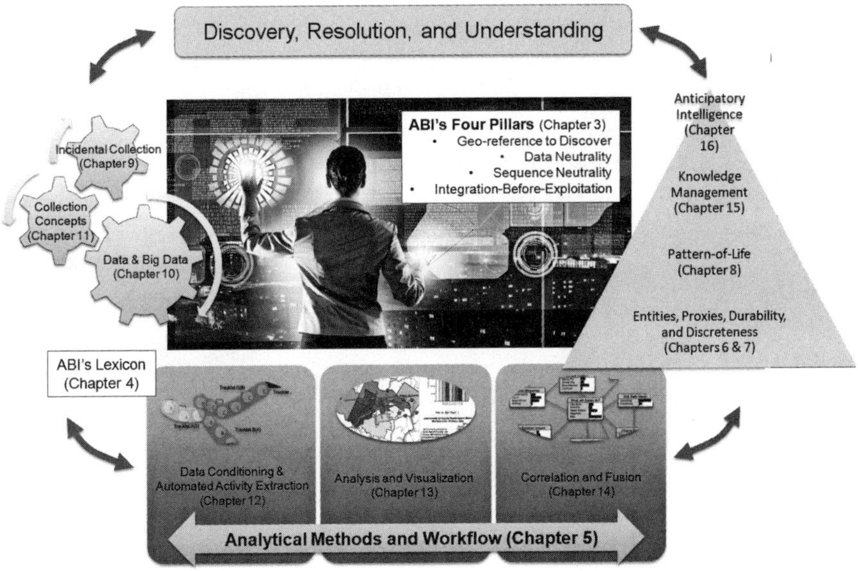

Figure 1.5 Organizing model for ABI. (Adapted from [29], [30–32].)

period of years and retroactively defined by a thorough analysis of the emergent methodology.

Chapters 4–9 introduces novice and experienced professionals to "what's different" about ABI. Chapter 4 exposes the student to the lexicon of ABI and introduces the core concept of entities, events, activities, and transactions. A focus on these fundamental terms is the primary shift from a traditional intelligence methodology to a new model based on activities. Chapter 5 defines the analytical methods and workflow required to understand ABI data and to reason through complex problems. Further, Chapter 5 expands on classical work on decision-making, hypothesis testing, and judgments in the context of ABI. It also introduces fundamental workflow changes required to give analysts the intellectual freedom to correlate data and develop valid conclusions from large and diverse data sets. In addition, Chapter 5 introduces the notion of a "nonlinear" workflow and describes the difference between inductive, deductive, and abductive reasoning. Chapter 6 explains the concept of entity resolution and proxies. This capability evolved from counterterrorism and manhunting but has since been expanded to a wide class of problems. Resolving unknowns—especially the identities and behaviors of entities—is required to truly realize ABI. Chapter 7 expands on these topics to include durability (of signatures and behaviors) and discreteness (of location and time). These two concepts go hand-in-hand to understand geospatial information and how it is exploited in the context of the human dimension. The seminal concept of "pattern of life" is introduced in Chapter 8. Chapter 8 exposes the nuances of "pattern of life" versus pattern analysis and describes how both concepts can be used to understand complex data and draw conclusions using the activities and transactions of entities. The final key concept, incidental collection, is the subject of Chapter 9. Incidental collection is a core mindset shift from target-based point collection to wide area activity-based surveillance.

Because tradecraft and technology for ABI have coevolved over the last decade, the book goes on to describe enabling technology for ABI. First, Chapter 10 introduces key concepts in data and big data. Because of the volume of open-source material on this topic, Chapter 10 is a primer and exposes students to key concepts in the context of ABI analysis. Chapter 11 describes some of the breakthrough collection technologies—especially increasingly capable commercial remote sensing capabilities—and how such collectors gather large-scale multi-INT data. Chapter 11 also introduces the concept of persistence. Readers will learn about long-duration aircraft, airships, ground-based sensors, closed-circuit TV cameras, and advanced persistent satellites that are used to gather enormous volumes of remote sensing data. Chapter 12 introduces machine learning, pattern matching, and artificial intelligence technologies for automated processing, data conditioning, and machine-assisted analysis that have the potential to revolutionize the way humans interact with data. Chapter 13 reviews some fundamental analysis and visualization technologies for ABI sense-making. Chapter 14 presents core concepts for knowledge management, collaboration, and sharing of ABI data and intelligence. Chapter 15 provides a high-level overview of data fusion and correlation methods as applied in this discipline. The technology section of the book closes with Chapter 16, which integrates all the previous concepts and describes a framework and methods for anticipatory intelligence, including modeling, pattern learning, and complex event processing.

A unique feature of this textbook is its focus on applications from the public domain. Chapters 17–24 review the concepts and technologies from the Chapters 1–16 in the context of contemporary problems from a wide range of disciplines including law enforcement, pattern analysis, transaction analysis, and the search for a missing airliner. These unclassified examples and case studies are rich with graphical examples of ABI analysis and provide the student with a broad foundation that can be applied across analytic disciplines.

1.4 Disclaimer About Sources and Methods

Protecting sources and methods is the most paramount and sacred duty of intelligence professionals. This central tenet will be carried throughout this book. The development of ABI was catalyzed by advances in commercial data management and analytics technology applied to unique sources of data. Practitioners deployed to the field have the benefit of on-the-job training and experience working with diverse and difficult data sets. A primary function of this textbook is to normalize understanding across the community and inform emerging intelligence professionals of the latest advances in data analysis and visual analytics.

All of the application examples in this textbook are derived from information entirely in the public domain. Some of these examples have corollaries to intelligence operations and intelligence functions. Some are merely interesting applications of the basic principles of ABI to other fields where multisource correlation, patterns of life, and anticipatory analytics are commonplace. Increasingly, commercial companies are using similar "big data analytics" to understand patterns, resolve unknowns, and anticipate what may happen.

Multiple government organizations have reviewed the content of this book, found it devoid of classified information, and approved this publication for release.

1.5 A Focus on Geospatial Intelligence (GEOINT)

Because ABI is an inherently spatial discipline and because GEOINT tradecraft is more available, more literal, and more understandable to entry-level professionals, many of the examples in this book apply GEOINT principles instead of more sensitive methodological examples from the human intelligence (HUMINT) and signals intelligence (SIGINT) domains. This is not to say that those domains—and the respective government agencies that act as functional managers for these "INTs"—do not have unique and enriched tradecraft for ABI, but many of the examples of their greatest successes remain highly classified.

1.6 Suggested Readings

Readers unfamiliar with intelligence analysis, the disciplines of intelligence, and the U.S. intelligence community are encouraged to review the following texts before delving deep into the world of ABI:

- Lowenthal, Mark M., *Intelligence: From Secrets to Policy.* Lowenthal's legendary text is the premier introduction to the U.S. intelligence community, the primary principles of intelligence, and the intelligence relationship to policy. The frequently updated text has been expanded to include Lowenthal's running commentary on various policy issues including the Obama administration, intelligence reform, and Wikileaks. Lowenthal, once the assistant director of analysis at the CIA and vice chairman of Evaluation for the National Intelligence Council, is the ideal intellectual mentor for an early intelligence professional.
- George, Roger Z., and James B. Bruce, *Analyzing Intelligence: Origins, Obstacles, and Innovations.* This excellent introductory text by two Georgetown University professors is the most comprehensive text on analysis currently in print. It provides an overview of analysis tradecraft and how analysis is used to produce intelligence, with a focus on all-source intelligence.
- Heuer, Richards J., *The Psychology of Intelligence Analysis.* This book is required reading for intelligence analysts and documents how analysts think. It introduces the method of analysis of competing hypotheses (ACH) and deductive reasoning, a core principle of ABI.
- Heuer, Richards J., and Randolph H. Pherson, *Structured Analytic Techniques for Intelligence Analysis.* An extension of Heuer's previous work, this is an excellent handbook of techniques for all-source analysts. Their techniques pair well with the spatiotemporal analytic methods discussed in this text.
- Waltz, Edward, *Quantitative Intelligence Analysis: Applied Analytic Models, Simulations, and Games.* Waltz's highly detailed book describes modern modeling techniques for intelligence analysis. It is an essential companion text to many of the analytic methods described in Chapters 12–16.

References

[1] "History of American Intelligence," Central Intelligence Agency, March 23, 2013, Web, June 15, 2014.

[2] Magnuson, S, "'Coin of the Realm': Military 'Swimming in Sensors and Drowning in Data,'" National Defense, 1 Jan. 2010, Web.

[3] Drew, C., "Military is Awash in Data from Drones," *New York Times,* January 10, 2010.

[4] Robinson, C. A., "Sensor, Listening Device Integration Provide Battlefield Intelligence Boon," *SIGNAL,* February 1, 2013, web.

[5] Ackerman, R. K., "Cultural Changes Drive Intelligence Analysis," *SIGNAL,* May 2007.

[6] Alsop, R., *The Trophy Kids Grow Up: How the Millennial Generation Is Shaking up the Workplace,* San Francisco: Jossey-Bass, 2008.

[7] Seffers, G., "War is Fought in Chat Rooms," *SIGNAL,* March 28, 2012.

[8] McWhorter, D. "Mandiant Exposes APT1—One of China's Cyber Espionage Units & Releases 3,000 Indicators," February 18, 2013, web https://www.fireeye.com/blog/threat-research/2013/02/mandiant-exposes-apt1-chinas-cyber-espionage-units.html, accessed July 5, 2014.

[9] Ziobro, P., "Target Earnings Slide 46% After Data Breach," *Wall Street Journal,* February 26, 2014.

[10] "Data Breach Chronology," Privacy Rights Clearinghouse, http://www.privacyrights.org/data-breach/.

[11] "Center for the Study of National Reconnaissance Classics: *Hexagon* KH-9 Imagery," National Reconnaissance Office, April 2012.

[12] "The *Hexagon* Story." National Reconnaissance Office, 1988, approved for public release September 17, 2011.

[13] "Satellite Database," Union of Concerned Scientists, 2014.

[14] Thomas A., "NGA Employs Emerging Commercial Space Radars," *Pathfinder,* Vol. 8, No. 1, September/October 2010.

[15] Simpson, T. "Information Dominance Trends and Strategies," presented at the NDIA Luncheon, October 6, 2010, p. 8, approved for public release and unlimited distribution.

[16] Bryson, S., et al., "Visually Exploring Gigabyte Data Sets in Real Time," *Commun. ACM,* Vol. 42, No. 8, August 1999, pp. 82–90.

[17] "Big Data: The Next Frontier for Innovation, Competition, and Productivity," McKinsey Global Institute, June 2011.

[18] "The World in 2014: ICT Facts and Figures," International Telecoms Union, 2014.

[19] Wieland, K., "Cisco Says Mobile Devices To Generate More IP Traffic Than PCs by 2018," *Mobile World Live,* accessed July 13, 2014.

[20] Thomas, J., and K. Cook, "Illuminating the Path: The R&D Agenda for Visual Analytics," National Visualization and Analytics Center, Pacific Northwest National Laboratory, 2004.

[21] Farhi, P., "Elephants Are Red, Donkeys Are Blue," *The Washington Post,* November 2, 2004, p. C01.

[22] Treverton, G. F., "Creatively Disrupting the Intelligence Paradigm," *The International Relations and Security Network,* August 13, 2014.

[23] Ackerman, R. K., "Digitization Brings Quantum Growth in Geospatial Products," SIGNAL, August 2004, accessed June 15, 2014.

[24] Moore, D. T., *Sensemaking: A Structure for an Intelligence Revolution,* Washington, D.C.: National Defense Intelligence College, 2011.

[25] "Discover the CIA with The Work of a Nation," Central Intelligence Agency, Central Intelligence Agency, April 30, 2013, web, accessed June 2, 2015.

[26] Committee on Homeland Security and Governmental Affairs, "Senate Permanent Subcommittee on Investigations Federal Support for Fusion Centers Report," October 3, 2012, web, accessed March 19, 2015.

[27] Cardillo, R., "Remarks as Prepared for Robert Cardillo, Director, National Geospatial-Intelligence Agency," AFCEA/NGA Industry Day, Springfield, VA, March 16, 2015, speech, approved for public release NGA Case #15-281.

[28] "Northern Fleet," Wikipedia, June 14, 2014.

[29] Gauthier, D., "Activity Based Intelligence: Finding Things That Don't Want to be Found," presented at the 2013* GEOINT Symposium, Tampa, FL, April 16, 2014, approved for public release NGA Case #14-233.

[30] McAuliffe, M. S., "CIA Documents on the Cuban Missile Crisis (1962)," CIA History Staff, October 1992.

[31] Waltz, E., *Knowledge Management in the Intelligence Enterprise,* Norwood, MA: Artech House, 2003.

[32] Gauthier, D., "Activity-Based Intelligence: NGA Initiatives," December 17, 2013, approved for public release NGA Case #13-509.

2

ABI History and Origins

Over the past 15 years, ABI has entered the intelligence vernacular. Former NGA Letitia Long, said it is "a new foundation for intelligence analysis, as basic and as important as photographic interpretation and imagery analysis became during World War II" [1]. The method and associated terminology have also evolved significantly since they were developed to find terrorists in Iraq and Afghanistan. This chapter introduces the origins of ABI and describes the evolution of the tradecraft to the present day.

2.1 Wartime Beginnings

ABI methods have been compared to many other disciplines including submarine hunting and policing, but the modern concepts of ABI trace their roots to the Global War on Terror. According to Long, "Special operations led the development of GEOINT-based multi-INT fusion techniques on which ABI is founded" [1, p. 8]. Robert Zitz, a former government executive and senior vice president at Leidos explains, "Special operations were not only melding SIGINT and GEOINT, but then bringing in HUMINT and OSINT" to evolve the methods of ABI [2].

GEOINT analysts were instrumental in this process, working with special operations forces in Iraq and Afghanistan. These analysts discovered that the units they supported had access to mountains of georeferenced intelligence data from various INTs, and that most of the data went unexamined by human eyes. If the data was exploited, it was done in a linear workflow with a single-INT focus, with SIGINT experts examining SIGINT, HUMINT experts HUMINT, and so on.

NGA's David Gauthier said that between 2004 and 2006, GEOINT analysts in Iraq and Afghanistan combined information from multiple sources into a single georeferenced database. "They… searched the database to identify adversary locations to the military operators could act against them," Gauthier said [1]. They initially called this method "geospatial multi-INT fusion" (GMIF) [2].

Analysts also worked with full-motion video (FMV) from unmanned aerial vehicles like the Air Force's MQ-1B *Predator*. The video would prove to be a rich data source, providing incredibly detailed surveillance information on potential adversaries. Biltgen and Tomes noted that FMV analysts were among the first experts reporting on "patterns of life" as they persistently tracked entities with high-resolution video in Iraq and Afghanistan [3, 4].

The analysts supporting special operations team would continue their work overseas. At the same time, however, geospatial multi-INT fusion would come to the attention of the Office of the Undersecretary of Defense for Intelligence [OUSD(I)] at the Pentagon.

2.2 OUSD(I) Studies and the Origin of the Term ABI

During the summer of 2008 the technical collection and analysis (TCA) branch within the OUSD(I) determined the need for a document defining "persistent surveillance" in support of irregular warfare. The initial concept was a "pamphlet" that would briefly define persistence and expose the reader to the various surveillance concepts that supported this persistence. U.S. Central Command, the combatant command with assigned responsibility throughout the Middle East, expressed interest in using the pamphlet as a training aid and as a means to get its components to use the same vocabulary. Mark Phillips, an OUSD(I) staff member, new to TCA, was assigned the task of writing this pamphlet.

Around the same time, OUSD(I) sent two staff members to the annual MIT Lincoln Laboratories Intelligence, Surveillance, and Reconnaissance (ISR) Symposium in October 2008. A U.S. government briefer summarized operational successes in GMIF, with a focus on the methods of employment of persistent surveillance to achieve success in irregular warfare. The OUSD(I) staff spent several weeks learning the tradecraft, interacting with analysts, and understanding the successes. This research transformed the overview pamphlet into a classified white paper, "Surveillance Employment Strategies for Irregular Warfare," released in December of 2009 [5]. The document not only contained government all-source analysts' perspectives, but also perspectives gleaned from discussions of the novel surveillance techniques used by the Nevada Gaming Commission to catch casino cheats as well as those used by various law enforcement gang task forces to unravel nefarious networks. The paper formally defined persistence and described various methods of surveillance that supported counterinsurgency and counterterrorism. One of these surveillance methods was called *activity-based surveillance* (ABS): the surveillance necessary to gather the data to support the GMIF method.

Although not universally accepted outside of small communities that had experienced its successes, this paper brought a new way of thinking to a broad audience under the then-Undersecretary of Defense (Intelligence) James Clapper.

In a briefing about the ABS principles, one senior government official stated: "I get activity-based surveillance, but I am much more interested in the intelligence resulting from that collection." TCA began a study focused on the analysis aspects associated with ABS. As the study participants dug deeper, they recognized a growing desire to understand and characterize the entities of interest to GMIF analysts and the intensely human element endemic to counterterrorism and counterinsurgency.

OUSD(I) published a second paper, "The Human Dimension: Analyzing the Role of the Human Element in Operational Environment," in September 2010. This paper had two distinct components: 1) it modeled people and outlined the information necessary to uniquely characterize an entity, and 2) it focused on the intelligence resulting from the GMIF analytical methods. Phillips described the analytical methodology of GMIF in the context of the intelligence community and introduced the term activity-based intelligence [6]. ABI was formally defined by the now widely circulated "USD(I) definition":

> A discipline of intelligence, where the analysis and subsequent collection is focused on the activity and transactions associated with an entity, a population, or an area of interest [6].

There are several key elements of this definition. First, OUSD(I) sought to define ABI as a separate discipline of intelligence like HUMINT or SIGINT: SIGINT is to the communications domain as activity-INT is to the human domain. Recognizing that the INTs are defined by an act of Congress, this definition was later softened into a "method" or "methodology."

The definition recognizes that ABI is focused on *activity* (composed of events and transactions, further explored in Chapter 4) rather than a specific target. It introduces the term *entity,* but also recognizes that analysis of the human domain could include populations or areas, as recognized by the related study called "human geography." Finally, the definition makes note of *analysis and subsequent collection,* also sometimes referred to as analysis driving collection. This emphasizes the importance of analysis over collection—a dramatic shift from the traditional collection-focused mindset of the intelligence community. To underscore the shift in focus from targets to entities, the paper introduced the topic of "human domain analytics."

2.3 Human Domain Analytics

Human domain analytics is the global understanding of anything associated with people. The human domain provides the context and understanding of the activities and transactions necessary to resolve entities in the ABI method. Based on its study of the counterterrorism, law enforcement, and counterfraud missions, TCA divided the human domain into four data categories that summarize what can be captured or collected about people (Figure 2.1). The first is biographical information, or "who they are." This includes information directly associated with an individual. The second data type is activities, or "what they do." This data category associates specific actions to an entity. The third data category is relational, or "who they know," the entities' family, friends, and associates. The final data category is contextual (metaknowledge), which is information about the context or the environment in which the entity is found. Examples include most of the information found within the sociocultural/human terrain studies. Taken in total, these data categories support ABI analysts in the analysis of entities, identity resolution of unknown entities, and placing the entities actions in a social context.

As military and intelligence organizations practiced ABS between 2006 and 2010, it became increasingly evident that such collection would generate

1. Biographical (Who They Are)	2. Activities (What They Do)
• Name • Address • Gender • DOB • Age • Height • Weight • Race • Religion • Financial Status • Languages • Occupation • Skills • Level of Education • Biometrics • Personality • Values • Beliefs • Marital Status • Passport Number • Email Address • Personal Attributes	• Travel (Flights, Trains, Cars, Hotels • Communication (Mail, Phones, Email, In-Person) • Financial Transactions (Purchases, Sales, Leases, Deposits, Transfers, Loans, Imports, Exports) • Movement of Physical Assets • Military Attacks • Coordinated Activities • Criminal Acts
3. Relational (Who They Know)	**4. Contextual (Meta-Knowledge)**
• Family • Friends • Employer • Co-workers • Colleagues • Associates • Partners • Enemies • Suppliers • Customers • Leaders • Followers • Neighbors • Superiors • Subordinates • Ties from Organizational Memberships • Ties from Community Involvement	• Demographics (i.e. Socio-cultural, Socio-economic) • Organizational Profiles (i.e. Size, Affiliations, Political Support, Business Directories) • Geography (i.e. Elevation, Weather, Resource Usage) • Anthropology • Psychology • Political Climate

Figure 2.1 Human dimension data categories and taxonomy. (Presented at the 2010 USGIF Symposium [7].)

an exponentially increasing amount of multisensor data. TCA tasked Phillips to study the increasing volume, velocity, and variety of data and how and ABI could be applied to make sense of it.

In "Activity-Based Intelligence Knowledge Management" (August 2011), OUSD(I) exposed new principles of knowledge management that would support ABI as well as the technical advances that needed to be made to make it a reality (see Chapter 15). TCA referred to the three papers collectively as the *Strategic Advantage Series.*

The reception of the papers was mixed across the community. Within the analytical corps, there were two distinct camps: 1) analysts who had direct experience with ABI and who felt their use of this methodology was vindicated and 2) the bulk of the traditional analytical corps who shunned the methodology for various reasons. Within industry and academia there was general enthusiasm for the concepts of the papers and the possibility of increased expenditure on enabling technologies, sensors, and processing systems, especially after the Secretary of Defense stood up a special task force to rush promising ISR capabilities to support the ongoing wars in Iraq and Afghanistan.

2.4 ABI Research and Development

One of the agencies involved in the rush to field new capabilities was NGA, headquartered in Springfield, Virginia. NGA has shared purview over the processing, exploitation, and dissemination (PED) of various sources of GEOINT data, from unmanned aerial vehicles (UAVs) like the Air Force's Predator. They also had a role in forensic exploitation of a new source, wide area motion imagery (WAMI) (see Chapter 11). There was a burgeoning demand for sensors that could provide tactical information across an entire area of interest—sometimes as large as an entire city.

These sensors generated enormous amounts of data, so much, in fact, that it would be impossible for traditional methods to keep up with the data stream; simply training thousands of additional analysts was an unsustainable position. NGA was also concerned with the storage and transmission of the data sets when a single mission could be comprised of nearly 100 terabytes of pixel data—100 times more data than produced by the most complex sensors in theater at the time. Figure 2.2 shows an early concept for shifting from "pixel-based PED" to "[activity-based] PED." The *x*-axis of Figure 2.2 shows the expansion of sensor coverage areas and notes that it must drive a change in exploitation methodology due to the massive size of new sensor data sets and challenges in applying pixel-based methodologies to WAMI sensor technology [8]. Although NGA's mandate focused on the PED, this first public reference

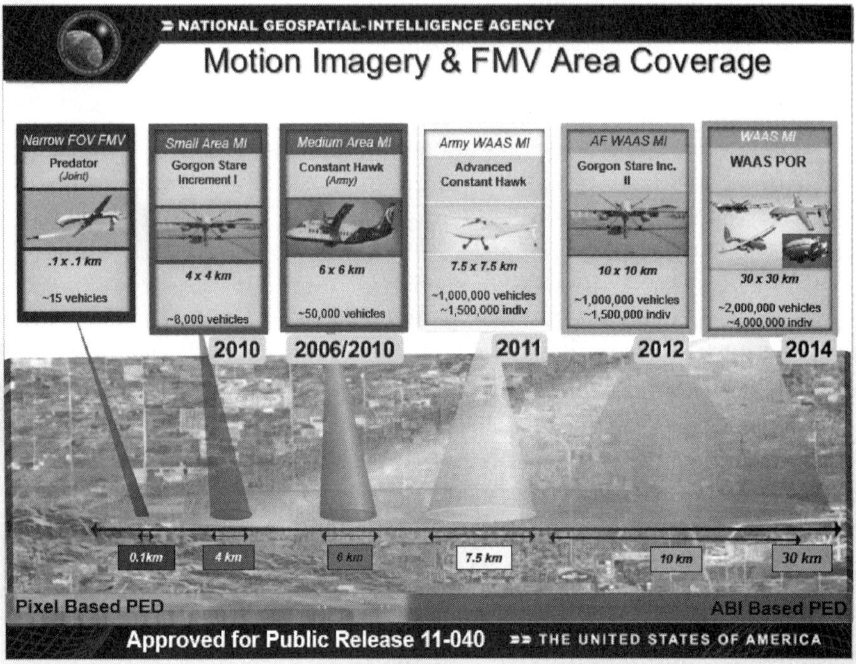

Figure 2.2 An early concept for "activity-based PED." (Source: NGA [8].)

to ABI still reinforces the traditional intelligence focus on the collection systems and their attributes.

NGA, under the Office of Military Readiness, received congressional funding in FY2010 to support risk reduction for the Air Force's wide area airborne surveillance (WAAS) program, Gorgon Stare Increment 2 [9]. Phillips, having recently authored the three papers and participated in an OUSD(I) ABI technology request for information, was brought into the office to scope requirements for a risk-reduction prototype. Around the same time, BAE Systems had conducted an internal research and development program on the concept of automated activity extraction from their ARGUS-IS WAMI sensor using an automated algorithm called WAVELIB (see Chapters 11 and 12). NGA's need for a prototype accelerated when the army asked NGA for PED concepts for a rapid overseas deployment of ARGUS-IS with the army's A-160 Hummingbird drone on a program called AAA [10]. NGA selected BAE Systems to develop the prototype—nicknamed M111—to support joint Army, Air Force, and NGA goals. This effort lead to a family of closely related systems designed from the "ground up" with activity extraction and the ABI pillars (described in Chapter 3) in mind.

2.5 ABI-Enabling Technology Accelerates

In January 2011, the Office of Military Readiness and representatives from NGA's analysis directorate demonstrated the concepts of ABI as implemented in M111 to NGA director Long. The team demonstrated how events and transactions could be automatically extracted from WAMI data while discarding the irrelevant background data—saving millions of dollars in storage costs and thousands of hours of analysis time. Long directed that the effort be accelerated as part of the NSG expeditionary architecture (NEA) [11]. In 2012, NEA deployed an "ABI quick-reaction capability, a web-based service to support U.S. military operations in theater" [12, p. 9]. Also in 2012, NGA began a technology effort to consolidate multiple ABI technology prototypes into a single, scalable, web-based architecture for advanced analytics. In December 2012, BAE Systems was awarded a multiyear $60-million contract to provide "ABI systems, tools, and support for mission priorities" under the agency's total application services for enterprise requirements (TASER) contract [13].

While these technology developments would bring new data sources to analysts, they also created confusion as the tools became conflated with the analytical methodology they were designed around. The phrase "ABI tool" would be attached to M111 and its successor program awarded under TASER.

For several years, the engineers (developers of M111 and subsequent tools) and analysts (supporting special operations) would continue their independent development paths until former NGA deputy director Lloyd Rowland directed that the two groups be unified. Leadership of this task ultimately fell to David Gauthier and the ABI Roundtable.

2.6 Evolution of the Terminology

The term ABI and the introduction of the four pillars was first mentioned to the unclassified community during an educational session hosted by the U.S. Geospatial Intelligence Foundation (USGIF) at the GEOINT Symposium in 2010, but the term was introduced broadly in comments by Director of National Intelligence (DNI) Clapper and NGA director Long in their remarks at the 2012 symposium [14, 15]. A breakout session at the symposium anchored by Gauthier included notable experts and senior leaders discussing multiple viewpoints on the emerging discipline of ABI.

Back at NGA, the ABI Roundtable began to study and normalize ABI concepts and technologies between 2012 and 2013, evolving the definition and terminology used to describe ABI. Most of the efforts centered on creating and mediating a common understanding of ABI between the analysts and technologists that originally held fundamentally divergent viewpoints. The analysts focused on the core methodology: correlating multi-INT data using spatiotemporal metadata in order to resolve entities (people). The technologists would begin to grapple with using spatiotemporal correlation to drive more timely and efficient collection, marrying it with ABI as an "end result" of the analytical process.

As wider intelligence community efforts to adapt ABI to multiple missions took shape, the definition of ABI became generalized and evolved to a broader perspective as shown in Table 2.1. NGA's Gauthier described it as "a set of methods for discovering patterns of behavior by correlating activity data at network speed and enormous scale" [16, p. 1]. It was also colloquially described by Gauthier and Long as, "finding things that don't want to be found."

The evolution of key definitions is summarized in Table 2.1. These definitions show the increase in scope but also have many consistent elements including a focus on multi-INT analysis, data correlation, analysis driving collection, and analysis focused on events, transactions, and patterns of life.

The military's initial reaction to ABI has been mixed. At the 2012 GEOINT Symposium, during a panel session on ABI, Army Maj. Gen. Steven Fogarty, the commanding general of the Intelligence and Security Command stunned the audience by saying, "The Army does not recognize ABI as a doctrinal term"[17]. Fogarty was not saying that the army does not practice ABI but rather that the term does not appear in army manuals or joint publications. He said that the army refers to the concepts in terms of "real-time intelligence collection and fusion. For us those are the two principle issues" [17]. Air Force Lieutenant General Robert P. Otto, in the Air Force's *ISR 2023: Delivering Decision Advantage* policy paper summed up the method by saying, "Whether it is labeled as 'big data,' data mining, activity-based intelligence (ABI), or object-based production (OBP), the vast amount of information that we collect demands a transformation in the way we process, organize, and present data"

Table 2.1
Evolving Community Definitions of ABI

Year	Definition	Source
2010	A discipline of intelligence where the analysis and subsequent collection is focused on the activity and transactions associated with an entity, population, or area of interest.	Undersecretary of Defense for Intelligence [6].
2013	A multi-INT approach to activity and transactional data analysis to resolve unknowns, develop object and network knowledge, and drive collection.	Office of the Director of National Intelligence (ODNI) ABI Community of Practice (CoP). [1][16, p. 1].
2014	A set of methods for discovering patterns of behavior by correlating activity data at network speed and enormous scale.	David Gauthier, NGA [16, p. 1].
2015	A set of spatio-temporal analytic methods to discover correlations, resolve unknowns, understand networks, develop knowledge, and drive collection using diverse multi-INT data sets.	Biltgen and Ryan

[18]. This book will describe each of these terms and provide insight into the methods developed to deal with this ongoing transformation.

2.7 Summary

Long described ABI as "the most important intelligence methodology of the first quarter of the 21st century," noting the convergence of cloud computing technology, advanced tracking algorithms, inexpensive data storage, and revolutionary tradecraft that drove adoption of the methods [1]. Chapters 3–9 introduce the basic terminology and principles of the ABI methods, which were documented through empirical observation of what worked for a challenging and dynamic set of intelligence problems.

References

[1] Long, L., "ABI: Activity-Based Intelligence, Understanding the Unknown," *The Intelligencer: Journal of U.S. Intelligence Studies,* Vol. 20, No. 2, fall/winter 2013, pp. 7–15.

[2] Quinn, K., "A Better Toolbox," *Trajectory,* winter 2012.

[3] Biltgen, P. and R. Tomes, "Rebalancing ISR," *Geospatial Intelligence Forum,* Vol. 8, No. 9, September 2010, pp. 14–16.

[4] Tomes, R., "Beyond Eyes on Target: Training the Next Generation of ISR Analysts," Presented at the FMV Conference for Defense and Intelligence Operations. Washington, D.C., 28 February 2011.

[5] "Surveillance Employment Strategies for Irregular Warfare." Undersecretary of Defense for Intelligence [USD(I)], 2009.

[6] "The Human Dimension: Analyzing the Role of the Human Element in the Operational Environment." Undersecretary of Defense for Intelligence [USD(I)],15 September 2010.

[7] Arbetter, R., "Understanding Activity-Based Intelligence and the Human Dimension," presented at the 2010 *GEOINT Symposium,* New Orleans, LA, November 1, 2010.

[8] Keene, K., "Wide Area Airborne Surveillance Activity Based Intelligence Processing, Exploitation and Dissemination Construct," presented at the 2010 GEOINT Symposium. New Orleans, LA, November 1, 2010. Approved for public release. NGA case #11-040.

[9] "RDT&E Budget Item Justification, Exhibit R-2, PE Number 0305206F, Airborne Reconnaissance Systems," May 2009.

[10] "Intelligence: Army and ARGUS Together At Last." *Strategy Page,* January 4, 2012, web, accessed August 20, 2014.

[11] Barber, K. L., "NSG Expeditionary Architecture Reshapes GEOINT," *Pathfinder,* Vol. 10, No. 3, May 2012.

[12] Barber, K. L., "NSG Expeditionary Architecture: Harnessing Big Data." *Pathfinder,* Vol. 10, No. 5, September/October 2012, pp. 8–10.

[13] "BAE Systems Selected to Provide Activity-Based Intelligence Support for National Geospatial-Intelligence Agency," *Business Wire,* December 19, 2012, accessed November 9, 2014.

[14] Clapper, J. R., keynote address at the 2012 GEOINT Symposium, Orlando, Florida, October 9, 2012.

[15] Long, L., remarks at the 2012 USGIF GEOINT Symposium, Orlando, Florida, October 9, 2012.

[16] Gauthier, D., "Activity Based Intelligence: Finding Things That Don't Want to be Found," presented at the 2013* GEOINT Symposium. Tampa, Florida, April 16, 2014, approved for Public Release, NGA Case #14-233.

[17] Fogarty, S., comments on the Activity-Based Intelligence Panel at the 2012 GEOINT Symposium. Orlando, Florida, October 10, 2012.

[18] "Air Force ISR 2023: Delivering Decision Advantage. A Strategic Vision for the AF ISR Enterprise." United States Air Force, 2013.

3

Discovering the Pillars of ABI

The basic principles of ABI have been categorized as four fundamental "pillars." These simple but powerful principles were developed by practitioners by cross-fertilizing best practices from other disciplines and applying them to intelligence problems in the field. They have evolved and solidified over the past five years as a community of interest developed around the topic. This chapter describes the origin and practice of the four pillars: georeference to discover, data neutrality, sequence neutrality, and integration before exploitation.

3.1 The First Day of a Different War

The U.S. intelligence community and most of the broader U.S. and western national security apparatus, was created to fight—and is expertly tuned for—the bipolar, state-centric conflict of the Cold War. Large states with vast bureaucracies and militaries molded in their image dominated the geopolitical landscape. One intelligence priority of the Cold War—one of the greatest priorities for the U.S. and its allies—became understanding what the Soviet Union was doing within space that was largely denied to many sources of U.S. and North Atlantic Treaty Organization (NATO) intelligence. The era of overhead reconnaissance changed this: First, briefly, the clandestine development of the U-2 spy plane and missions over the USSR until Francis Gary Powers was shot down on an overflight mission effectively ending the use of the U-2 over the Soviet Union; second, the CORONA program, leading to the development and launch of the United States' first overhead imaging satellite as Discoverer XIV in August 1960 [1, 2].

With the discovery of Soviet MRBMs in Cuba in 1961, the value of overhead reconnaissance became clear: No longer could closed borders prevent a high-resolution camera from seeing an adversary's movements inside its territory. Figure 3.1 shows a team of photographic interpreters at the National Photographic Interpretation Center (NPIC)[1] identifying Soviet missiles in Cuba in 1961.

Fast-forward to the War on Terror after September 11, 2001. Far from a slow-moving state, the adversary had changed: It was nimble and fast, possessed no doctrine or signature, and blended into the population where the United States had theoretical freedom of movement [3]. A clear mismatch between the technology of the last war and the adversary of the next war emerged. The tools of the Cold War, overhead satellites, collectively referred to as national technical means (NTMs), and analysis tools aimed at state actors were ill-suited for this new age of intelligence.

Aided by technological advances described in Chapters 10–17, analysts on the front lines of this new war developed concepts that were refined into the pillars of ABI: a grassroots, evidence-based methodology instead of a top-down, academic redefinition of intelligence.

3.2 Georeference to Discover: "Everything Happens Somewhere"

Georeference to discover is the foundational pillar of ABI. It was derived from the simplest of notions but proves that simple concepts have tremendous power in their application. Imagine for a moment you are an intelligence analyst, trained in the interpretation of satellite imagery and equipped with some basic cartography and GIS skills. You arrive in a dusty room, far from headquarters

Figure 3.1 The NPIC team of imagery analysts that identified Soviet missiles in Cuba in 1961. (Source: NGA [1].)

1. NPIC would later be merged with the Defense Mapping Agency (DMA) and evolve into NGA [1].

and your heritage. Instead, you are in an auditorium with rows and rows of people sitting at computer workstations, all trying to determine friend from foe and to dismantle a deadly network of committed radicals. You realize your toolkit, tuned for the last war, is woefully equipped to find these men and women, hiding in plain sight in the city mere minutes from your desk. However, as you settle in, you begin to realize something: Everyone, each row in front of the other, is creating and receiving mounds of information at the most granular level.

Awash in data, few are sharing it, and the reason why is readily apparent. In one corner, spreadsheets are carefully crafted, line-by-line, capturing data from technical collection systems. Further down the row, others write text reports, capturing in narrative the events of the day or information from an informant, called in to a tip line. But all of this data is about "activity"—things that happen at a location on Earth. "Everything happens somewhere," said then Undersecretary of Defense James Clapper in 2004 [4]. Where activity happens—the spatial component—is the one aspect of these diverse data that is (potentially) common. The advent of the global positioning system (GPS)—and perhaps most importantly for the commercial realm, the de-activation of a mode called "selective availability"—has made precisely capturing "where things happen" move from the realm of science fiction to the reality of day-to-day living. With technological advances, location has become knowable. Georeferencing refers to the ability to add location information to otherwise untagged data. Often, this includes the addition of spatial coordinates (sometimes called geocoding). General spatial referencing to a geographic place name is accomplished by correlation with a gazetteer. The simple—but powerful—concept of adding location information and discovering correlations using this information is the pillar of "georeference to discover."

This approach is not without difficulty: Some information is naturally easier to "pin to the map" than other, depending on what kind of "activity" it represents. This process breaks out into several distinct classes and subclasses of data, explained in Table 3.1 and detailed below in Sections 3.2.1–3.2.3.

3.2.1 First-Degree Direct Georeference

The most straightforward of these is direct georeferencing, which is where machine-readable geospatial content in the form of a coordinate system or known cadastral system is present in the metadata of a type of information. An example of this is metadata (simply, "data about data") of a photo a GPS-enabled handheld camera or cell phone, for example, might give a series of GPS coordinates in degrees-minutes-seconds format. Regardless of the format, the presence of a coordinate system readable by a GIS is what defines direct georeferences.

Table 3.1
Categories of Georeferenced Information

Degree	Type	Basis	Example
First Degree	Direct	Metadata	GPS location "tag" on a still image
	Indirect	Content	Text document stating individual's residence
	Indirect	Metadata	Biographical profile with a metadata tag for residence
Second Degree	Indirect	Metadata/ Context	Synthesized from both content and context of data (e.g. georeferenced poem)

3.2.2 First-Degree Indirect Georeference

By contrast, indirect georeferencing contains spatial information in non-machine-readable content, not ready for ingestion into a GIS. Indirect georeferences further break down into content georeferences and metadata georeferences, depending on the part of the data in which they are found. An example of a content-based georeference is a text document with no metadata that states, in part "John Smith lives in Nome, Alaska." By contrast, an example of a metadata-based georeference in the same context would be a biographical profile of John Smith with the metadata tag "RESIDENCE: NOME, ALASKA."

3.2.3 Second-Degree Georeference

Further down the georeferencing rabbit hole is the concept of a second-degree georeference. This is a special case of georeferencing where the content and metadata contain no first-degree georeferences, but analysis of the data in its context can provide a georeference. For example, a poem about a beautiful summer day might not contain any first-degree georeferences, as it describes only a generic location. By reconsidering the poem as the "event" of "poem composition, a georeference can be derived. Because the poet lived at a known location, and the date of the poem's composition is also known, the "poem composition event" occurred at "the poet's house" on "the date of composition," creating a second-degree georeference for a poem [5].

The concept of second-degree georeferencing is how we solve the vexing problem of data that does not appear, at first glance, to be "georeferenceable." The above example shows how, by deriving events from data, we can identify activity that is more easily georeferenceable. This is one of the strongest responses to critics of the ABI methodology who argue that much, if not most, data does not lend itself to the georeferencing and overall data-conditioning process.

With this georeferenced data, you now have a *geospatial discovery environment.* Some, perhaps much of the data is also temporally referenced, allowing the introduction of this additional unique element that allows another way for the user to smartly filter through the data. The concept of time-filtering will be revisited in Section 3.6 during discussion of *sequence neutrality*, and the importance of not being bound in the search for potential correlations by the linear-forward path of time (see Chapters 13 and 22 for examples).

3.3 Discover to Georeference Versus Georeference to Discover

It is also important to contrast the philosophy of georeference to discover with the more traditional mindset of discover to georeference. Discover to georeference is a concept often not given a name but aptly describes the more traditional approach to geographically referencing information. This traditional process, based on keyword, relational, or Boolean-type queries, is illustrated in Figure 3.2. Often, the georeferencing that occurs in this process is manual, done via copy-paste from free text documents accessible to analysts.

With discover to georeference, the first question that is asked, often unconsciously, is, "This is an interesting piece of information; I should find out where it happened." It can also be described as "pin-mapping," based on the process of placing pins in a map to describe events of interest. The key difference is the a priori decision that a given event is relevant or irrelevant before the process of georeferencing begins.

Using the pillar of georeference to discover, the act of georeferencing is an integral part of the act of processing data, through either first- or second-degree attributes. It is the first step of the ABI analytic process and begins before the analyst ever looks at the data. In the "pin-mapping" approach, one cannot spatially discover any information other than what has already been labeled "relevant" or "of interest" to a given problem (see Chapter 17 for examples of crime

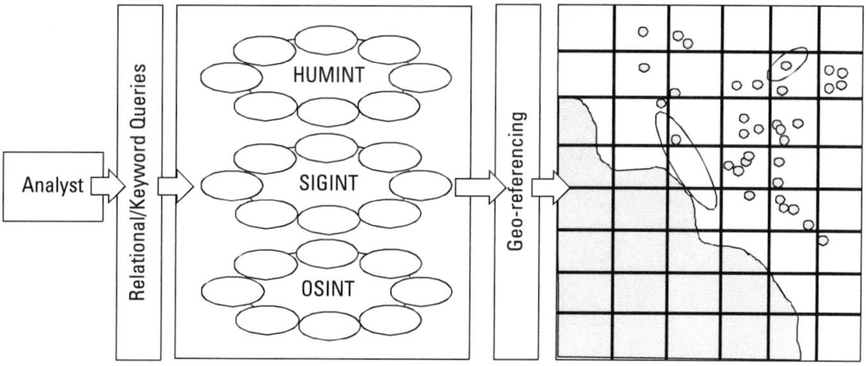

Figure 3.2 The discover to georeference workflow, moving from left to right.

mapping, which limits pins to "crimes" instead of "everything that happened"). This intellectual decision, many times made at the subconscious level, has the effect of drastically limiting the data set available to an analyst. It also destroys any possibility of an effective environment for geospatial discovery of information relevant to a potentially unknown problem or entity. The georeference to discover workflow used in ABI is illustrated in Figure 3.3. Note that, unlike in Figure 3.2, the georeferencing step occurs first in the process, and spatial/temporal queries looking for potential correlations in various domains drive the analytic process.

The act of georeferencing creates an inherently spatial and temporal data environment in which ABI analysts spend the bulk of their time, identifying spatial and temporal co-occurrences and examining said co-occurrences to identify correlations. This environment naturally leads the analyst to seek more sources of data to improve correlations and subsequent discovery. Our entry-level analyst eventually georeferenced everything is his shoebox and all the data in his dusty room. The quest for more data inspired him to leave the confines of INT-specific analysis and led directly to the development of the second pillar: *data neutrality.*

3.4 Data Neutrality: Seeding the Multi-INT Spatial Data Environment

Data neutrality is the premise that all data may be relevant regardless of the source from which it was obtained. This is perhaps the most overlooked of the pillars of ABI because it is so simple as to be obvious. Some may dismiss this pillar as not important to the overall process of ABI, but it is central to the need to break down the cultural and institutional barriers between INT-specific

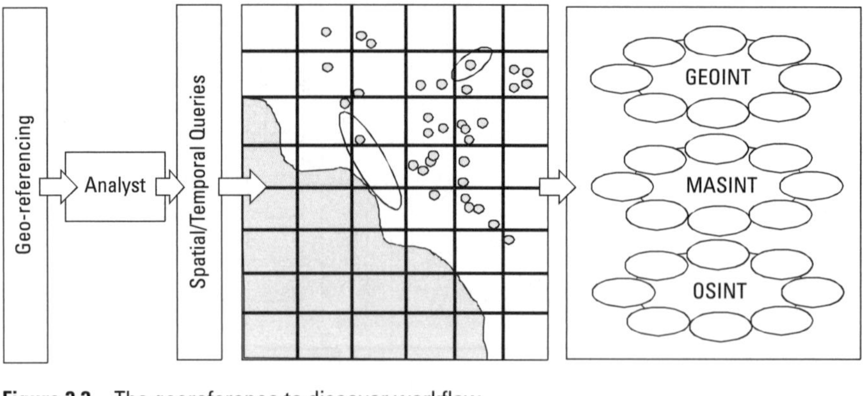

Figure 3.3 The georeference to discover workflow.

"stovepipes" and consider all possible sources for understanding entities and their activities.

In addition, as the pillars were being developed, the practitioners who helped to write much of ABI's lexicon spoke of data neutrality as a goal instead of a consequence. The importance of this distinction will be explored below, as it relates to the first pillar of georeference to discover.

Imagine again you are the analyst described in the prior section. In front of you is a spatial data environment in your GIS consisting of data obtained from many different sources of information, everything from reports from the lowest level of foot patrols to data collected from exquisite national assets. This data is represented as vectors: dots and lines (events and transactions) on your map. As you begin to correlate data via spatial and temporal attributes, you realize that data is data, and no one data source is necessarily favored over the others. The second pillar of ABI serves to then reinforce the importance of the first and naturally follows as a logical consequence.

It is important, however, to not dismiss the idea of data neutrality as a goal or an attitude on the part of an analyst. This begins our first significant exploration of a concept discussed at length later in Chapter 5, which is the *mindset of discovery.* This is an important concept because it is an attitude that helped to create and maintain ABI as a methodology long before it was even called ABI.

Wearing your analyst hat again, imagine you are back in front of your computer, awash in georeferenced data. You have so much data, in fact, that it is tempting to say, "I have enough," or perhaps, "I have too much." But these concepts bear further exploration. What is "enough" data in the context of ABI—and does such a state even exist? Repeat the same question for "too much" data.

Given that *the act of data correlation* is a *core function of ABI,* the conclusion that there can never be "too much" data is inevitable. "Too much," in the inexact terms of an analyst, often means "more than I have the time, inclination, or capacity to understand," but more often than that means "data that is not in a format conducive to examination in a single environment." This becomes an important feature in understanding the data discovery mindset.

Returning to your analyst workstation, you notice that another person in the far corner of your room is working with a novel data type from an information source you are unfamiliar with. You see this and ask, "What is this? Where does it come from?" While he or she is initially hesitant, you are able to start a conversation about this data source and you realize, through the course of conversation, that you can georeference this information, too. You ask how to obtain the data and begin to chart a course of action for how to integrate this new data type manually into your spatial environment, even as you simultaneously begin to think about how, as a technologist, the process might be automated so that the data seamlessly arrives on your desktop.

As the density of data increases, the necessity for smart technology for attribute correlation becomes a key component of the technical aspects of ABI. This challenge is exacerbated by the fact that some data sources include inherent uncertainty and must be represented by fuzzy boundaries, confidence bands, spatial polygons, ellipses, or circles representing circular error probability (CEP). Chapters 6 and 14 explore this problem in greater depth and introduce methods for data correlation under these conditions.

The spatial and temporal environment provides two of the three primary data filters for the ABI methodology: correlation on location and correlation on attributes. Attribute-based correlation becomes important to rule out false-positive correlations that have occurred solely based on space and time. Although automated algorithms show promise, the nature of many data sources almost always requires human judgment regarding correlation across multiple domains or sources of information. Machine learning continues to struggle with these especially as it is difficult to describe the intangible context in which potential correlations occur. For example, in Washington, D.C., "the mall" could refer to the large green space between the Washington monument and the Capitol building or a shopping center in nearby Crystal City. These issues are discussed in Chapter 7 as the ABI concepts of *discreteness* and *durability*.

The concept of mindset applies to data neutrality. Most analysts are trained to perform analysis on a given type of data (or collection of like data types). Technical analysts are trained on the specific outputs of sensors and collection systems, while traditional all-source analysts are trained to read and extract information of value from preliminary intelligence reports, often the very same reports that technical analysts themselves prepare. Part of the importance of the data neutrality mindset is realizing the unique perspective that analysts bring to data analysis; moreover, this perspective cannot be easily realized in one type of analyst but is at its core *the product of different perspectives collaborating on a similar problem set* [6]. This syncretic approach to analysis was central to the revolution of ABI, with technical analysts from two distinct intelligence disciplines collaborating and bringing their unique perspectives to their counterparts' data sets.

The act of georeferencing to discover creates a data-neutral environment of spatially and temporally referenced data ready for discovery. More importantly, the single environment now allows analysts to consider the concept of data integration much further "upstream" than the traditional workflow applied to intelligence analysis in single domains. This leads to the third pillar of ABI: *integration before exploitation.*

3.5 Integration Before Exploitation: From Correlation to Discovery

The traditional intelligence cycle is a process often referred to as tasking, collection, processing, exploitation, and dissemination (TCPED).[2] TCPED is a familiar concept to intelligence professionals working in various technical disciplines who are responsible for making sense out of data in domains such as SIGINT and IMINT. Although often depicted as a cycle as shown in Figure 3.4, the process is also described as linear.

From a philosophical standpoint, TCPED makes several key assumptions:

- The ability to collect data is the scarcest resource, which implies that tasking is the most critical part of the data exploitation process. The first step of the process begins with tasking against a target, which assumes the target is known a priori.
- The most efficient way to derive knowledge in a single domain is through focused analysis of data, generally to the exclusion of specific contextual data.

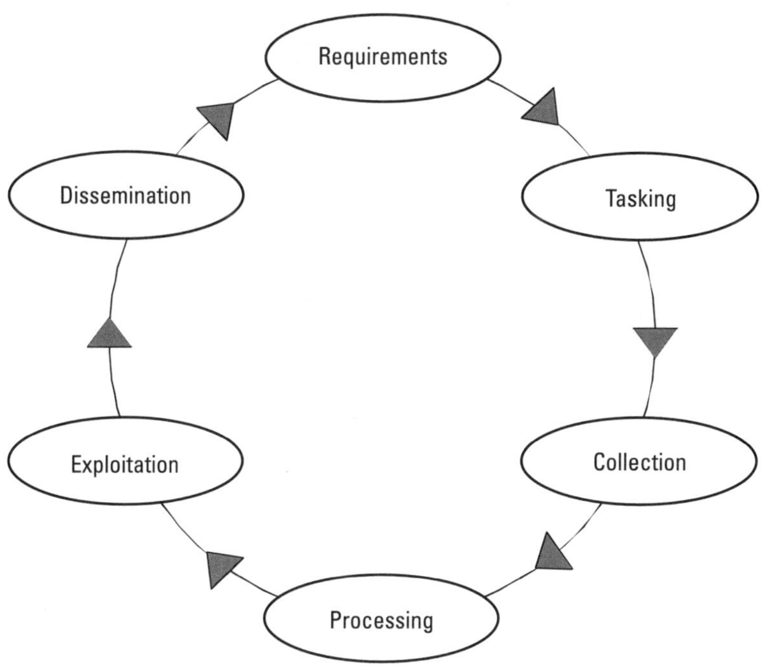

Figure 3.4 The TCPED process or "cycle."

2. Note that "requirements," while not part of the acronym TCPED, ultimately drive the entire process.

- All data that is collected should be exploited and disseminated.

Understanding these key assumptions is critical to understanding TCPED. This is also critical to understanding why ABI challenges many of these notions, themselves product of Cold War–era intelligence gathering and analysis. For example, let us go back to the era of *CORONA,* the United States' first overhead reconnaissance satellite. *CORONA* operated using film canisters that were deorbited from space and snagged, midair by specially equipped cargo aircraft. The process used TCPED: identify targets, launch satellite on flight path over targets, image targets until the film was expended, drop the canister, retrieve the canister, develop the film, analyze the film, and report on what was observed on the film. The limiting factor for *CORONA* missions was the number of images that could be taken by the satellite. In this model, tasking becomes supremely important: There are many more potential targets that can be imaged on a single roll of film. However, since satellite imaging in the *CORONA* era was a constrained exercise, processes were put in place to vet, validate, and rank-order tasking through an elaborate bureaucracy.

With this much emphasis on deciding what is collected, it follows logically that each piece of the highly limited collection—each *CORONA* image—must be exploited for full intelligence value. In fact, because collection was so limited and considered so valuable, a phased exploitation process (Figure 3.5) was implemented to ensure maximum value by sequentially exploiting the same data multiple times.

The early phases are directed at immediate warning on known targets in known locations. During the Cold War, analysts were looking for large military force buildups presaging an invasion through the Fulda Gap or heightened levels of "activity" at known missile test ranges. Later phases of the process in Figure 3.5 focused on in-depth exploitation and deep learning about the target. These assessments took days, weeks, months, or years depending on the intelligence requirements.

The other reality of phased exploitation is that it was a product of an adversary with signature and doctrine that, while not necessarily known, could be

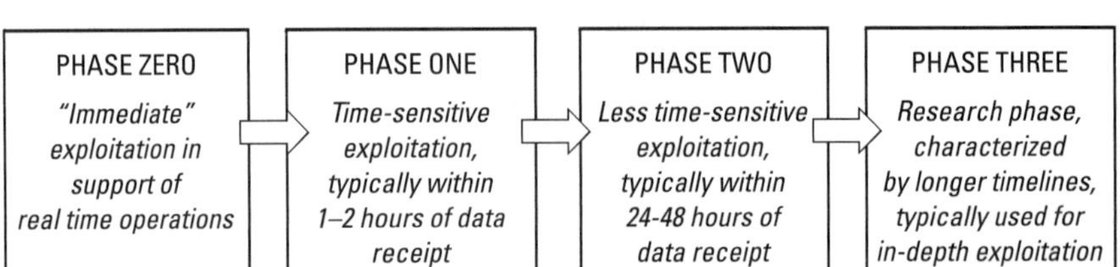

Figure 3.5 The phased exploitation process in traditional intelligence.

deduced or inferred over repeated observations. Large, conventional, doctrine-driven adversaries like the Soviet Union not only had large signatures, but their observable activities played out over a time scale that was easily captured by infrequent, scheduled revisit with satellites like CORONA. Although they developed advanced denial and deception techniques employed against imaging systems, both airborne and national, their large, observable activities were hard to hide [7].

Once analysts examined the satellite data and interpreted it—annotating it with various equipment types and indicating "what" was present on the imagery—reports could be prepared along with the images themselves, and the information, now exploited, was disseminated through to policymakers and others with an interest in the data. This would naturally generate new requirements based on information needs and restart the TCPED process from the beginning.

But where is integration in this process? There is no "I," big or small, in TCPED. Rather, integration was a subsequent step conducted very often by completely different analysts. In the case of imagery, returning to our *CORONA* example, the reports prepared by imagery analysts would then flow to all-source analysts at the CIA's Directorate of Intelligence, who would then combine those reports with other information—open sources, but most often reports generated from clandestine HUMINT sources operating in areas of concern to the U.S. government. Only then would integration truly occur, and then only as the final step in the process before assessments were made and most often, incorporated into narrative judgments. The effectiveness of this process (at its maximum) would not last long, as adversaries gradually understood more and more regarding overhead capabilities and were able to successfully use denial and deception measures to reduce the amount of information gained from U.S. capabilities [8]. In today's era of reduced observable signatures, fleeting enemies, and rapidly changing threat environments, integration after exploitation is seldom timely enough to provide decision advantage. The traditional concept of integration after exploitation, where finished reporting is only released when it exceeds the single-INT reporting threshold is shown in Figure 3.6. This approach not only suffers from a lack of timeliness but also is limited by the fact that only information deemed significant within a single-INT domain (without the contextual information provided by other INTs) is available for integration. For this reason, the single-INT workflows are often derisively referred to by intelligence professionals as "stovepipes" or as "stovepiped exploitation".

Returning to our operations center, you as the analyst are discovering the pillars of ABI. Rather than rely on written assessments generated by other analysts further "upstream" in the data process, you realize that you have access to what some might call "raw" intelligence. While "raw" is a loaded term with specific meanings in certain disciplines and collection modalities, the theory is

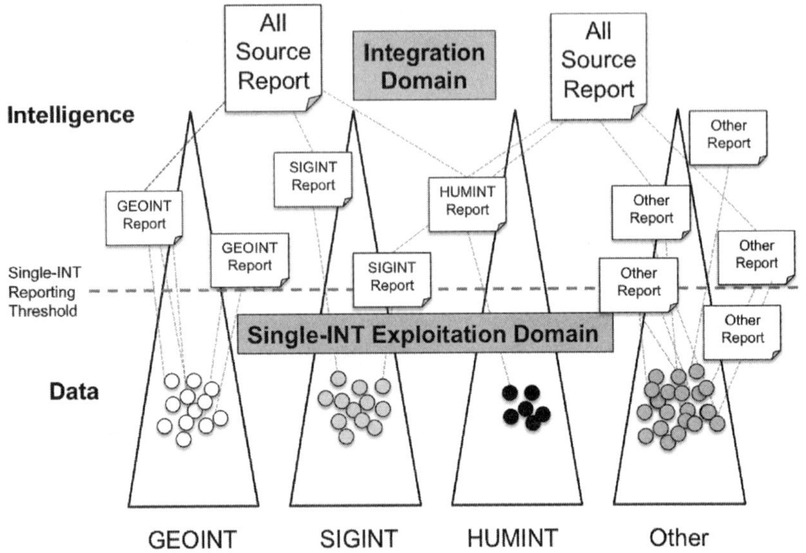

Figure 3.6 The traditional concept of integration after exploitation.

the same: The data you find yourself georeferencing, from any source you can get your hands on, is data that very often, has not made it into the formal intelligence report preparation and dissemination process. It is a very different kind of data, one for which the existing processes of TCPED and the intelligence cycle are inexactly tuned. Much of this information is well below the single-INT reporting threshold in Figure 3.6, but data neutrality tells us that while the individual pieces of information may not exceed the domain thresholds, the combined value of several pieces in an integrated review may not only exceed reporting thresholds but could reveal unique insight to a problem that would be otherwise undiscoverable to the analyst.

Spatially and temporally correlating data is a powerful form of integration. In ABI, integration is performed before the individual data streams are exploited in their individual workflows. But why integrate prior to exploitation? There are many who argue that the best way to obtain maximum value from an information source is to exploit it in a vacuum, looking to pull as much as possible. This approach has corollaries in scientific research, wherein the idea of controlling certain variables weighs heavily on the overall process in order to determine results. What you find, as the analyst looking at spatial correlations, is that these correlations are ones that would not be visible in any other way. This is one-half of the key value proposition of ABI as a methodology. (We will subsequently discuss entity resolution through incidental collection, a compound concept that is the other half of ABI's unique value proposition.)

Another key result of spatiotemporal data integration, executed well prior to exploitation, is a distinct change in the emphasis of the TCPED cycle. TCPED is a dated concept because of its inherent emphasis on the tasking and collection functions. The mindset that collection is a limited commodity influences and biases the gathering of information by requiring such analysts to decide a priori what is important. This is inconsistent with the goals of the ABI methodology. Instead, ABI offers a paradigm more suited to a world in which data has become not a scarcity, but a commodity: the relative de-emphasis of tasking collection versus a new emphasis on the tasking of analysis and exploitation (Figure 3.7).

This change in emphasis was inevitable: More and more of the world around us is being collected and represented as data. Simultaneously, technology has advanced to the point wherein the ability to collect and represent the variations of everyday life as data is cheaply and widely available. Cukier and Mayer-Schoenberg, the authors of a widely read journal article on "big data" even have a name for this process: "datafication" [9]. Chapter 10 describes the datafication of intelligence in more detail.

Re-examining the process of georeference to discover, it is clearly a specific form and view of datafication: a particular method and view to represent,

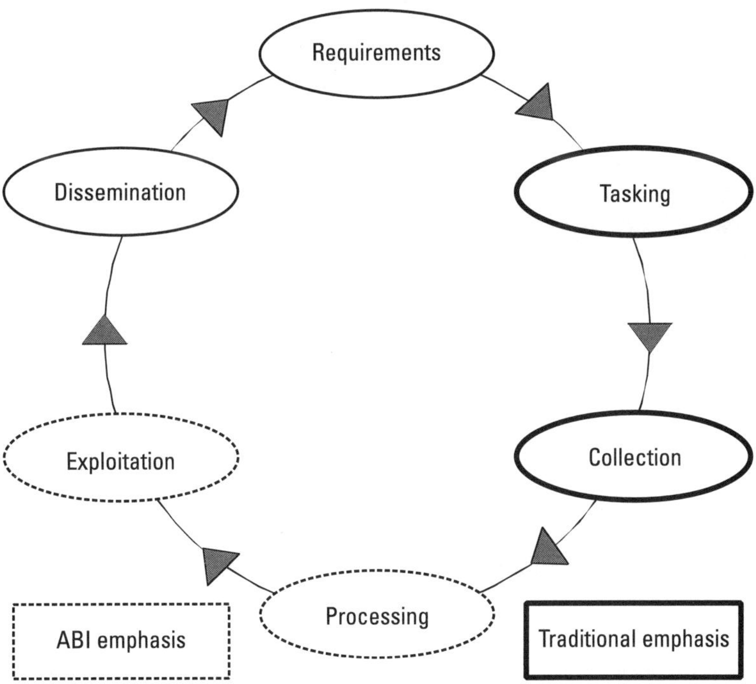

Figure 3.7 Contrasting the points of emphasis in TCPED between the traditional approach and ABI approach.

discover, and explore information. The result of being awash in data is that no longer can manual exploitation processes scale. New advances in collection systems like the constellation of small satellites proposed by Google's Skybox will offer far more data than even a legion of trained imagery analysts could possibly exploit. There are several solutions to this problem of "drowning in data":

- Collect less data (or perhaps, less irrelevant data and more relevant data);
- Integrate data earlier, using correlations to guide labor-intensive exploitation processes;
- Use smart technology to move techniques traditionally deemed "exploitation" into the "processing" stage.

These three solutions are not mutually exclusive, though note that the first two represent philosophically divergent viewpoints on the problem of data. ABI naturally chooses both the second and third solution. In fact, ABI is one of a small handful of approaches that actually becomes far more powerful as the represented data volume of activity increases because of the increased probability of correlations.

The analytic process emphasis in ABI also bears resemblance to the structured geospatial analytic method (SGAM), first posited by researchers at Penn State University [10]. Of particular importance is the concept of two major iterative loops consisting of "foraging" and "sensemaking" as illustrated in Figure 3.8.

Mapped to ABI, reflections of foraging are evident in the constructs of data neutrality as well as sequence neutrality. Foraging, then, is not only a process that analysts use but also a type of attitude that seeks to be embedded in the analytical mindset: The process of foraging is a continual one spanning not only specific lines of inquiry but also evolves beyond the bounderies of specfic questions, turning the "foraging process" into a consistent quest for more data. As we saw from Figure 3.4, more data creates more opportunities for spatiotemporal correlation and therefore, possibly more new discoveries and chances to resolve entities within the data sets available to an analyst. Another implication is precisely where in the data acquisition chain an ABI analyst should ideally be placed. Rather than putting integrated analysis at the end of the TCPED process, this concept argues for placing the analyst as close to the data collection point (or point of operational integration) as possible. While this differs greatly for tactical missions versus strategic missions, the result of placing the analyst as close to the data acquisition and processing components is clear: The analyst has additional opportunities not only to acquire new data but affect the acquisition and processing of data from the ground up, making more data available to the entire enterprise through his or her individual efforts.

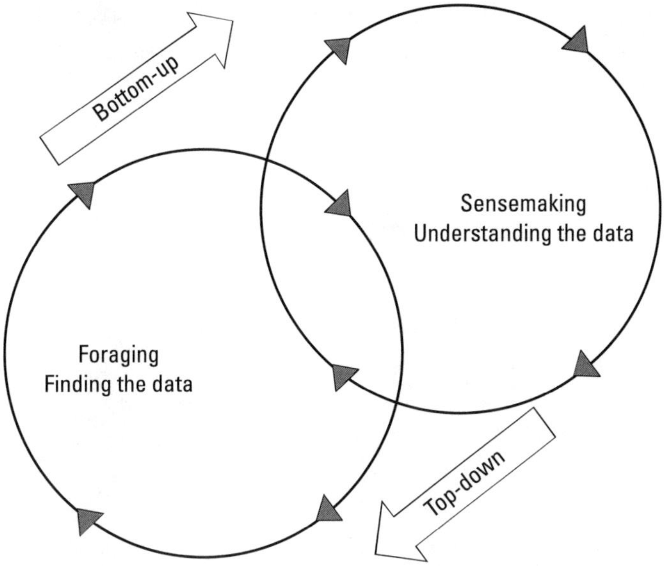

Figure 3.8 The "foraging" and "sensemaking" processes of the SGAM. (Adapted from [10].)

With the increased probability of correlations over space, the implications of time as a filtering mechanism for data exploration as well as a component of the iterative process of analytical question-and-answer must be discussed. This discussion leads directly to the final pillar of ABI.

3.6 Sequence Neutrality: Temporal Implications for Data Correlation

Sequence neutrality is perhaps the least understood and most complex of the pillars of ABI. The first three pillars are generally easily understood after a sentence or two of explanation (though they have deeper implications for the analytical process as we continually explore their meaning). Sequence neutrality, on the other hand, forces us to consider—and in many ways, reconsider—the implications of temporality with regard to causality and causal reasoning. As ABI moves data analysis to a world governed by correlation rather than causation, the specter of causation must be addressed.

"Post hoc, ergo, propter hoc." The direct translation of this Latin phrase is "After it, therefore, because of it." Converted into more understandable English, it is a logical proposition: "Because event B happens after event A, event B must therefore have been caused by event A." (See Figure 3.9.)

The problem with this statement is evident to most: it is rarely true. Many events are followed by many other events, yet the former events do not

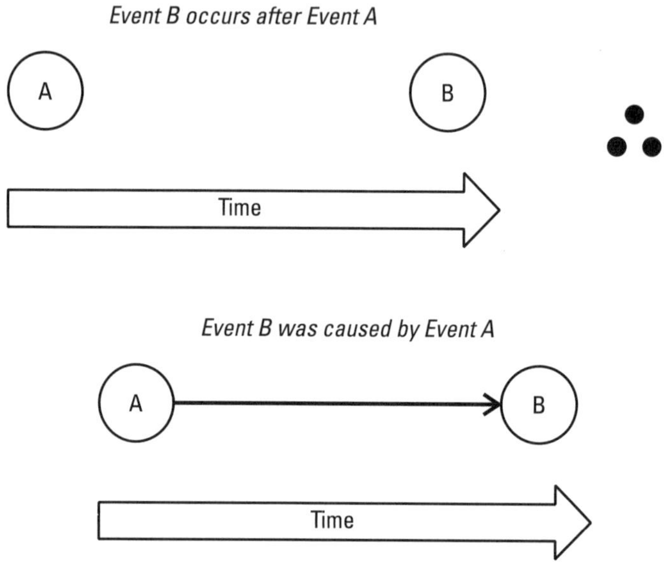

Figure 3.9 A basic illustration of assumed temporal causality: "Because event B occurred after event A, event B was caused by event A."

necessarily cause the latter events. Historians often confront this issue as they look to reconstruct the narrative surrounding a sequence of events. Moving from "the rise of Adolf Hitler" to "Germany" to "France and England's response" and ultimately to the complex chain of events leading to the Second World War illustrates this problem well [11].

In epistemology, this concept is described as narrative fallacy. Naseem Taleb, in his 2007 work *Black Swan,* explains it as "[addressing] our limited ability to look at sequences of facts without weaving an explanation into them, or, equivalently, forcing a logical link, an arrow of relationship upon them. Explanations bind facts together. They make them all the more easily remembered; they help them make more sense" [12]. What is important in Taleb's statement is the concept of sequence: Events occur in order, and we weave a logical relationship around them.

In the context of ABI, events can be understood as activity data: the "grist of the mill" for ABI. As events happen in sequence, we chain them together even given our limited perspective on the accuracy with which those events represent reality. When assessing patterns—true patterns, not correlations—in single-source data sets, time proves to be a useful filter presuming that the percentage of the "full" data set represented remains relatively consistent. As we introduce additional datasets, the potential gaps multiply causing uncertainty to exponentially increase. In intelligence, as many data sets are acquired in an adversarial rather than cooperative fashion (as opposed to in traditional civil

GIS approaches, or even crime mapping approaches), this concept becomes so important that it is given a name: *sparse data*.

Uncertainty—caused by incomplete or sparse data—poses a major challenged for a narrative, sequenced understanding of the world. However, if we re-examine our approach, focusing on correlation rather than causation, we begin to see that the sequence in which events occur becomes fundamentally irrelevant—our only goal is to examine whether those data points can be correlated in space and time and only then begin to contextualize the meaning of those potential correlations.

Revisiting our scene: You are back again, sitting in the operations center. You have georeferenced every bit of available data (and are continuing to forage for more). You are integrating the data well before stovepiped exploitation and have created a data-neutral environment in which you can ask complex questions of the data. This enables and illuminates a key concept of sequence neutrality: The data itself drives the kinds of questions that you ask. In this way, we express a key component of sequence neutrality as "understanding that we have the answers to many questions we do not yet know to ask."

The corollary to this realization is the importance of forensic correlation versus linear-forward correlation. If we have the answers to many questions in our spatial-temporal data environment, it then follows logically that the first place to search for answers—to search for correlations—is in the data environment we have already created. Since the data environment is based on what has already been collected, the resultant approach is necessarily forensic. Look backward, before looking forward.

Forensic has often been conflated incorrectly with the speed of analysis rather than the approach. This is largely a legacy of the pre-big data era (remnants of which still dominate in some fields today) in which it is extraordinarily difficult to quickly understand the sum of data already collected. Here, we see a concrete example of technology driving tradecraft evolutions. From card catalogs and libraries we have moved to search algorithms and metadata, allowing us as analysts to quickly and efficiently employ a forensic, research-based approach to seeking correlations.

This also mandates a discussion on the use of sequenced, temporal filtering of data sets. Early in the history of ABI, analysts were leveraging GIS-based technology that did not easily filter based on time (though later evolutions of the software package have added the ability to perform more interactive time-based filtering). Indeed, separate layers, or data files, would represent the sum total of a year's worth of collected data, and an analyst could filter by activating or de-activating certain years of data. While it was a valuable process particularly when large data sets were present or processing power was limited, it was also highly cumbersome.

As software platforms evolved, more intuitive time-based filtering was employed, allowing analysts to easily "scroll through time." As with many technological developments, however, there was also a less obvious downside related to narrative fallacy and event sequencing: The time slider allowed analysts to see temporally referenced data occur in sequence, reinforcing the belief that because certain events happened after other events, they may have been caused by them. It also made it easy to temporally search for patterns in data sets: useful again in single data sets, but potentially highly misleading in multisourced data sets due to the previously discussed sparse data problem. Sequence neutrality, then, is not only an expression of the forensic mindset but a statement of warning to the analyst to consider the value of sequenced versus nonsequenced approaches to analysis. Humans have an intuitive bias to see causality when there is only correlation. We caution against use of advanced analytic tools without the proper training and mindset adjustment.

The manual approach, by forcing analysts to jump back and forth across the timeline—from 2004 data to 2007 data to 2006 data—took temporal causality out of the analytic process, without analysts realizing it. You realize, scrolling through metadata, that you might not have seen the correlations if you had employed a "replay" approach in which you scrolled through data based on when it occurred. By first exploring spatial correlation and only then examining temporal attribution, you are able to find correlations across vast chunks of time that other analysts, viewing data in a linear-forward construct, would likely have missed. While neither approach is wrong, in the context of ABI, the sequence neutral approach is more right.

3.6.1 Sequence Neutrality's Focus on Metadata: Section 215 and the Bulk Telephony Metadata Program Under the USA Patriot Act

Sequence neutrality also has important implications for how we collect, store, and tag data. By positing that all data represents answers to certain questions, it implores us to collect and preserve the maximum amount of data as possible, limited only by storage space and cost. It also begs the creation of indexes within supermassive data sets, allowing us to zero in on key attributes of data that may only represent a fraction of the total data size. By preserving at least the metadata, we are able to preserve the potential for correlation even if the content becomes unavailable at a later date. (Naturally, the availability of content increases our understanding of the context of correlation thereby enabling us to better understand whether a correlation is valid or not, and what it means in a broader context.) One example of the importance of metadata is the now declassified Section 215 bulk telephony metadata collection program under the USA PATRIOT Act, which was acknowledged by the Office of the Director of

National Intelligence on December 20, 2013, and multiple subsequent declarations available on ICONTHERECORD.tumblr.com [13].

A controversial provision of the USA PATRIOT act, Section 215, allows the director of the Federal Bureau of Investigation (or designee) to seek access to "certain business records" which may include "any tangible things (including books, records, papers, documents, and other items) for an investigation to protect against international terrorism or clandestine intelligence activities, provided that such investigation of a United States person is not conducted sole upon the basis of activities protected by the first amendment to the Constitution" [14].

In support of the provisions of Section 215 of the Act, the Foreign Intelligence Surveillance Court (FISC) ruled that telephony metadata included "comprehensive communications routing information, including but not limited to session routing information (e.g. originating and terminating telephone number, International Mobile Subscriber Identity (IMSI) number, International Mobile station Equipment Identifier (IMEI number), etc.), trunk identifier, telephone calling card numbers, and time and duration of call)" could be collected [15]. Under the now declassified program, metadata was archived and indexed in case future investigations against targeted identifiers for which there was a reasonable, articulable suspicion of participation in terrorist activities. If the metadata was not indexed, sequence neutral discovery of associates through forensic investigation would have been impossible.

3.7 After Next: From Pillars, to Concepts, to Practical Applications

The pillars of ABI represent the core concepts, as derived by the first practitioners of ABI. Rather than a framework invented in a classroom, the pillars were based on the actual experiences of analysts in the field, working with real data against real mission sets. It was in this environment, forged by the demands of asymmetric warfare and enabled by accelerating technology, in which ABI emerged as one of the first examples of data-driven intelligence analysis approaches, focused primarily on spatial and temporal correlation as a means to discover.

The foraging-sensemaking, data-centric, sequence-neutral analysis paradigm of ABI conflicts with the linear-forward TCPED-centric approaches used in "tipping-cueing" constructs. The tip/cue concept slews (moves) sensors to observed or predicted activity based on forecasted sensor collection, accelerating warning on known-knowns. This concept ignores the wealth of known data about the world in favor of simple additive constructs that if not carefully understood, risk biasing analysts with predetermined conclusions from arrayed collection systems. A heavy focus on tipping and cueing risks devolving the deep spatiotemporal analytic environment created through the first three pillars into a discussion about speeding up the linear TCPED process. This invariably drives the focus away from integrating activity to discover unknowns and toward monitoring known targets for change.

Sequence neutrality leads to the important concept of incidental collection of data (as in the controversial NSA metadata program). This has implications for designs not only for data processing and storage but the very way in which sensors and platforms are built. By using systems that are consistently acquiring and processing data—and by reprocessing targeted data in an intelligent fashion—all data types can be transformed into incidentally collected data. In the parlance of Cukier and Mayer-Schönberger, this approach leads to an early consideration of how we might maximize the reuse value of data, changing our entire approach to the process of datafication. This concept will be explored further in Chapter 10.

The pillars, called the foundation of the ABI method, also form the foundation of a much needed tradecraft revolution that integrates all sources of data and improves the timeliness of results. Figure 3.10 illustrates the data-to-intelligence process using the four pillars of ABI. Integration before exploitation lowers the threshold for "reporting" below the single-INT, finished reporting threshold. It allows multi-INT correlations to be created much earlier in the processing cycle and enables discovery of that which would otherwise be ignored.

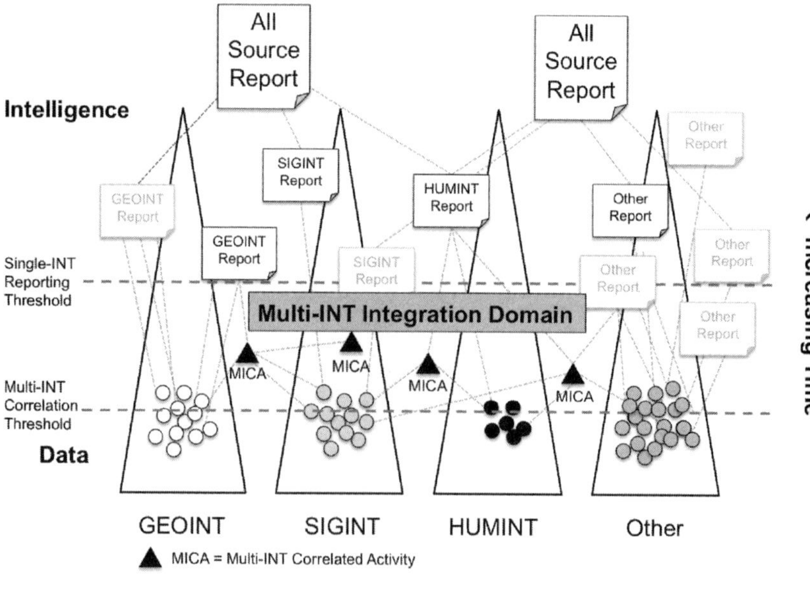

Figure 3.10 Improving the quality and timeliness of intelligence through integration before exploitation.

While some traditional practitioners are uncomfortable with the prospect of "releasing unfinished intelligence," the ABI paradigm—awash in data—leverages the power of "everything happens somewhere" to discover the unknown. As a corollary, when many things happen in the same place and time, this is generally an indication of activity of interest. Correlation across multiple data sources improves our confidence in true positives and eliminates false positives.

3.8 Summary

The four pillars of ABI are the conceptual foundation of this new approach and philosophy for intelligence. We will discuss the more advanced principles of ABI, as well as ABI's unique top-level data ontology necessitated by the diverse data types with which this chapter's hypothetical analyst is confronted, in Chapter 4.

References

[1] "The 1 May U-2 Incident and Powers' Fate," Central Intelligence Agency, 1961.

[2] "National Space Science Data Center - Discoverer 14 Spacecraft Details," National Space Science Data Center.

[3] Treverton, G., Cambridge, England: *Intelligence for an Age of Terror,* Cambridge University Press, 2009.

[4] Quoted in Phillips, Mark, "A Brief Overview of ABI and Human Domain Analytics," *Trajectory,* http://trajectorymagazine.com/web-exclusives/item/1369-human-domain-analytics.html. 28 Sep. 2012. Accessed 28 May 2014. Approved for Public Release. NGA Case #12-463.

[5] Frost, R., "The Road Not Taken," *Mountain Interval,* New York: Henry Holt and Company, 1916.

[6] Matheny, J., "Tournaments for Geopolitical Forecasting," presented at the SAP NS2 Solutions Summit, October 29, 2013.

[7] Richelson, J. T.,"Intelligence: The Imagery Dimension," in *The Intelligence Cycle: From Spies to Policymakers,* Vol. 2, Santa Barbara, CA: Praeger, 2006, p. 71–72.

[8] Bennett, M., and E. Waltz, *Counterdeception Principles and Applications for National Security,* Norwood, MA: Artech House, 2007.

[9] Mayer-Schonberger, V., and K. Cukier, *Big Data: A Revolution That Will Transform How We Live, Work, and Think,* New York: Houghton Mifflin Harcourt, 2013, p. 15.

[10] Penn State University Department of Geography, "Base Theory of Structured Geospatial Analytic Model," *The Learner's Guide to Geospatial Analysis,* available: https://www.e-education.psu.edu/sgam/node/173, accessed August 22, 2014.

[11] Overy, R. J., *The Origins of the Second World War,* 2nd ed., London: Longman. 1998, pp. 1–3.

[12] Taleb, N., *The Black Swan: The Impact of the Highly Improbable,* 2nd ed., New York: Random House, 2010, p. 303.

[13] "IC ON THE RECORD" Office of the Director of National Intelligence, http://icontherecord.tumblr.com/tagged/declassified, web, 2014.

[14] UNITING AND STRENGTHENING AMERICA BY PROVIDING APPROPRIATE TOOLS REQUIRED TO INTERCEPT AND OBSTRUCT TERRORISM (USA PATRIOT ACT) ACT OF 2001. Public Law 107-56, October 26, 2001.

[15] IN RE APPLICATION OF THE FEDERAL BUREAU OF INVESTIGATION FOR AN ORDER REQUIRING THE PRODUCTION OF TANGIBLE THINGS FROM [REDACTED]. Foreign Intelligence Surveillance Court, United States of America, 04–12-2013. [Online]. Available: http://www.dni.gov/files/documents/PrimaryOrder_Collection_215.pdf.

4

The Lexicon of ABI

The development of ABI also included the development of a unique lexicon, terminology, and ontology to accompany it. Activity data "comprises physical actions, behaviors, and information received about entities. The focus of analysis in ABI, activity is the overarching term used for 'what entities do.' Activity can be subdivided into two types based on its accompanying metadata and analytical use: events and transactions" [1]. A community of interest developed this definition to be somewhat all-encompassing. As with any bureaucratic definition, it serves the reader well to do some unpacking and truly understand what activity data is, and why we care about it.

4.1 Ontology for ABI

One of the challenges of intelligence approaches for the data-rich world that we now live in is integration of data. In Chapter 3, the reader was introduced to the integration before exploitation pillar and the idea of integration using spatial and temporal referencing. The principle of data neutrality drove analysts to consider new data sources that sometimes only had spatial and temporal metadata in common. As the diversity of data increased, analysts were confronted with the problem that most human analysts deal with today: How does one represent diverse data in a common way?

An *ontology* is the formal naming and documentation of interrelationships between concepts and terms in a discipline. Established fields like biology and telecommunications have well-established standards and ontologies. As the diversity of data and the scope of a discipline increases, so does the complexity of the ontology. If the ontology becomes too rigid and requires too many committee approvals to adapt to change, it cannot easily account for new data types that emerge as technology advances. Moreover, with complex ontologies for data, complex environments are required for analysis, and it becomes extraordinarily difficult to correlate and connect data (to say nothing of conclusions derived from data correlations) in any environment other than a pen-to-paper notebook or a person's mind.

As ABI developed, there was a need to create a simple, yet appropriately flexible data ontology based on metadata elements. Because of ABI's emphasis on metadata before content (a consequence of the pillar of georeference to discover) any data ontology would have to be based around a handful of data elements. When the OUSD(I) defined the human dimension in the *Strategic Advantage* series of papers, they offered four categories of data that form the basic ontology to support the ABI method (Table 4.1). The focus of ABI is on the use of activity and context data, but biographical and relational information about specific entities is often vital in the overall analytic process. An analytic workflow can often involve discovery of pieces of information in each of the four categories.

The interplay between these four data types, and the nonlinear process by which the information is obtained, is perhaps the most important goal of ABI. Chapters 5–9 explore how to use activity data, as well as other data types, in analytical process, while the remaining sections of this chapter describe in detail the features of these important data types.

4.2 Activity Data: "Things People Do"

The first core concept that "activity" data reinforces is the idea that ABI is ultimately about people, which, in ABI, we primarily refer to as "entities."

This is an important distinction for a number of reasons, but is particularly important because of ABI's heritage in the imagery and geospatial worlds. The pre-9/11 national security strategy focused heavily on counterproliferation and deterrence of conventional military forces, with lesser focus on terrorism, non-state actors, and transnational organizations. IMINT requirements in this era tended to focus on assessing military order of battle (OOB), because the priority intelligence questions related to state actors and their respective military forces. This draws a sharp contrast with ABI, where the focus originated on human beings (terrorists and insurgents) who blended into their surrounding populations. Accordingly, still images did not offer the same kind of advantage in the context of person-level activities.

Thus, activity in ABI is information relating to "things people do." While this is perhaps a simplistic explanation, it is important to the role of ABI. In ABI parlance, activities are not about places or equipment or objects. Activity

Table 4.1
The Four Elements of Data in the Human Domain Supporting the ABI Methodology

Data type	Focus	Primary purpose
Activity	Discrete activities conducted by individual entities	Entity resolution/identification, pattern-of-life assessment
Context	Aggregated data of any type	Provides context for observed activities of entities
Biographical	Attribute information of an entity	Provide information pertaining to a specific entity such as age or name
Relational	Information describing relationships between entities	Understand and visualize the formal and informal social networks to which an entity belongs

data tells us about what people have done, even when we do not know their identities.

This ultimately leads us to more complex concepts about how we represent activity and how it is sensed in the environment around us. In differentiating between activity and contextual data, it becomes clear that part of defining data is understanding potential uses and granularity.

4.2.1 "Activity" Versus "Activities"

The vernacular and book title use the term "activity-based intelligence," but in early discussions, the phrase was "activities-based intelligence." Activities are the differentiated, atomic, individual activities of entities (people). Activity is a broad construct to describe aggregated activities over space and time.

To illustrate the difference, consider the heat maps of traffic speed provided in Google Maps. The underlying math is ultimately based on aggregation. The speed of individual cars is irrelevant; the goal is to produce a general understanding of the average speed over a given stretch of road in order to provide an estimate of travel time[1]. If one particular criminal flees police by weaving through cars on a motorcycle at 100 mph, this doesn't impact the aggregate "activity" of the highway.

In another example, traditional imagery analysts are fond of saying: "During the Cuban Missile Crisis, we saw activity at missile sites in Cuba!" This was a slow, aggregate change in a fixed site over time. If individual humans were observed unloading crates, driving trucks, setting up equipment, or communicating with the Russians, these "activities," if detected would serve as early indicators of missile proliferation [2].

The goal of ABI focuses more on the use of individual activities in order to understand the identities and patterns of life specific to individual entities

1. This integrative traffic data is ultimately useful (at times) as context data in ABI, but is not the focus of analysis because the aggregation removes the granularity necessary to discern specific activities conducted by individual entities.

(combinations of specific behaviors performed by entities conducting everyday activities, discussed further in Chapter 8). While activity data as context or backdrop is useful, this kind of data is often difficult to use to disambiguate one entity from another due to the aggregation. This is slightly different when using anonymized but individual data, where research has shown that the anonymization process can be effectively undone with a surprisingly small number of unique data points [3]. Interestingly, commercial companies perform this de-anonymization of personal data by correlating activities across multiple locations using methods familiar to ABI practitioners.

4.2.2 Events and Transactions

The definition in the introduction to this chapter defined activity data as "physical actions, behaviors, and information received about entities" but also divided activity data into two categories: events and transactions. These types are distinguished based on their metadata and utility for analysis. To limit the scope of the ABI ontology (translation: to avoid making an ontology that describes every possible action that could be performed by every possible type of entity), we specifically categorize all activity data into either an event or transaction based on the metadata that accompanies the data of interest.

An event is "a recognizable movement or change conducted by an entity that has a specific meaning when viewed within a relevant context. Analysts use events to characterize locations and entities. Events are defined by their spatial metadata components." Some examples include an IED attack, a missile test, and a person living in a residence [1].

The provided examples are extraordinarily illustrative. The first two examples—an IED attack and a missile test—provide us examples of singular events, that is, events that occur once for a very distinct period of time. The third example—a person living in a residence—provides a very different kind of event, one that is far less specific. While a residential address or location can also be considered biographical data, the fact of a person living in a specific place is treated as an event because of its spatial metadata component.

In all three examples, spatial metadata is the most important component. While both the IED attack and missile test have accompanying temporal registration, the person living in a residence is far more open-ended; thus, it is not precisely registered temporally. An example of a georeferenced event/situation report is shown in Figure 4.1.

The concept of analyzing georeferenced events is not specific to military or intelligence analysis. The GDELT project maintains a 100% free and open database of 300 kinds of events using data in over 100 languages with daily updates from January 1, 1979, to the present. The database contains over 400 million georeferenced data points [4]. Analysts use events to spatially characterize

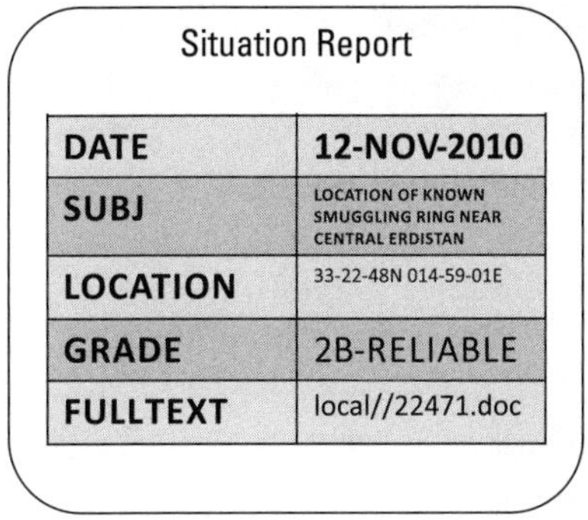

Situation Report

DATE	12-NOV-2010
SUBJ	LOCATION OF KNOWN SMUGGLING RING NEAR CENTRAL ERDISTAN
LOCATION	33-22-48N 014-59-01E
GRADE	2B-RELIABLE
FULLTEXT	local//22471.doc

Figure 4.1 Example of an event report.

locations and entities. In a GIS, events are usually represented one-dimensionally as "dots," although some formulations also consider measures of spatial and temporal uncertainty. Note the precise spatial registration in the form of degrees-minutes-seconds (DMS) as well as a link to the full text of the document. Chapter 13 describes event analysis using GIS tools in more detail.

Characterization is an important concept because it can sometimes appear as if we are using events as a type of context. In this way, activities can characterize other activities. This is important because most activity conducted by entities does not occur in a vacuum; it occurs simultaneously with activities conducted by different entities that occur in either the same place or time—and sometimes both. This can be true whether the events are related or unrelated. (A *co-occurrence,* describing two events that take place in close proximity spatially and temporally, is one kind of noncausal relationship.)

Events that occur in close proximity provide us an indirect way to relate entities together based on individual data points. There is, however, a more direct way to relate entities together through the second type of activity data: *transactions.*

4.2.3 Transactions: Temporal Registration

Transactions in ABI provide us with our first form of data that directly relates entities. A transaction is defined as "an exchange of information between entities (through the observation of proxies) and has a finite beginning and end" [1]. This exchange of information is essentially the instantaneous expression of a relationship between two entities. This relationship can take many forms, but it exists for at least the duration of the transaction. Whereas spatial registration defines event data, transaction data is defined by its temporal registration: All transactions have a precise beginning and end.

Transactions are of supreme importance in ABI because they represent relationships between entities. Transactions are typically observed between proxies, or representations of entities, and are therefore indirect representations of the entities themselves. For example, police performing a stakeout of a suspect's home may not observe the entity of interest, but they may follow his or her vehicle. The vehicle is a proxy. The departure from the home is an event. The origin-destination motion of the vehicle is a transaction. Analysts use transactions to connect entities and locations together, depending on the type of transaction. Chapter 6 details the concept of proxies.

Transactions come in two major subtypes: *physical transactions* and *logical transactions*. Physical transactions are exchanges that occur primarily in physical space, or, in other words, the real world. An example of a physical transaction is a car driving between location A and location B. The connection and exchange takes place between the entity or entities driving the car, and the entity or entities at each location. Subconnections that are possible include the driver of the car to the entity at location A, the driver to the entity at location B, and the entities at location A to the entities at location B via the driver. Often, additional information is needed to discern which of the entities involved represents the exchange in the case of a physical transaction, which we will discuss in Chapter 7 as part of ABI's concept of discreteness. Figure 4.2 illustrates an example of a physical transaction based on the path of a red sedan between two points. Note the precise time given for both the start and the stop of the vehicle, while the line represents the physical path taken by the vehicle.

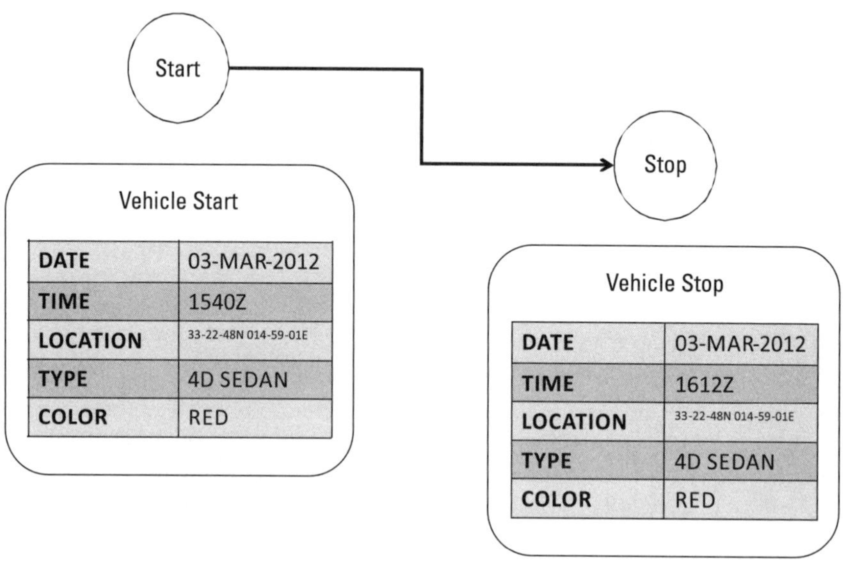

Figure 4.2 Example of a physical transaction and associated metadata.

Logical transactions represent the other major subtype of transaction. These types of transactions are easier to join directly to proxies for entities (and therefore, the entities themselves) because the actual transaction occurs in cyberspace as opposed to physical space. Despite this, the endpoints—or vertices—of the transaction do occur in physical space, as they are conducted by the individuals in question.

A good example of this is a financial transaction between two individually numbered bank accounts. Each account is a proxy for an individual, or entity, and the flow of funds into one account and out of another represents an exchange. Even with this data, context is king in evaluating the meaning of the transactions: is there an employee-employer relationship between the accounts? Or does, perhaps, the transfer represent a small loan given from one friend to another? With only the single transaction itself, it is difficult to provide an accurate assessment.

Defining events and transactions by their respective critical metadata components—spatial and temporal registration—leads to an obvious question: What happens when a piece of data has both spatial and temporal registration? Can something be both an event and a transaction?

4.2.4 Event or Transaction? The Answer is (Sometimes) Yes

Defining data as either an event or transaction is as much a function of understanding its role in the analytical process as much as it is about recognizing present metadata fields and "binning" it into one of two large groups. Consequently, there are certain data types that can be treated as both events and transactions depending on the circumstances and analytical use.

Because analytical use—spatial characterization versus logical connection—is part of determining whether a piece of data is considered an event or transaction, we must consider the possibility that some data types fit into both categories. In most cases, this type of data will be fundamentally a transaction. In a transaction, there are two important parts: the relationship or connection between the entities in question, and the vertices—the entities (or proxy observations thereof).

In considering data that might potentially fall into both the "event" and "transaction" bins, it is helpful to revisit the metadata fields required to make a datum an event or transaction: spatial or temporal registration. Considering Figure 4.2, we see that the vertices of the transaction are spatially registered. We also know that at each of the vertices is a different entity, represented by a proxy. Therefore, we have an entity with specific spatial registration, meeting our original definition of an event. Thus, analytically, the vertices of transactions with spatial registration can be treated as events in the analytical process.

4.3 Contextual Data: Providing the Backdrop to Understand Activity

One of the important points to understand with regard to activity data is that its full meaning is often unintelligible without understanding the context in which observed activity is occurring. This is true from both a conceptual point of view as well as a data point of view. Here, we focus on the importance of contextual data, which helps us understand the environment in which activity is occurring. This offers us another parallel to commercial big data, where the importance of placing data in context is clear. Alissa Lorentz from Augify writes, "Contextualization is crucial in transforming senseless data into real information" [5]. Activity data in ABI is the same: To understand it fully, we must understand the context in which it occurs, and context is a kind of data all unto itself.

There are many different kinds of contextual data. One easy example of contextual data is weather data. By understanding weather patterns during the time of an activity of interest, an analyst can begin to draw the kinds of conclusions that put him or her in the mind of the entities of interest: The neighbor was expected to cut his lawn. Did the weather have an impact on the activity in question? The fusion of activity and context allows us to begin to ask intelligent questions of our spatial data environments and understand relationships that are otherwise hidden in a mess of activity data.

Another kind of contextual data is foundational imagery or maps. An example of this is controlled image base (CIB), produced by the Defense Logistics Agency (DLA). The agency's product web site states that CIB is "an unclassified seamless dataset of orthophotos, made from rectified grayscale aerial images" [6]. CIB provides planners with a worldwide data set on which they can see landscape and structures, telling us if activity is happening in cities, near houses, or in rural areas; it is imagery as context data to activity. The spatial location of the data anchors the context, but an image fully defines the relevant context of an activity. Did it occur in Baghdad, or in the Indian Ocean? The quickest and easiest way to see this is on an image. Digital map overlays, similarly, can provide contextual information about the locations where activity occurs. Since the concept of place is inherent to the georeference-to-discover technique, a strong foundation of geospatial contextual data is important to understand activities and transactions.

Consider the case of a car approaching a four-way intersection. Will the car go straight, turn left, or turn right? There is a 33% chance of each event[2]. Contextual information might show that the left path is a dead-end road and the right path leads to a popular department store. Who is the driver? Do they live on the dead-end road? Have they shopped at that store before? Are they

2. Or the car could stop indefinitely; or it could turn around; or it could be crushed by a falling meteorite. (See how the question biases the possible answers?)

likely to? Activity data and contextual data help understand the nature of events and transactions—and sometimes even to anticipate what might happen. However, this example also illustrates the power of two additional data types identified by OUSD(I): biographical data and relational data [7].

4.4 Biographical Data: Attributes of Entities

Biographical data provides information about an entity: name, age, date of birth, and other similar attributes. Because ABI is as much about entities as it is about activity, considering the types of data that apply specifically to entities is extremely important. Biographical data provides analysts with context to understand the meaning of activity conducted between entities.

Biographical data is perhaps the most intuitive data type in ABI, because it is something we easily relate to as people. This data is information about entities (e.g., given names, aliases, dates of birth, and bank account numbers). Some of these data types are proxies, while some are merely contextual. All help provide analysts with an understanding of what specific events and transactions might mean and how different events and transactions may or may not be related to specific entities.

Biographical data is also important to consider in the context of entity resolution. While we will explore entity resolution in greater context in Chapter 6, it is important to briefly discuss this concept here, because the process of entity resolution (fundamentally, disambiguation) enables us to understand additional biographical information about entities. If we envision each entity as having a kind of "baseball card," providing information like name, telephone, e-mail, and bank account number, then the process of entity resolution is about determining which proxies (things like e-mails and bank account numbers, for example) belong to which entities, all of which has the effect of filling out biographical data about particular entities.

Reading the above paragraph, it is easy to think that ABI is just new terminology around old approaches to analysis. Police departments, intelligence agencies, and even private organizations have long desired to understand specific details about individuals; therefore, what is it that makes ABI a fundamentally different analytic methodology? The answer is in the relationship of this biographical data to events and transactions described in Sections 4.2.2–4.2.4 and the fusion of different data types across the ABI ontology at speed and scale.

Unlike in more traditional approaches, wherein analysts might start with an individual of interest and attempt to "fill out" the baseball card, ABI starts with the events and transactions (activity) of many entities, ultimately attempting to narrow down to specific individuals of interest. This is one of the techniques that ABI uses to conquer the problem of unknown individuals in a network, which guards against the possibility that the most important entities might be ones that are completely unknown to individual analysts. Even in instances where an individual practicing ABI might be looking for an individual in particular, the analytic approach of beginning with the activity in question rather than the entity helps set ABI apart from other methodologies. This will be explored further in practice in Chapter 5.

Figure 4.3 contrasts the ABI entity research flow with the traditional entity research flow, like that practiced by law enforcement, for example. In both, common pieces of information about an entity are the values: "name," "residence," "phone number," "e-mail address." All are aspects of understanding an entity's behavior, as described in detail in Chapter 6 regarding the relationship of proxies to entities. In the traditional research flow, the biographical information of a known entity is used to identify accesses to that individual. With a court order, law enforcement personnel can wiretap a telephone or search a residence for evidence. They can also follow a vehicle and monitor an entity's transactions to attempt to identify associated locations and other entities. The small ovals in the ABI workflow represent the many potential phones, vehicles,

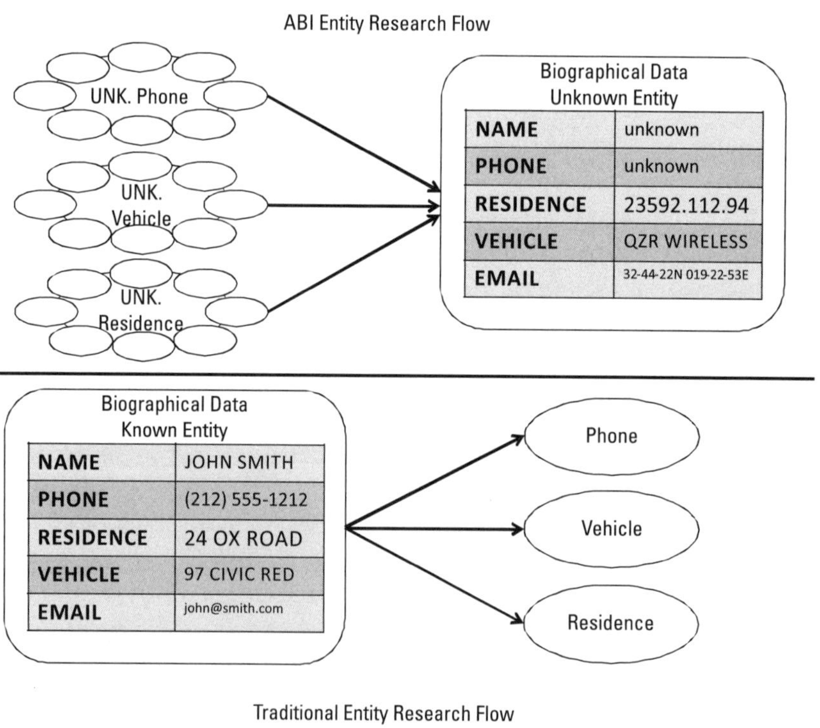

Figure 4.3 Comparison of entity research flows across traditional analysis and with the ABI process.

and residences conducting activity, out of which entities can be deduced. The difference in the process is that ABI analysts consider all of the sources of data as data (the principle of data neutrality in action) and resolve the unknown entity through correlation of different sources.

The final piece of the "puzzle" of ABI's data ontology is relating entities to each other—but unlike transactions, we begin to understand generalized links and macronetworks. Fundamentally, this is relational data.

4.5 Relational Data: Networks of Entities

Entities do not exist in vacuums. Even Osama bin Laden, perhaps one of the world's most reclusive entities until his death in 2011, maintained links to the outside world, attempting to influence entities leading the various al-Qaeda affiliates [8]. Therefore, considering the context of relationships between entities is also of extreme importance in ABI. Relational data tells us about the entity's relationships to other entities, through formal and informal institutions, social networks, and other means.

Initially, it is difficult to differentiate relational data from transaction data. Both data types are fundamentally about relating entities together; what, therefore, is the difference between the two? The answer is that one type—transactions—represents specific expressions of a relationship, while the other type—relational data—is the generalized data based on both biographical data and activity data relevant to specific entities. This concept is depicted in Figure 4.4.

The importance of understanding general relationships between entities cannot be overstated; it is one of several effective ways to contextualize specific expressions of relationships in the form of transactions. Traditionally, this process would be to simply use specific data to form general conclusions (an

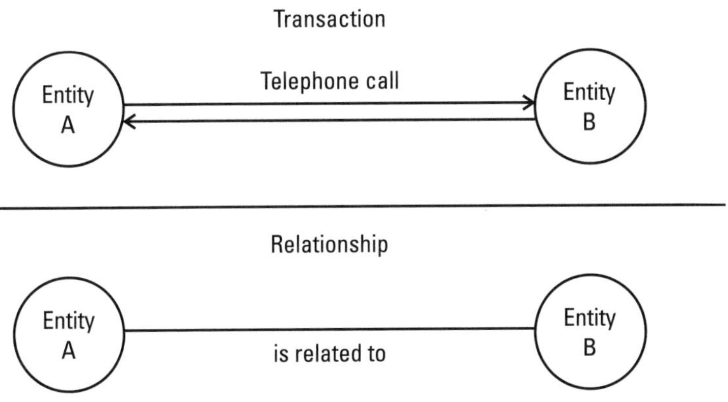

Figure 4.4 Contrast between transactions and relationships.

inductive process, explored in Chapter 5). In ABI, however, deductive and abductive processes are preferred (whereby the general informs our evaluation of the specific). In the context of events and transactions, our understanding of the relational networks pertinent to two or more entities can help us determine whether connections between events and transactions are the product of mere coincidence (or density of activity in a given environment) or the product of a relationship between individuals or networks.

This type of data is often used in social network analysis (SNA), which is a particular approach to analyze relational data in the context of social networks (i.e., networks of entities) [9]. In this way, SNA can be an important complementary approach to ABI, but each focuses on different aspects of data and seeks a fundamentally different outcome, indicating that the two are not duplicative approaches. What ABI and SNA share, however, is an appreciation for the importance of understanding entities and relationships as a method for answering particular types of questions.

4.6 Analytical and Technological Implications

The consequence of ABI's broad approach to data ontology—focused on activity—is a means of understanding, analytically and technologically, how we deal with data. By considering the spatiotemporal discovery environment as a primary interface, and understanding the spatial and temporal aspects of events and transactions respectively, we can begin to design technological solutions to transform and apply the event-transaction structure to data. In some cases, this technology can be built into sensor data processing systems; in others, especially with repurposed data (discussed further in Chapter 9 concerning incidental collection), downstream data conditioning must take data outputs and convert data into the event-transaction ontology.

Relational and biographical information regarding entities is supremely important for contextualizing events and transactions, but unlike earlier approaches to analysis and traditional manhunting, focusing on specific entities from the outset is not the hallmark innovation of ABI. Rather, in ABI the analysis of the activity of many entities helps us unlock the keys to understanding and discovering information about important entities. This is true even if we were unaware the entities even existed; in other words, activity helps us unlock unknowns and potentially prevent analytic surprise.

4.7 Summary

With an understanding of ABI's approach to data, the next step is to understand ABI's unique approach to analysis. Chapter 5 will explore ABI's place in intelligence analytic methodologies, and Chapters 6–9 will outline and explain ABI's core set of concepts and definitions and their implications for the analyst and technologist.

References

[1] Ryan, S., "ABI Draft Lexicon Discussion," National Geospatial-Intelligence Agency, approved for public release, NGA Case #13-437, August 15, 2013.

[2] McAuliffe, M. S., "CIA Documents on the Cuban Missile Crisis (1962)," CIA History Staff, October 1992.

[3] de Montjoye, Y.A., et al., "Unique in the Crowd: The Privacy Bounds of Human Mobility," *Scientific Reports,* March 23, 2013.

[4] "The GDELT Project," web, available: http://gdeltproject.org/.

[5] Lorentz, A., "With Big Data, Context is a Big Issue," *Wired Innovation Insights,* April 23, 2013, web.

[6] "Mapping Customer Operations: Digital Products," Defense Logistics Agency, web, available: http://www.aviation.dla.mil/rmf/products_digital.htm.

[7] Phillips, M., "A Brief Overview of ABI and Human Domain Analytics." *Trajectory,* 28 Sep 2012, approved for public release, NGA Case #12-463.

[8] bin Laden, U., "Letter to Nasir al-Wuhayshi (English translation)," SOCOM-2012-0000016-HT, Combating Terrorism Center, United States Military Academy, 2012, web, available: https://www.ctc.usma.edu/posts/letter-to-nasir-al-wuhayshi-english-translation-2.

[9] Johnson, J., et al., "Social Network Analysis: A Systematic Approach for Investigating," *Law Enforcement Bulletin,* March 2013, Federal Bureau of Investigation. web, available: http://www.fbi.gov/stats-services/publications/law-enforcement-bulletin/2013/March/social-network-analysis.

5

Analytical Methods and ABI

Over the past five years, the intelligence community and the analytic corps have adopted the term ABI and ABI-like principles into their analytic workflows. While the methods have easily been adapted by those new to the field—especially those "digital natives" with strong analytic credentials from their everyday lives—traditionalists have been confused about the nature of this intelligence revolution. This chapter describes some of the fundamental methodological advances and a new framework for analysis, integration, and exploitation in a dynamically changing world.

5.1 Revisiting the Modern Intelligence Framework

Sherman Kent is one of a small group of individuals who has to date had a profound impact on the field of intelligence analysis. Along with men like Robert Gates, Douglas MacEachin, and Richard Heuer, Kent took the field of intelligence and argued—eloquently—for the concept of professionalization of analysis and the methods used therein [1]. While words such as tradecraft were and will continue to be used to describe ABI and other analytical methods, these esteemed observers of the intelligence analysis profession speak to the idea of a set of methods: consistent, repeatable, and, most importantly, teachable.

ABI shares in this intellectual heritage while contributing something fundamentally unique to the pantheon of analytic methods: the use of spatiotemporal correlations to disambiguate and identify entities of interest from large, incidentally collected data sets.

This poses a challenge to the current literature of intelligence. John Hollister Hedley, a long-serving CIA officer and editor of the President's Daily Brief (PDB) outlines three broad categories of intelligence: 1) strategic or "estimative" intelligence; 2) current intelligence, and 3) basic intelligence [2]. The difficulty with these three categories is that they, like most of the literature of intelligence analysis, are focused on the work conducted by all-source[1] intelligence analysts, primarily within the CIA's directorate of intelligence (DI) but also within the Defense Intelligence Agency (DIA) and military services. Comparably less has been written on the analytic methods of other, technically focused analytic occupations such as GEOINT and SIGINT because of the close coupling of the methodology to collection modalities. As a result, these agencies consider (with good reason) much of the tradecraft tied to "sources and methods of intelligence collection" to be itself classified.

Assessments or conclusions derived from ABI's methodology could very well be strategic in nature, but considering ABI's origins on the modern battlefields of Iraq and Afghanistan, strategic intelligence is insufficient to capture ABI. It is equally likely (or unlikely) to be current intelligence, as ABI's forensic and sequence-neutral approaches encourage analysts to consider links between data points from the past as well as present. ABI is also not necessarily predictive, lending itself poorly to warning, contained within current intelligence. Nor is it basic intelligence, laying the groundwork for other types of intelligence, though it indeed depends on basic intelligence as the other categories do. While Hedley writes of these categories in the context of finished intelligence, "finished" intelligence continues to be the frame around which much of today's intelligence literature is constructed.

From this brief exposition of intelligence categories, it is clear that our existing intelligence framework needs expansion to account for ABI and other methodologies sharing similar intellectual approaches. These approaches comprise a category labeled "discovery" as shown in Figure 5.1.

5.2 The Case for Discovery

In an increasingly data-driven world, the possibility of analytical methods that do not square with our existing categories of intelligence seems inevitable. The authors argue that ABI is the first of potentially many methods that belong in this category, which can be broadly labeled as "discovery," sitting equally alongside current intelligence, strategic intelligence, and basic intelligence [3].

What characterizes discovery? Most intelligence analysts, many of whom are naturally inquisitive, already conduct aspects of discovery instinctively as they go about their day-to-day jobs. But there has been a growing chorus of concerns from both the analytical community and IC leadership that intelligence

1. All-source intelligence is defined by the Department of Defense as "intelligence products and/or organizations and activities that incorporate all sources of information in the production of finished intelligence" [4].

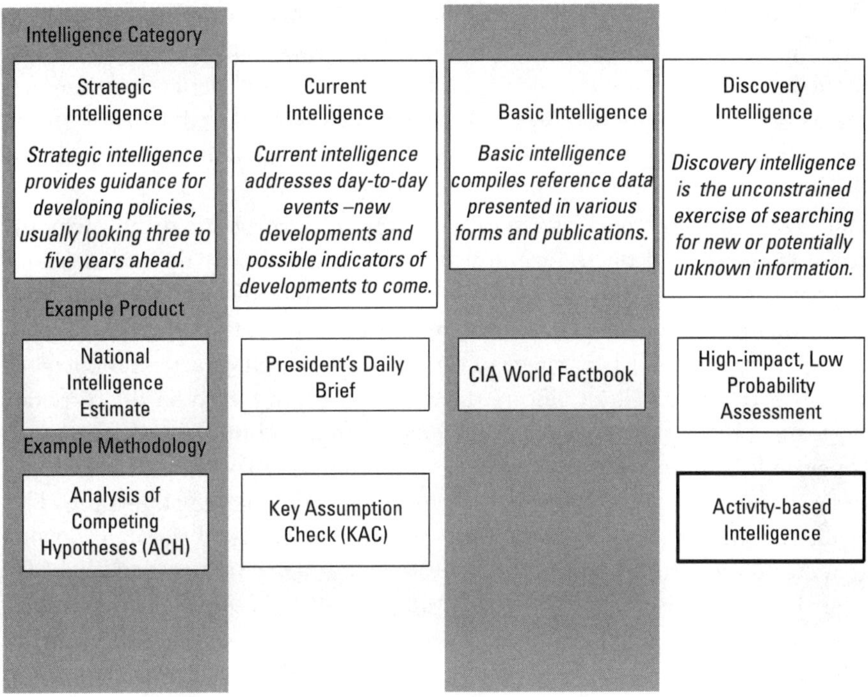

Figure 5.1 Four major categories of intelligence, including "discovery intelligence." (Adapted in part from Hedley in Johnson, pp. 126–127.)

production has become increasingly driven by specific tasks and questions posed by policymakers and warfighters. In part, this is understandable: If policymakers and warfighters are the two major customer sets served by intelligence agencies, then it is natural for these agencies to be responsive to the perceived or articulated needs of those customers. However, need responsiveness does not encompass the potential to identify correlations and issues previously unknown or poorly understood by consumers of intelligence production. This is where discovery comes in: *the category of intelligence primarily focused on identifying relevant and previously unknown potential information to provide decision advantage in the absence of specific requirements to do so.* Bruce and George observe:

> ...we believe there is still a notably thin professional literature on intelligence analysis. Part of this glaring absence is the result of management imperatives that are driven by current intelligence demands (as opposed to more in-depth research and less time pressure analysis) and do not permit sufficient time to reflect on the intelligence community's past performance or to record the lessons learned, from which subsequent intelligence analysts can benefit [5].

Even the title "production" implies an assembly-line process, unsuited to the complexities of the real world and foreshadowing a grave potential for failing to acknowledge, understand, or find "unknowns." Answering these requirements (which are growing faster and faster while timelines and demands shrink just as fast) can begin to crowd out the intelligence analyst's ability to simply discover new and potentially useful information.

Here there is a parallel between innovation in corporate firms, including technology-oriented ones. Technical personnel, with clearly-defined tasks, often have little to no time specifically set aside for innovation. While some leaders in the field specifically carve out time for innovation (Google, for example, allows its employees a percentage of time to work on self-directed products) all too often the potential for perceived "waste" or lack of return on the corporate investment of employee time can be enough to squelch institutional attempts to innovate [6]. In addition, institutional innovation often assumes (implicitly) a desire to innovate equally distributed across a given employee population. This egalitarian model of innovation, however, is belied by actual research showing that creativity is more concentrated in certain segments of the population [7].

If "discovery" in intelligence is similar to "innovation" in technology, one consequence is that the desire to perform—and success at performing—"discovery" is unequally distributed across the population of intelligence analysts, and that different analysts will want to (and be able to) spend different amounts of time on "discovery." Innovation is about finding new things based on a broad understanding of needs but lacking specific subtasks or requirements; no one asked Thomas Edison to invent the incandescent filament that made the electric light bulb possible. It is not the police detective working a crime, but more akin to using new bits of information gleaned from current investigations to add context and knowledge (and potentially solve) "cold" cases outside the scope of current responsibility.

ABI is one set of methods under the broad heading of discovery, but other methods—some familiar to the world of big data—also fit in the heading. ABI's focus on spatial and temporal correlation for entity resolution through disambiguation is a specific set of methods designed for the specific problem of human networks. The concept of discovery, however, as a category of intelligence, deserves a full exposition and the authors anticipate many other methods—some existing and some yet undiscovered—to join ABI under the heading of "discovery."

Beyond our four categories of intelligence, ABI's next challenge is its tenuous place between single-INT and all-source intelligence as well as between exploitation and "finished" intelligence. Identifying a place for multi-INT, problem-focused methodologies like ABI is a necessity in the modern world, and the ambiguity of current intelligence constructs forces either an identification of a new "type" of intelligence or an expansion of the existing types.

5.3 The Spectrum of "INTS" and Exploitation Versus Finished Intelligence

While ABI originated within the realm of GEOINT analysis, its inherently multi-INT approach makes it difficult to reconcile from either a methodological or policy standpoint with the traditional work of GEOINT analysts—or for that matter, any other kind of single-INT technical analysis. Joint Publication 1-02 defines GEOINT as "the exploitation and analysis of imagery and geospatial information to describe, assess, and visually depict physical features and geographically referenced activities on the Earth. Geospatial intelligence consists of imagery, imagery intelligence, and geospatial information" [3]. While debate continues about whether ABI belongs as a part of GEOINT, it is clear that ABI is distinct from "traditional" approaches to GEOINT. However, it also does not seem to fit well in Kent's world of DI-style analysis, where reports and "finished" products dominate all with an eye toward writing for the penultimate U.S. intelligence publication, the PDB, which has existed in one form or another since the establishment of CIA's forerunner the Central Intelligence Group in 1946 [8].

It is in between these approaches that ABI finds its home alongside other data-driven approaches to analysis. Traditionally, single-INT approaches to analysis focus on what intelligence analysts call "exploitation," a set of techniques used to extract maximum value from a single source or common set of sources of intelligence collection systems. These sources are often groups of similar technical collection systems: SIGINT; measurement and signatures intelligence (MASINT); and GEOINT.

The contrast between ABI and traditional exploitation approaches is apparent. Two of ABI's four pillars specifically reject principles of single-INT exploitation: integrate before exploitation, which seeks to minimize exploitation of individual domains and data sources, and data neutrality, which through georeferencing sets the various INTs on equal footing. In addition, data neutrality's application puts information gathered from open sources and social media up against information collected from clandestine and technical means. Rather than biasing analysis in favor of traditional sources of intelligence data, social media data is brought into the fold without establishing a separate exploitation workflow. One of the criticisms of the Director of National Intelligence (DNI) Open Source Center, and the creation of OSINT as another domain of intelligence, was that it effectively served to create another stovepipe within the intelligence world, a point made by former chairman of the National Intelligence Council Dr. John Gannon during congressional testimony in 2005 [9].

Yet the relationship between ABI and single-INT exploitation is symbiotic, not antagonistic. The individual domains have developed unique processing—conditioning—approaches that produce, at times, the kind of data that an ABI analyst needs. By instilling the principles of georeference to discover within the individual domain workflows, the requisite georeferenced, granular intelligence can be produced as a function of existing exploitation workflows while placing minimal burdens on single-INT analysts. ABI's successes came from partnering, not replacing, single-INT analysts in battlefield tactical operations centers (TOCs).

At the same time, the pillar of integrate before exploitation relieves some of the burden placed on individual domain exploitation personnel: By focusing on exploiting only small pieces of time-consuming data types, ABI helps make exploiters more effective, particularly in forensic exploitation approaches that occur well after the point of collection. (See Figure 5.2.)

Assessing the line between ABI and all-source analysis is not quite as easy at first glance. Data neutrality indicates that ABI, like all-source analysis, may, and, in fact, does, consider the potential of all sources of intelligence and non-intelligence data. This indicates a blurring of traditional lines, most often under the heading of "multi-INT." Yet the use of many different types of data is only part of ABI; it is important to remember here, as always, ABI's core goal of disambiguation and ultimately entity resolution from large data sets. This is and remains the key distinction between ABI and all-source analysis. The all-source analysis field is more typically (though not always) focused on higher-order judgments and adversary intentions; it effectively operates at a level of abstraction above both ABI and single-INT exploitation. This is most evident in approaches to strategic issues dealing with state actors; all-source analysis seeks to provide a comprehensive understanding of current issues enabling intelligent forecasting of future events, while ABI focuses on entity resolution through disambiguation (using identical methodological approaches found on the counterinsurgency/counterterrorism battlefield) relevant to the very same

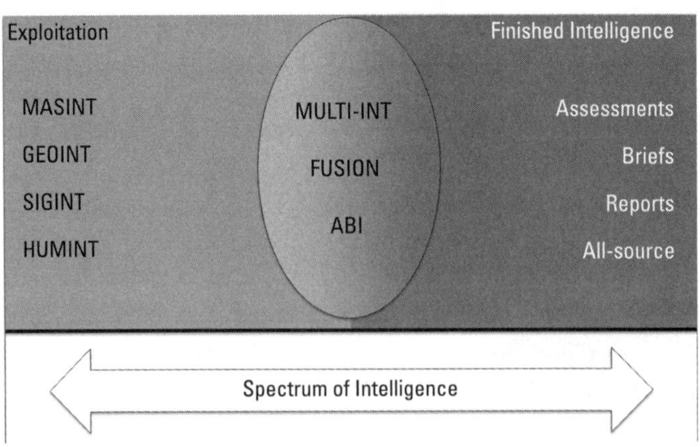

Figure 5.2 An example spectrum from exploitation to finished intelligence.

state actors. It is easy to envision the utility of understanding the precise activity of an adversary's leadership, particularly if that leadership is poorly understood or somewhat ambiguous in terms of composition or power structure. Here ABI can provide a distinct, data-driven edge over more anecdotal assessments based on human observation and assessment.

This difference in echelons, or levels of analysis, is where the most important contrast between ABI and all-source, "finished" intelligence is found. ABI is not finished intelligence, nor is it intended to be. Rather, ABI is a more investigative approach that identifies leads—spatiotemporal data correlations—for both single-INT and all-source analysts to consider. It informs both exploitation processes and all-source assessments without supplanting either with a series of unique core assumptions and methodological approaches centered on entity resolution through disambiguation. One of the most unique core assumptions of ABI is its search for members of human networks based on activity rather than perceived or actual hierarchies. This approach allows it to identify network members who may not be known to the entire network (making it particularly effective against compartmentalized or cell-based networks) without necessarily focusing on them as individuals. Rather than filling out a network diagram, ABI lets human activity "do the talking."

One of the most important strengths of ABI is its application in finding "unknowns." This term, though often overused, deals with the incomplete or fuzzy aspects of knowledge. With respect to networks, "unknowns" in practice often refers to unknown entities connected to known entities. ABI's two major methodological approaches, outlined in Section 5.5, provide two different but related approaches to conquering this problem. Before addressing these approaches, however, an intelligence problem must be decomposed to the point where ABI becomes a useful methodology.

5.4 Decomposing an Intelligence Problem for ABI

One of the critical aspects of properly applying ABI is about asking the "right" questions. In essence, the challenge is to decompose a high-level intelligence problem into a series of subproblems, often posed as questions, that can potentially be answered using ABI methods.

While ABI's roots in counterterrorism are well-documented, a more interesting question is to examine a simple, state actor–based problem and identify where ABI might lend unique insight. Heuer, in Chapter 7 of "Psychology of Intelligence Analysis," discusses structuring analytical problems and notes that problem decomposition is a key tool for all forms of analysis [10].

In this notional example (Figure 5.3), consider a near-peer competitor to the United States. As ABI focuses on disambiguation of entities, the problem

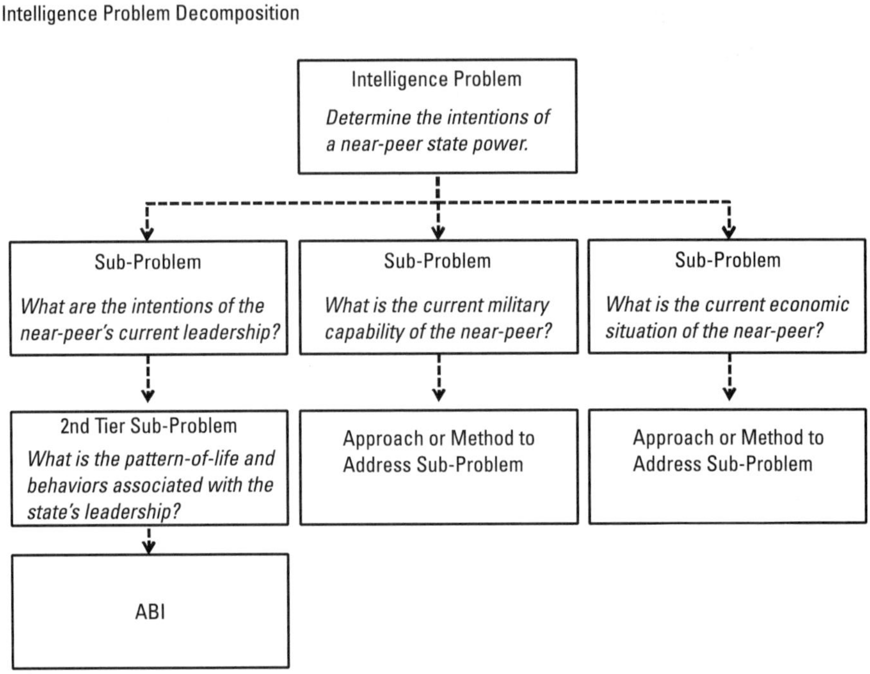

Figure 5.3 A notional decomposition of intentions of a near-peer state power.

must be decomposed to a level where disambiguation of particular entities helps fill intelligence gaps relating to the near-peer state power. As subproblems are identified, approaches or methods to address the specific subproblems are aligned to each subproblem in turn, creating an overall approach for tackling the larger intelligence problem. In this case, ABI does not become directly applicable to the overall intelligence problem until the subproblem specifically dealing with the pattern of life of a group of entities is extracted from the larger problem.

Note that ABI is used to address the second-tier subproblem of specific patterns of life associated with state leadership and that the problem has been decomposed two levels before ABI is directly applicable.

Another example problem to which ABI would be applicable is identifying unknown entities outside of formal leadership structures who may be key influencers outside of the given hierarchy through analyzing entities present at a location known to be associated with high-level leadership of the near-peer state.

Once the problem has been decomposed to the level where ABI is clearly applicable, understanding and applying the two major analytical methods resident within ABI is the next step. Collectively, these are referred to as the "W3" approaches.

5.5 The W3 Approaches: Locations Connected Through People and People Connected Through Locations

Once immersed in a multi-INT spatial data environment, there are two major approaches used in ABI to establish network knowledge and connect entities (and to various proxies, covered in detail in Chapter 6). These two approaches are summarized below, both dealing with connecting entities and locations. Together they are known as "W3" approaches, combining "who" and "where" to extend analyst knowledge of social and physical networks.

5.5.1 Relating Entities Through Common Locations

This approach focuses on connecting entities based on presence at common locations. Analysis begins with a known entity and then moves to identifying other entities present at the same location. Thus, the location potentially connects disparate entities. The process for evaluating strength of relationship based on locational proximity and type of location relies on the concepts of durability and discreteness, a concept further explored in Chapter 7. Colloquially, this process is known as "who-where-who," and it is primarily focused on building out logical networks. (See Figure 5.4.)

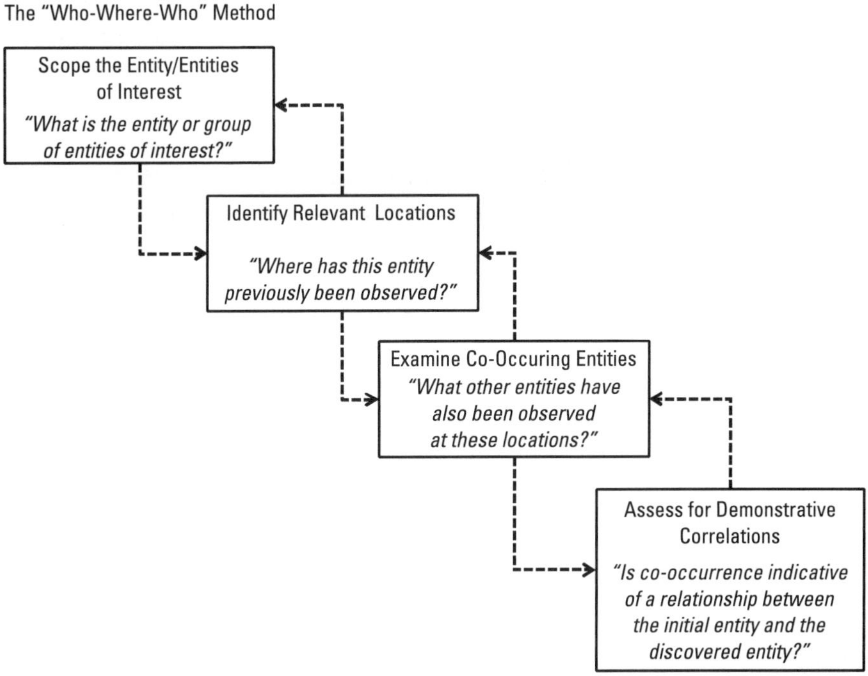

Figure 5.4 The sequential, iterative method and questions posed for "who-where-who."

A perfect example of building out logical networks through locations begins with two entities—people, unique individuals—observed at a private residence on multiple occasions. In a spatial data environment, the presence of two entities at the same location at multiple points in time might bear investigation into the various attributes of those entities. The research process initially might show no apparent connection between them, but by continuing to understand various aspects of the entities, the locational connection may be corroborated and "confirmed" via the respective attributes of the entities. This could take many forms, including common social contacts, family members, and employers.

The easiest way to step through "who-where-who" is through a series of four questions. These questions offer an analyst the ability to logically step through a potential relationship through the colocation of individual entities.

The first question is: "What is the entity or group of entities of interest?" This is often expressed as a simple "who" in shorthand, but the focus here is in identifying a specific entity or group of entities that are of interest to the analyst. Note that while ABI's origins are in counterterrorism and thus, the search for "hostile entities," the entities of interest could also be neutral or friendly entities, depending on what kind of organization the analyst is a part of.

These techniques have applications across a broad range of missions. In practice, this phase will consist of using known entities of interest and examining locations where the entities have been present. This process can often lead to constructing a full "pattern of life" for one or more specific entities, but it can also be as simple as identifying locations where entities were located on one or more specific occasions (N.B. the amount of detail required to establish an entity's pattern of life far exceeds a handful of specific observations). Of note, answering this question in the analytical process occurs primarily outside of the spatial-temporal environment and instead focuses on entities of interest, information (or attributes) of said entities, and relationships between entities that compose networks. As a result, effective tools for this stage can range from a whiteboard or piece of paper to more complex computer software that focuses on links between entities, such as i2 Analyst's Notebook [11].

The second question is: "Where has this entity been observed?" At this point, focus is on the spatial-temporal data environment. The goal here is to establish various locations where the entity was present along with as precise a time as possible. By focusing on specific locations of interest—as specific as a building on a city block or as general as a rural village—potential points of entity co-occurrence are established. The factors that aid in determining the validity of a co-occurrence will be discussed in depth in Chapter 7.

The third question is: "What other entities have also been observed at these locations?" This is perhaps the most important of the four questions in this process. Here, the goal is to identify entities co-occurring with the entity or entities of interest. The focus is on spatial co-occurrence, ideally over multiple locations. This intuitive point—more co-occurrences increases the likelihood of a true correlation—is present in the math used to describe a linear correlation function:

$$r = \frac{n\sum xy - (\sum x)(\sum y)}{\sqrt{n(\sum x^2) - (\sum x)^2}\sqrt{n(\sum y^2) - (\sum y)^2}}$$

Because the equation is designed for linear correlation in a two-dimensional graph, applying the concept behind the equation is the goal. In this equation, n is the number of pairs of data, and x and y are the two variables considered. With respect to co-occurrence between entities over multiple locations, as n—the number of locations where known entity x and unknown entity y co-occur—increases, overall value r draws closer to +1, indicating a correlation and thus a potential relationship between the known and unknown entities. Again, the characteristics of each location considered must be evaluated in order to separate out "chance co-occurrences" versus "demonstrative co-occurrences." In addition, referring back to the pillar of sequence neutrality, it is vitally important to consider the potential for co-occurrences that are temporally separated. This often occurs when networks of entities change their membership but use consistent locations for their activities, as is the case with many clubs and societies. Despite temporal separation, spatial co-occurrence can in fact be indicative of a potential relationship.

The fourth and final question is: "Is locational proximity indicative of some kind of relationship between the initial entity and the discovered entity?" Considering ABI's two primary data types—events and transactions—the goal is to evaluate a potential relationship between entities using two (or more) events occurring in close spatial or spatiotemporal proximity. Note that this is possible even when events are separated in time significantly, though as temporal distance increases between events, the possibility of the two events being fundamentally unconnected increases greatly. This concept is extremely intuitive. Consider the following example: Entity A is located at a private residence in 2007. Entity B is located at the same private residence in 2008. In order to determine whether the presence of entity A and entity B at the same location indicates the two are somehow connected, more information—about the use of the residence, the presence of other entities, and numerous other factors—must be evaluated as part of the analytical process.

The application of this technique varies from its sister technique. Here, the goal is to take an existing network of entities and identify additional entities that may have been partially known or completely unknown. The overwhelming majority of entities must interact with each other, particularly to achieve common goals, and this analytic technique helps identify entities that are related based on common locations before metadata or attribute-based explicit relationships. This technique can be applied even in areas that are "dense" with other entities because it narrows in on a specific entity or network before actively considering location. It helps grapple with potential noise posed by public areas such as shopping malls, markets, and dense urban environments.

5.5.2 Relating Locations Through Common Entities

This approach is the inverse of the previous approach and focuses on connecting locations based on the presence of common entities. By tracking entities to multiple locations, connections between locations can be revealed. This begins with a known location, identifies entities present at the known location, and then examines other locations where the same entities are or have been present. Colloquially, this process is known as "where-who-where," and it is primarily focused on building out physical networks. (See Figure 5.5.)

The process of relating locations through common entities also involves stepping through a sequence of four questions. In many ways, these questions mirror the questions of the previous process, simply substituting "locations" for entities. Where the previous process is focused on building out logical networks

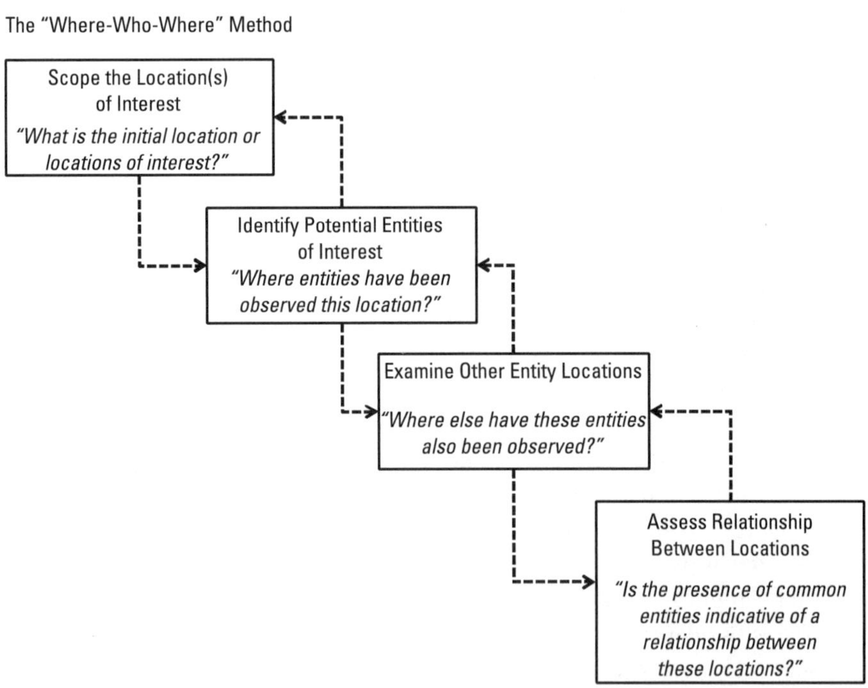

Figure 5.5 The sequential, iterative method and questions posed for "where-who-where."

where entities are the nodes, this process focuses on building out either logical or physical networks where locations are the nodes. While at first this can seem less relevant to a methodology focused on understanding networks of entities, understanding the physical network of locations helps indirectly reveal information about entities who use physical locations for various means (nefarious and nonnefarious alike).

The first question asked in this process is, "What is the initial location or locations of interest?" This is the most deceptively difficult question to answer, because it involves bounding the initial area of interest. Yet there is no hard and fast formula for the techniques to bound the area of interest. They are dependent upon data density as well as the type of entities of concern (despite starting with locations, entities remain of paramount relevance to this intellectual process). Population density (and therefore the data collected about this population) is an important consideration because it effectively limits the amount of information that a human analyst can consider.

Here, smartly applied machine correlation techniques and automated search tools can greatly increase the amount of data considered, but this sometimes comes at the expense of the most nuanced correlations.

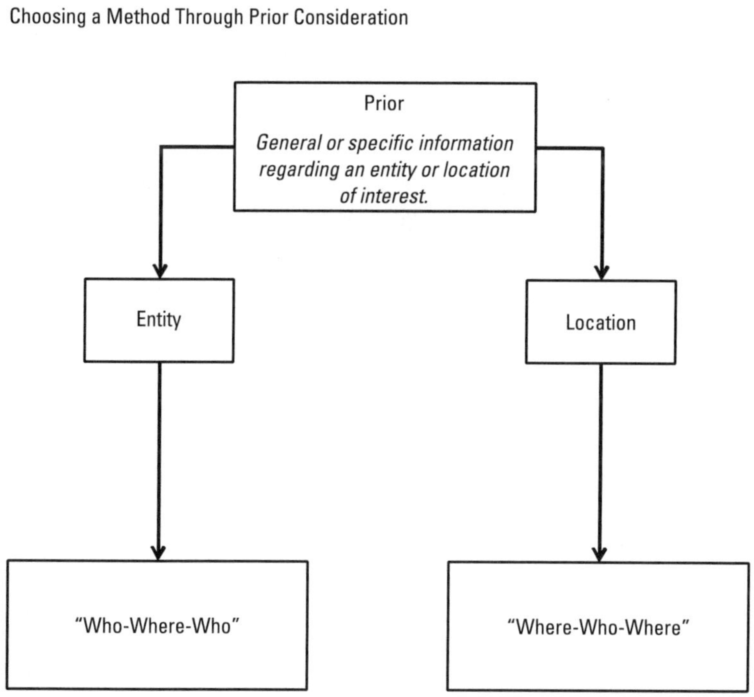

Figure 5.6 A notional workflow consisting of a prior, or piece of information, and then the applied workflows based on the presence of entities or locations within that prior.

Accordingly, a small rural village with mostly private residences may have roughly the same amount of data as a single high-rise apartment building in a busy downtown area, assuming the rate of collection across modalities has been generally consistent. In most cases, this process begins with a "prior"—a piece of information that occurs at a specific location, focusing attention and further analysis (see Figure 5.6). Priors can be explicit—such as a confidential informant's report of a drug distribution ring at a specific address—or implicit, such as a colleague's query about a particular marketplace outside of a city center. In both cases, data drives the prior's existence and sets the stage for "where-who-where."

The next question brings us back to entities: "What entities have been observed at this location?" Whether considering one or more locations, this is where specific entities can be identified or partially known entities can be identified for further research. This is one of the core differences between the two approaches, in that there is no explicit a priori assumption regarding the entities of interest. This question is where pure "entity discovery" occurs, as focusing on locations allows entities not discovered through traditional, relational searches to emerge as potentially relevant players in multiple networks of interest.

The third question is, "Where else have these entities been observed?" This is where a network of related locations is principally constructed. Based on the entities—or networks—discovered in the previous phase of research, the goal is now to associate additional, previously unknown locations based on common entities. In this step, portions of patterns of life are used and examined against contextual knowledge and other information. One of the principal uses of this information is to identify locations that share a common group of entities. In limited cases, this approach can be predictive, indicating locations that entities may be associated with even if they have not yet been observed at a given location.

The final question is thus, "Is the presence of common entities indicative of a relationship between locations?" The obvious problem of public locations (discussed further in Chapter 7) is only one of the difficulties that must be overcome in order to assess that locations are connected based on common entities. Here again, contextual data reigns supreme: Discovering correlation between entities and locations is only the first step, as subsequently contextual information must be examined dispassionately to support or refute the hypothesis suggested by entity commonality.

At this point, the assessment aspect of both methods must be discussed. By separating what is "known" to be true versus what is "believed" to be true, analysts can attempt to provide maximum value to intelligence customers. Maximizing the utility of said assessments is highly dependent on the ability to distinguish types of information from each other.

5.6 Assessments: What Is Known Versus What Is Believed

At the end of both methods is an assessment question: Has the process of moving from vast amounts of data to specific data about entities and locations provided correlations that demonstrate actual relationships between entities and/or locations? Correlation versus causation can quickly become a problem in the assessment phase, as well as the role of chance in spatial or temporal correlation of data. The assessment phase of each method is designed to help analysts separate out random chance from relevant relationships in the data.

ABI adapts new terminology from a classic problem of intelligence assessments, which is separating *fact* from *belief*. Particularly with assessments that rest on correlations present across several degrees of data, the potential for alternative explanations must always be considered. While the concepts themselves are common across intelligence methodologies, these are of paramount importance in properly understanding and assessing the "information" created through assessment of correlated data. Intelligence failures often occur when longstanding assessments are allowed to masquerade as facts, and the same problem occurs in ABI. Equally important in understanding assessments is understanding gaps—pieces of knowledge that are definitely unknown and if filled, would enhance overall understanding. The process of separating fact from belief is achieved through logical reasoning processes. The three most common that analysts use when making assessments are inductive, deductive, and abductive reasoning.

Induction represents perhaps the most common form of reasoning employed by analysts: deriving a general rule from specific observations. In the context of analysis, specific observations can be derived from any sources, and analysts attempt to identify broader conclusions from said information. This can be expressed as a pyramid: going from a "small" observation to a "larger" general rule.

Deduction represents the opposite process: Conclusions are reached from specific information informed by general rules. Most importantly, these conclusions are *guaranteed* by the premises on which they are based. As a result, the conclusions "must" be true as long as the premises are true.

Abduction, perhaps the least known in popular culture, represents the most relevant form of inferential reasoning for the ABI analyst. It is also the form of reasoning most commonly employed by Sir Arthur Conan Doyle's Sherlock Holmes, despite references to Holmes as the master of deduction. Abduction can be thought of as "inference to the best explanation," where rather than a conclusion guaranteed by the premises, the conclusion is expressed as a "best guess" based on background knowledge and specific observations. The differences between the three forms of reasoning, using similar information, are outlined in Figure 5.7. [12, 13]. All three forms of inferential reasoning, however, are informed by and must involve facts.

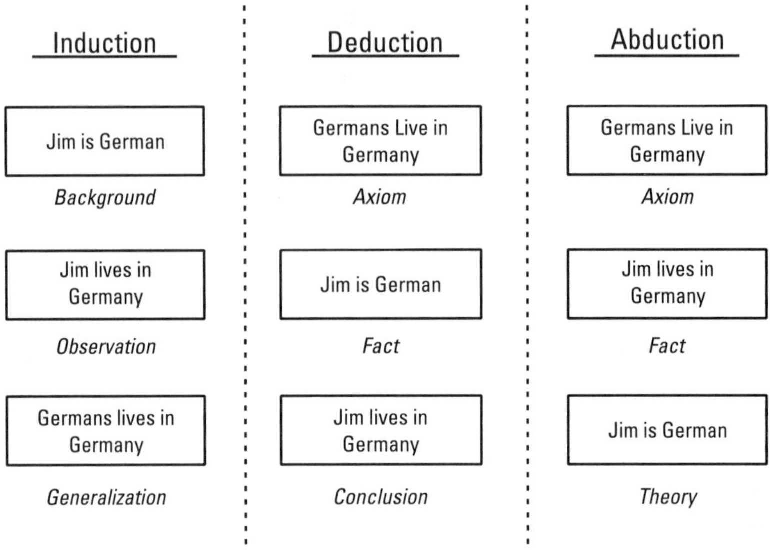

Figure 5.7 Three types of inferential reasoning. ABI analysts most commonly employ abduction to generate "best explanations" for spatial correlations.

5.7 Facts: What Is Known

First and most important is stating what is true. One might state, "I know person X is a member of criminal organization Y." This is a perfect example of assessment masquerading as fact. A good rule of thumb is the more general the statement, the more likely that it is an assessment instead of a fact. In order to identify facts, more granular pieces of data must be consulted.

A better example of a fact is "On 24 July at 2000L, sensor A made detection B." A fact is thus the minimum statement that must be true. Of course, allowance must be made for uncertainty even in the identification of facts; even narrowly scoped, facts can turn out to be untrue for a variety of reasons. Despite this tension, distinguishing between facts and assessments is a useful mental exercise. It also serves to introduce the concept of a key assumption check (KAC) into ABI, as what ABI terms "facts" overlaps some with what other intelligence methodologies term "assumptions."

Another useful way to conceptualize facts is "information as reported from the primary source." A confidential informant, for example, might state, "Last month, person A lived in location B, but location B is owned by person C." The informant may believe that both facts reported (person A living in location B, person C owning location B) are true, but one, both, or neither may be true due to no deliberate intent on the part of the informant. And as with technical error, human sources can introduce bad information into the intelligence process deliberately.

5.8 Assessments: What Is Believed or "Thought"

Assessment is where the specific becomes general. Assessment is one of the key functions performed by intelligence analysts, and it is one of very few common attributes across functions, echelons, and types of analysts. It is also not, strictly speaking, the sole province of ABI.

This may seem counterintuitive. Why mention assessment in a book on ABI if it is not exclusive to the method? The answer is that ABI, like other methods, is one of many ways of generating intelligence assessments. Many assessments are the product of multiple methods, sometimes including ABI, sometimes leaving it out, all depending on the type of data and type of intelligence problem at hand.

Rather than a bright line, the point at which ABI stops and assessment begins is rather blurry. ABI identifies correlated data based on spatial and temporal co-occurrence, but it does not explicitly seek to assign meaning to the correlation or place it in a larger context. That is where the "assessment" process takes the baton and contextualizes what ABI provides along with other sources of information. There are times, however, when the method cannot even reach assessment level due to "getting stuck" during research of spatial and temporal correlations. This is where the concept of "unfinished threads' becomes vitally important.

5.9 Gaps: What Is Unknown

The last piece of the assessment puzzle is "gaps." This is in many ways the inverse of "facts" and can inform assessments as much as "facts" can. Gaps, like facts, must be stated as narrowly and explicitly as possible in order to identify areas for further research or where the collection of additional data is required.

Gap identification is a crucial skill for most analytic methods because of natural human tendencies to either ignore contradictory information or construct narratives that explain incomplete or missing information. Figure 5.8 contrasts facts, assessments, and gaps with a simple case dealing with the relationship between Jim, Sandy, company Y, and project X.

5.10 Unfinished Threads

Every time a prior initiates one or both of the principal methods discussed earlier in this chapter, an investigation begins. In ABI, these are often referred to as "threads," tying together the pieces of information an analyst discovers relevant to a particular line of inquiry (regarding a given location, a given network, or at times both). True to its place in "discovery intelligence," ABI not only makes

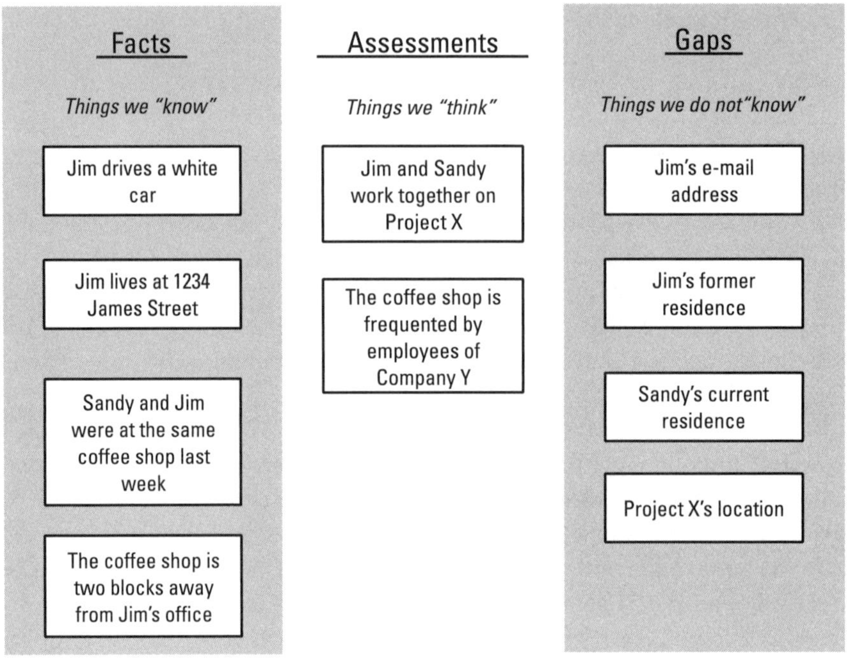

Figure 5.8 An example chart contrasting facts, assessments, and gaps using a simple case regarding the relationship between two individuals, a project, and a company.

allowances for the existence of these unfinished threads, it explicitly generates techniques to address these threads and uses them for the maximum benefit of the analytical process.

Unfinished threads are important for several reasons. First, they represent the institutionalization of the discovery process within ABI. Rather than force a process by which a finished product must be generated, ABI allows for the analyst to pause and even walk away from a particular line of inquiry for any number of reasons. Second, unfinished threads can at times lead an analyst into parallel or even completely unrelated threads that are as important, or even more important, than the initial thread. This process, called "thread hopping," is one expression of a nonlinear workflow inside of ABI. An example of this would be searching for a particular type of entity network at a group of locations, but finding a completely different network in answer to a different kind

One of the most challenging problems presented by unfinished threads is preserving threads for later investigation. Methods for doing so are both technical (computer software designed to preserve these threads, discussed further in Chapter 15) and nontechnical, such as scrap paper, whiteboards, and pen-and-paper notebooks. This is particularly important when new information arrives, especially when the investigating analyst did not specifically request the new information. As discussed in earlier in Chapter 4, and in detail in Chapter 9, the concept of repurposing information from many sources collected for multiple varying reasons is of paramount importance to ABI.

The concept of multiple threads pursued both sequentially and simultaneously is also important to place both unsuccessful threads and successful threads in the proper context. Entities cannot always be resolved, and not every data point leads to a groundbreaking intelligence "find." Nonetheless, the act of working through individual threads helps train individuals and teaches them to explore and work with real data on a daily basis, which is of critical importance. By maintaining a discovery mindset and continuing to explore threads from various different sources of information, the full power of ABI—combined with the art and intuition present in the best analysts—can be realized.

5.11 Leaving Room for Art And Intuition

One of the hardest challenges for structured approaches to intelligence analysis is carving out a place for human intuition and, indeed, a bit of artistry. The difficulty of describing and near impossibility of teaching intuition make it tempting to omit it from any discussion of analytic methods in an effort to focus on that which is teachable. To do so, however, would be both unrealistic as well as a disservice to the critical role that intuition—properly understood and subject to appropriate —can play in the analytic process.

The importance of intuition extends far beyond the field of intelligence analysis. Judge Richard Posner, in his analysis of the thought processes of appellate judges, notes:

> A hunch sounds like a guess, a shot in the dark, and there is that element in judging. But 'hunch' is a misleading as well as a belittling description of interpretation and appellate review. Both are areas where intuition reigns, but not in the form of guesswork. Interpretation is an innate, universal, and quintessentially intuitive human faculty. It is field-specific, in the sense that one's being good at interpreting, say, faces or pictures or modern poetry does not guarantee success at interpreting contracts or statues. It is not a rule-bound activity, and the reason a judge is likely to be a better interpreter of a statute than of a poem, and a literary critic a better interpreter of a poem than a statute, is the experience creates a repository of buried knowledge on which intuition can draw when one is faced with a new interpretandum [14].

Posner explicitly identifies experience as creating a "buried repository of knowledge" that intuition draws upon, a construct congruent with decision-making expert Daniel Kahneman's body of research. In his 2011 work

Thinking, Fast and Slow, he describes the interplay between two different modes of thinking: "System 1, which operates automatically and quickly, with little to no effort and no sense of voluntary control," and "System 2, [which] allocates attention to the effortful mental activities that demand it, including complex computations" [15, p. 20–21]. Kahneman notes that his particular exposure to decision-making has made him a skeptic with regard to so-called expert intuition, but at multiple points in the book he notes the potential for rapid interplay between System 1 and System 2 thinking, particularly with regard to his collaboration with leading naturalistic decision making proponent Gary Klein. "The situation has provided a cue," Kahneman writes, "This cue has given the expert access to information stored in memory, and the information provides the answer. Intuition is nothing more and nothing less than recognition" [15, p. 237].

Kahneman goes on to describe in detail conditions under which intuition can be useful, and through his collaboration with Klein settles on two basic principles in what they refer to as the "environment of skill":

- An environment that is sufficiently regular to be predictable;
- An opportunity to learn those regularities through prolonged practice [15, p. 240].

It is easy to see the problem with Kahneman's two principles and the environment faced by intelligence analysts, particularly those attempting the delicate business of long-range forecasting. (This problem is noted directly in Frank Babetski's review of Kahneman for the Center for the Study of Intelligence [16].) While many experienced analysts have achieved the second principle, the first—that an environment be sufficiently regular and therefore predictable—demolishes the notion of intuitive forecasts drawing upon a body of experience built from regular exposure and, more importantly, feedback.

This same weakness, however, does not apply in quite the same way to ABI's methods. Because ABI functions at the subproblem level (and sometimes even addresses second- or third-tier subproblems in intelligence), the environment in many ways becomes more regular. Whether it is sufficiently regular so as to be predictable is a matter for debate elsewhere, but the limitations applied by the physical world vice the world of intentions give intuition about co-occurrences and correlations an important place in ABI. At all times, however, these intuitions must be subject to rigorous scrutiny and cross-checking, to ensure their validity is supported by evidence and that alternative or "chance" explanations cannot also account for the spatial or temporal connections in data.

Fundamentally, there is a role for structured thinking about problems, application of documented techniques, and artistry and intuition when examining correlations in spatial and temporal data. Practice in these techniques and practical application that builds experience are both equally valuable in developing the skills of an ABI practitioner. With this general knowledge of analytical principles, it is now time to delve into the set of specific concepts used in ABI: proxies, entities, durability and discreteness, pattern of life, and incidental collection. Each one of these will be discussed in turn in Chapters 6–9.

References

[1] Kent, S., "The Need for an Intelligence Literature," in *Sherman Kent and the Board of National Estimates: Collected Essays* (D. Steury, ed.), Center for the Study of Intelligence, Washington, DC, University of Michigan Press, 1994, p.15–16.

[2] Hedley, J. H., "The Challenges of Intelligence Analysis," in *Strategic Intelligence: Understanding the Hidden Side of Government* (L. Johnson, ed.), vol. 1, 2006, p. 126-127.

[3] Kimminau, J., Analysis Mission Technical Advisor, Deputy Chief of Staff (ISR), U.S. Air Force, remarks at USGIF/ATIA ABI Forum, 24 July 2014.

[4] "Joint Publication 1-02: Department of Defense Dictionary of Military and Associated Terms," U.S. Department of Defense, 14 July 2014, web.

[5] Bruce, J. B., and J. B. George, "Intelligence Analysis—The Emergence of a Discipline," in *Analyzing Intelligence: Origins, Obstacles, and Innovations,* Washington, DC: Georgetown University Press, 2008, p. 5.

[6] Mims, C., "20% Time is Officially Alive and Well, says Google," *Quartz,* 21 August 2013, web.

[7] Martinsen, Ø. L., "The Creative Personality: A Synthesis and Development of the Creative Person Profile," *Creativity Research Journal,* Vol. 23, No. 3,, 2011, pp. 185–202.

[8] "A Look Back: The President's First Daily Brief," Central Intelligence Agency: News and Information, 6 February 2008, web, available: https://www.cia.gov/news-information/featured-story-archive/2008-featured-story-archive/the-presidents-first-daily-brief.html.

[9] Gannon, J., Statement for the Record, Subcommittee on Intelligence, Information Sharing, and Terrorism Risk Assessment, Committee on Homeland Security, U.S. House of Representatives. 109th Congress, First Session, 21 June 2005, web.

[10] Heuer, R. J., *The Psychology of Intelligence Analysis,* Center for the Study of Intelligence, 1999. pp. 84–97.

[11] "Data analysis—i2 Analyst's Notebook," IBM, web.

[12] Patokorpi, E., "Logic of Sherlock Holmes in Technology Enhanced Learning," *Educational Technology & Society,* Vol. 10, No. 1, 2007, pp. 171–185.

[13] Gust, H., and K. Kuhnberger, "Computational Logic and Cognitive Science: An Overview," presented at the *ICCL Summer School 2008,* Technical University of Dresden, 25 August 2008.

[14] Posner, R., *How Judges Think,* Cambridge, MA: Harvard University Press, 2008, p. 113.

[15] Kahneman, D., *Thinking, Fast and Slow,* New York: Farrar, Straus, and Giroux, 2011, pp. 20–21.

[16] Babetski, F. J., "Intelligence in Public Literature: *Thinking Fast and Slow* (book review)," *Studies In Intelligence,* Vol. 56, No. 2, July 2012, web.

6

Disambiguation and Entity Resolution

Previous chapters introduced the pillars of ABI and explained their importance for discovery, shifting the intelligence process from a linear one that monitors knowns to a more dynamic process that resolves unknowns. One of the most significant unknowns is often the identity of entities conducting activities. Entity resolution or disambiguation through multi-INT correlation is a primary function of ABI. Entities and their activities, however, are rarely directly observable across multiple phenomenologies. Thus, we need an approach that considers proxies—indirect representations of entities—which are often directly observable through various means.

6.1 A World of Proxies

Consider for a moment today's average business card. On it, one finds a remarkable amount of information about an individual: a name, for certain; likely a job title; perhaps an address or place of business; an e-mail address, most definitely; one or more phone numbers, probably. If the individual in question is an entity, the pieces of information describe the entity's attributes (Figure 6.1).

As entities are a central focus of ABI, all of an entity's attributes are potentially relevant to the analytical process. That said, a subset of attributes called proxies is the focus of analysis as described in Chapter 5. A *proxy* "is an observable identifier used as a substitute for an entity, limited by its durability (i.e., influenced by the entity's ability to change/alter proxies)" [1]. Based on this definition, a simple rule of thumb for distinguishing proxies from other types of attributes is that *proxies can be used to disambiguate entities*—essentially, proxies help distinguish one entity from another. Not all attributes are proxies; for

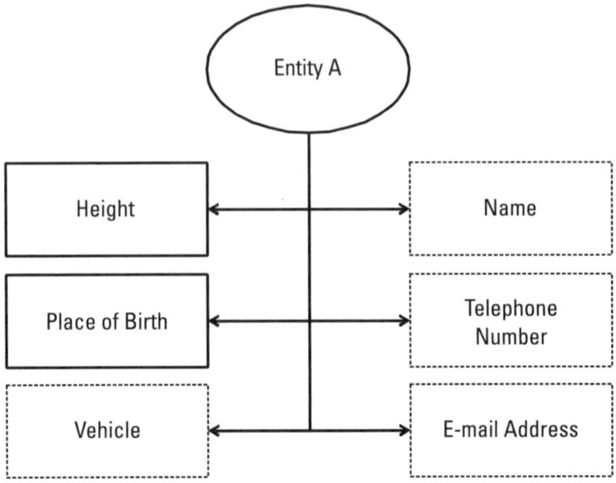

Figure 6.1 Attributes of a notional entity. The dotted boxes represent some of the entity's proxies.

instance, an entity's height (5'11," for example) is certainly an attribute, but not a proxy by itself.

Why do proxies matter? *In the epistemology of ABI, entities cannot be directly observed.* From this statement, it naturally follows that any observations made are actually observations of one or more types of proxies for an entity. This statement seems unnecessarily abstract at first. Naturally, if people are entities, and one person watches another, this is a direct observation of an entity.[1]

Communication technology, for instance, is, in essence, communication between entities, but the communication is routed to the proper entity—from sender to receiver—based on the relevant proxy. An e-mail address is a common example in everyday life. Whether a person's name—John.Smith@EmailService.com—or more generic/descriptive "handle"—ScubaDiver32394@EmailService.net—the unique combination of letters and numbers ensures that e-mails arrive where they are supposed to. Chapter 4 defines this type of activity as a transaction—the exchange of information between entities. Also noted in Chapter 4 is the fact that the exchange is based on the observation of proxies, rather than direct observation of entities.

Focusing on any particular, "average" entity results in a manageable number of proxies.[2] However, beginning with a given entity is fundamentally a problem of "knowns." How can an analyst identify an "unknown entity?" Now the problem becomes more acute. Without using a given entity to filter

1. Direct observation like that in this example is a special case not generally discussed within the bounds of ABI because it is outside the scope of the methodology: There is no spatial or temporal correlation, and ABI focuses on activity at a more general level of analysis
2. Though various regional and cultural factors can alter the number of proxies associated with a given entity.

potential proxies, all proxies must be considered; this number is likely very large and for the purposes of this chapter is n. The challenge that ABI's spatiotemporal methodology confronts is going from n, or all proxies, to a subset of n that relates to an individual or group of individuals. In some cases, n can be as limited as a single proxy. The process of moving from n to the subset of n is called disambiguation.

6.2 Disambiguation

Disambiguation is not a concept unique to ABI. Indeed, it is something most people do every day in a variety of settings, for a variety of different reasons. A simple example of disambiguation is using facial features to disambiguate between two different people. This basic staple of human interaction is so important that an inability to do so is a named disorder—prosopagnosia [2].

In ABI, the focus is on spatiotemporal disambiguation, that is, disambiguating between proxies of entities using spatial and temporally referenced information. This approach is what makes ABI a unique analytic methodology. The spatiotemporal methods inherent to ABI also cannot work without disambiguation based on attributes or proxies. As the population density of Earth continues to increase, there are more and more entities concentrated in the same places. The W3 approaches from Chapter 5—"who-where-who" and "where-who-where"—both use spatiotemporal disambiguation (the hallmark method of ABI) in the "where" phases to identify potentially relevant entities and increase the overall understanding of entity networks.

Disambiguation is a conceptually simple process; accordingly, the actual process of disambiguation is severely complicated by incomplete, erroneous, misleading, or insufficiently specific data. Moreover, certain types of proxies for individuals are less specific, or unique to a single individual. How specific a proxy is for a given individual can also vary based on a number of factors specifically related to the proxy as well as other, external factors.

Figure 6.2 shows a representative spectrum of proxies, with the best or most "unique" on the right side of the chart.

With some proxies, certain "values" are more useful for disambiguation than others. Names are a good example of this concept. The name John Smith (in the United States) does take one from n entities to a subset of n, but because of a large number of entities sharing the name "John Smith," the overall utility is diminished. When compared to "Stanley James Xavier Stellenbosch," the increased utility of the latter name-as-proxy is clear.

In other cases, the "format" of certain proxies can lend more naturally to disambiguation through uniqueness. An easy example is a telephone number, which uses a unique combination of numbers to identify endpoints. By adding

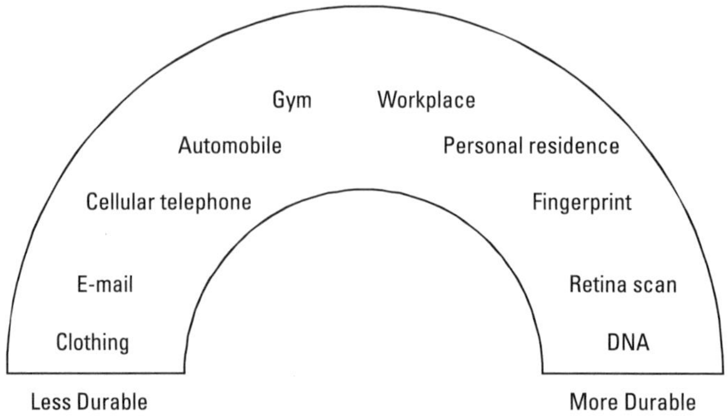

Figure 6.2 Proxies for an individual entity arrayed on a spectrum of relative durability.

digits, the number of possible combinations increases, and adding prefixes or subdividing existing prefixes also increases the number of unique combinations. In the United States, area codes are used to denote a physical region of telephone numbers. As the potential seven-digit telephone numbers are assigned within a given area code and the possibility of "running out" of numbers emerges, area codes are often split, resulting in the creation of a new area code and a new group of unique proxies.

In today's digital world, many aspects associated with our electronic life fall into this category: proxies that serve to more uniquely identify entities of interest, such as e-mail addresses, telephone numbers, MAC addresses, IP addresses, and radio frequency identification (RFID) chips. In Figure 6.3, two telephone numbers represent proxies for two entities—"A" and "B." This shows how transactions are actually indirect observations of entities, made through observation of activity of a given proxy (in this case, telephone calls).

As more and more discussion ensues of the "Internet of Things," wearable technology, and an interconnected ecosystem of electronics, the need for an analytic method focused on taking advantage of this type of data is clearly evident (see Chapter 27). This technological trend line has run in parallel with ABI's emergence and development. Without discounting the utility of more "general" proxies like appearance and clothing and vehicle types, it is the "unique" identifiers that offer the most probative value in the process of disambiguation and that, ultimately, are most useful in achieving the ultimate goal: entity resolution.

6.3 Unique Identifiers—"Better" Proxies

To understand fully why unique identifiers are of such importance to the analytical process in ABI, a concept extending the data ontology of "events"

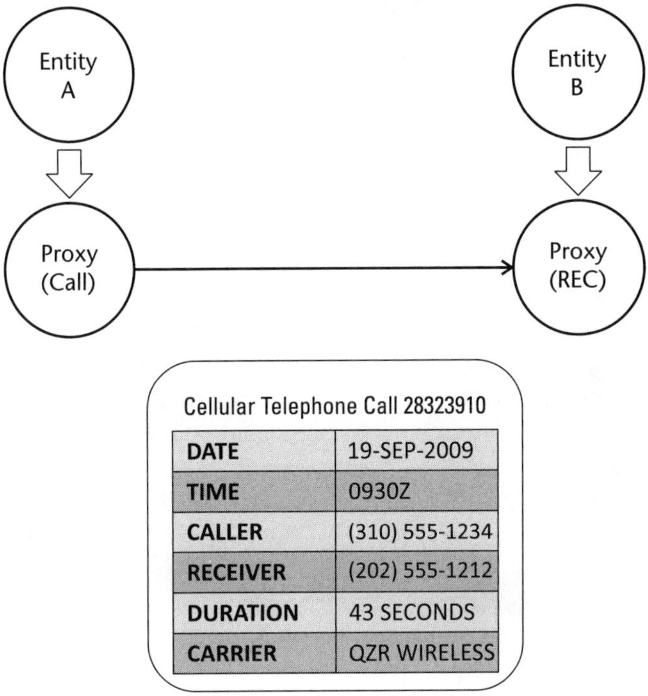

DATE	19-SEP-2009
TIME	0930Z
CALLER	(310) 555-1234
RECEIVER	(202) 555-1212
DURATION	43 SECONDS
CARRIER	QZR WIRELESS

Figure 6.3 Proxies (representing entities) form the logical endpoints of all types of transactions.

and "transactions" from Chapter 4 must be introduced. This concept is called certainty of identity. When two different "name" proxies were introduced in Section 6.2—"John Smith" and "Stanley James Xavier Stellenbosch"—the importance of disambiguating among instances of like classes of proxies became immediately obvious. Another intuitive example comes from the law enforcement world. When a police officer submits a "be on the look-out" (BOLO) report, a vehicle description of a suspect is often given. If the report is of a red, late-model four-door sedan, it immediately creates a disambiguation problem: Which of the potentially thousands of sedans on the road is the correct sedan? With a license plate number and state—a type of unique identifier—the disambiguation process becomes far easier. Consider, for example, the widespread use of the AMBER alert system, sending text message notifications with unique identifiers (license plates) for vehicles involved in suspected child abductions.

This concept has a direct analog in the computing world—the universal unique identifier (UUID) or globally unique identifier (GUID) [3, 4]. In distributed computing environments—linking together disparate databases—UUIDs or GUIDs are the mechanism to disambiguate objects in the computing environments [4]. This is done against the backdrop of massive data stores from various different sources in the computing and database world.

In ABI, the same concept is applied to the "world's" spatiotemporal data store: Space and time provide the functions to associate unique identifiers (proxies) with each other and with entities. The proxies can then be used to identify the same entity across multiple data sources, allowing for a highly accurate understanding of an entity's movement and therefore behavior. There are, however, enduring challenges to ABI's methodology of associating unique identifiers with entities known to be present in a given area at a given time; the most acute of these is a relative lack of spatial accuracy for unique identifiers.

An example of this is the error inherent in Internet protocol (IP)-based geolocation. An IP provides one of two ends of a transaction designed to send pieces of information from one connected computer to another (though the computer can also be a phone or tablet, among other possibilities). This protocol was specifically designed to facilitate networked communication for the Defense Advanced Research Projects Agency (DARPA) [5]. In ABI, we would characterize this as a transaction—conducted between two entities (the users of the computers)—observed via proxies (IP addresses).

But where, spatially, are the entities? IP-based geolocation struggles to provide an exact answer. In fact, in 2011, an enhanced technique for IP-based geolocation boasted of the ability to reduce error to less than 1 km. The technique, however, does not work in all instances [6]. Even in an area less than 1 km, the number of potential entities to connect the IP address to—the proxy—remains prohibitive for interrogation from either automated or manual testing, particularly as multiple observations over space and time are required to confirm or deny a potential "match." In order to get to a useful pairing, the second concept of certainty of presence must be addressed.

When examining information sources that provide either identity or presence, it becomes evident that very few data sources are capable of providing both types of data. This becomes especially true with noncooperative data collection, which is the most common data received by intelligence officers. Amazon.com and Target's big data successes, while impressive, are facilitated by the fact that data is collected cooperatively, and the end-to-end collection mechanisms are wholly owned by their respective companies [7]. Such is not true with intelligence agencies, where adversaries actively try to deny or mask relevant data and collection is often incomplete.

Moving from proxies into the process of entity resolution requires some degree of disambiguation prior to associating proxies with entities. Whether performed by a human being or a machine algorithm, however, this process is anything but straightforward. Section 6.4 begins to explore the complexity (and art) of resolving entities.

6.4 Resolving the Entity

As the core of ABI's analytic methodology revolves around discovering entities through spatial and temporal correlations in large data sets across multiple INTs, the process of entity resolution principally through spatial and temporal attributes is the defining attribute of ABI's analytical methodology and represents ABI's enduring contribution to the overall discipline of intelligence analysis.

Entity resolution is "the iterative and additive process of uniquely identifying and characterizing an [entity], known or unknown, through the process of correlating event/transaction data generated by proxies to the [entity]" [1, slide 4].

Entity resolution itself is not unique to ABI. Data mining and database efforts in computer science focus intense amounts of effort on entity resolution. These efforts are known by a number of different terms (e.g., record linkage, de-duplication, and co-reference resolution), but all focus on "the problem of extracting, matching, and resolving entity mentions in structured and unstructured data" [8]. In ABI, "entity mentions" are contained within activity data. This encompasses both events and transactions, as both can involve a specific detection of a proxy. As shown in Figure 6.4, transactions always involve proxies at the endpoints, or "vertices" of the transaction. Events also provide proxies, but these can range from general (example, a georeferenced report stating that a particular house is an entity's residence) to highly specific (a time-stamped detection of a radio-frequency identification tag at a given location).

6.5 Two Basic Types of Entity Resolution

Ultimately, the process of entity resolution can be broken into two categories: proxy-to-entity resolution and proxy-to-proxy resolution. Both types have spe-

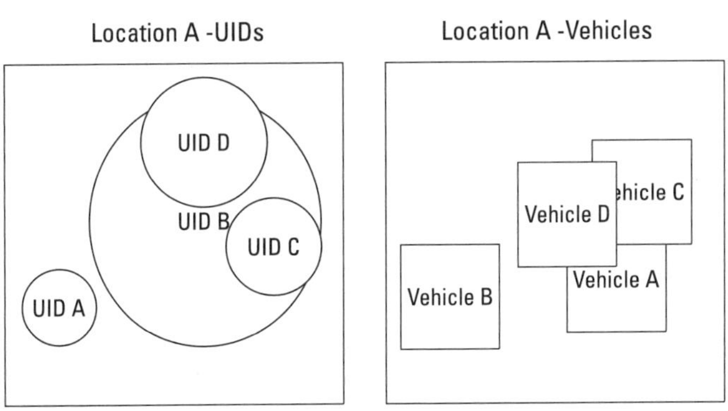

Figure 6.4 UIDs and vehicles (two types of proxies) in a bounded area at time Z.

cific use cases in ABI and can provide valuable information pertinent to an entity of interest, ultimately helping answer intelligence questions.

6.5.1 Proxy-to-Proxy Resolution

Proxy-to-proxy resolution through spatiotemporal correlation is not just an important aspect of ABI; it is one of the defining concepts of ABI. But why is this? At face value, entity resolution is ultimately the goal of ABI. Therefore, how does resolving one proxy to another proxy help advance understanding of an entity and relate it to its relevant proxies?

The answer is found at the beginning of this chapter: entities cannot be directly observed. Therefore, any kind of resolution must by definition be relating one proxy to another proxy, through space and time and across multiple domains of information. Figure 6.4 illustrates an example problem of proxy-to-proxy resolution: Four UIDs are present in the search area, and four vehicles are also present in the search area; both "snapshots" were taken at exactly the same time.

Proxy-to-proxy resolution is the attempt to identify which UID belongs with which vehicle. (For the sake of this exercise, it is assumed that each vehicle has exactly one UID. In the real world, some vehicles might have more than one while other vehicles might have none.) At first glance, this problem appears to be somewhat easy. UID "A" and vehicle "B" are quickly assessed to be the same based on their spatial positioning at the same time. Beyond this, however, difficulty emerges. Vehicles "A," "C," and "D" are all in very close proximity, and UIDs "B," "C," and "D" are also in the same rough cluster. Adding complexity is that the collection modality for UIDs seems to introduce some error (as is common with many types of technical data collection in intelligence, law enforcement, and even commercial applications): UID "B's" possible location completely contains all three vehicles, while UIDs "C" and "D" have smaller areas that still overlap more than one potential vehicle.

What the various sizes of circles introduce is the concept of CEP (Figure 6.5). CEP was originally introduced as a measure of accuracy in ballistics, representing the radius of the circle within which 50% of "rounds" or "warheads" were expected to fall. A smaller CEP indicated a more accurate weapon system. This concept has been expanded to represent the accuracy of geolocation of any item (not just a shell or round from a weapons system), particularly with the proliferation of GPS-based locations [9]. Even systems such as GPS, which are designed to provide accurate geolocations, have some degree of error.

Given that CEP is a reality of many kinds of technical collection, what mechanisms can an ABI analyst use to disambiguate and ultimately resolve the UID proxy to the vehicle proxy? With only a single point in time, as in Figure 6.4, complete disambiguation is unlikely, even impossible. With a single

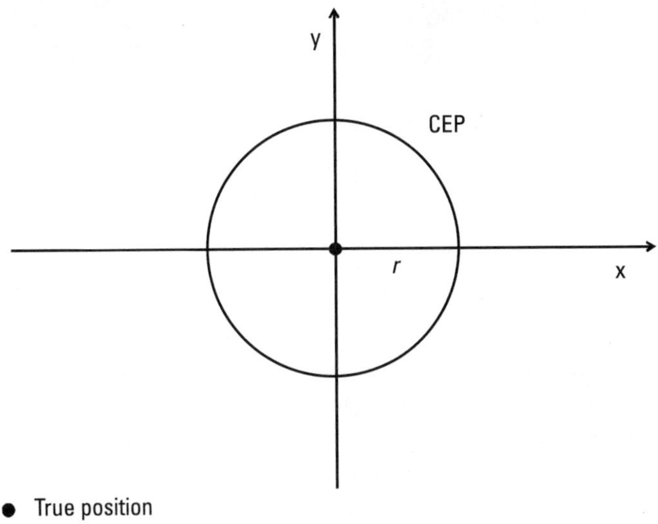

Figure 6.5 A circle with radius r, representing the CEP value of a measurement from true position located at (0,0) in a given coordinate plane.

point in time UID "A" was correlated to vehicle "B," but the validity of this correlation bears questioning. The solution to confirming this correlation, and identifying the proper vehicle-UID pairs for the remaining three items, rests in multiple observations over space and time. This can be at the same location or different locations depending on the behavior of the relevant proxies (and their underlying entities).

Figure 6.6 shows a second observation at a new location of the same four UIDs and vehicles. By examining the spatial characteristics of the UIDs and the vehicles, the correlation between UID "A" and vehicle "B" now gains an additional data point, adding strength to the potential correlation. By comparing

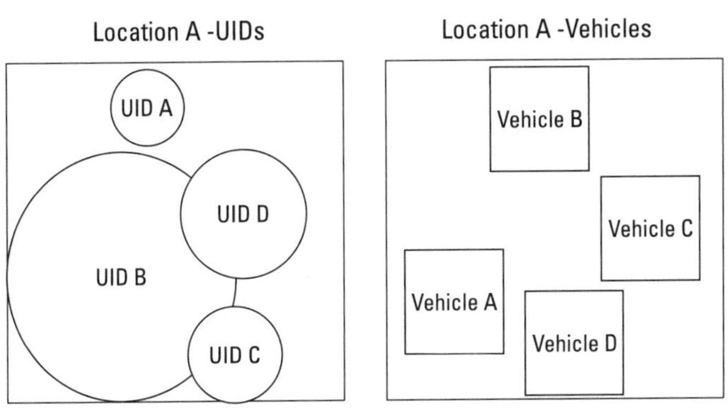

Figure 6.6 A second observation at time Z + 1 at a new location.

the overlap in the other UIDs and vehicle locations, tentative correlations between UID "B" and vehicle "A," UID "C" and vehicle "D", and UID "D" and vehicle "C" can be made.

This simple example illustrates the power of multiple observations over space and time for proper disambiguation and resolving proxies from one data source to proxies from another data source. This was a simplistic thought experiment. The bounds were clearly defined, and there was a 1:1 ratio of Vehicles:Unique Identifiers, both of which were of a known quantity (four each). Real-world conditions and problems will rarely present such clean results for an analyst or for a computer algorithm. The methods and techniques for entity disambiguation over space and time have been extensively researched over the past 30 years by the multisensor data fusion community. Some of these techniques are discussed in Chapter 15.

This process of proxy-to-proxy resolution is most closely analogous to "entity resolution" in the more traditional context of databases. John Talburt offers that "[entity resolution] is the process of determining whether two references to real-world objects are referring to the same, or to different, objects" [10]. This definition has both elements of disambiguation and proxy-to-proxy resolution: emphasis is on "two references to real-world objects." In the context of ABI, Talburt's references are proxies, and the latter half of his definition covers the use of multiple spatiotemporal observations to disambiguate proxies from one another and ultimately discover their associated entities.

6.5.2 Proxy-to-Entity Resolution: Indexing

While proxy-to-proxy resolution is at the heart of ABI, the importance of proxy-to-entity resolution, or indexing, cannot be overstated. Indexing is a broad term used for various processes, most outside the strict bounds of ABI, that help link proxies to entities through a variety of technical and nontechnical means. Indexing takes placed based on values within single information sources (rather than across them) and is often done in the process of focused exploitation on a single source or type of data.

An easy example of an index, or proxy-to-entity key, is a database of vehicle license plate registration and ownership information. Most states in the United States maintain such a database through their respective motor vehicle bureaus, and these are commonly used by law enforcement officers during traffic stops (see Chapter 23). In law enforcement, license plate information (a proxy) is used to retrieve information about the owner of the vehicle. Such indexes, however, are not perfect: Despite the fact that a vehicle might be registered to one person, it may in fact be used (routinely, even) by another person. This illustrates a key limitation of proxy-to-entity indexes that are common features of life in the 21st century. Hackers often use malware-infected personal computers to conduct cyberattacks on third parties, using the infected machines as "false" proxies to hide their true identity and location.

Indexing is essentially an automated way of helping analysts build up an understanding of attributes of an entity. In intelligence, this often focuses around a particular collection mechanism or phenomenology; the same is true with regard to law enforcement and public records, where vehicle registration databases, RFID toll road passes, and other useful information is binned according to data class and searchable using relational keys. While not directly a part of the ABI analytic process, access to these databases provides analysts with an important advantage in determining potential entities to resolve to proxies in the environment.

6.6 Iterative Resolution and Limitations on Entity Resolution

Even the best proxies, however, have limitations. This is why we refer to the relevant items as proxies instead of signatures in ABI. A signature is something characteristic, indicative of identity. Most importantly, signatures have inherent meaning, typically detectable in a single phenomenology or domain of information. Proxies, however, lack the same inherent meaning, though in everyday use, the two are often conflated. A phone, for example, discussed earlier in this chapter as a proxy, might be present at a certain location. Many would then state, "John Doe is present at the location," implicitly assuming that where John Doe's phone is located is also where John Doe is located.

This, however, is not always the case. Consider an individual's car. Most often, that car represents—is a proxy for, in the parlance of this chapter and of ABI—an individual, potentially of interest. On certain days, however, that car is loaned to the primary entity's best friend for use. How would an analyst determine the validity of car-as-proxy for the primary entity, and for what duration would this hold true?

These challenges necessitate the key concept of iterative resolution in ABI; in essence, practitioners must consistently re-examine proxies to determine whether they are still valid for entities of interest. By revisiting Figure 6.2, it is intuitively clear that certain proxies are easier to change, while others are far more difficult. When deliberate operations security (OPSEC) practices are introduced from terrorists, insurgents, intelligence officers, and other entities who are trained in countersurveillance and counterintelligence efforts, it can be even more challenging to evaluate the validity of a given proxy for an individual at an individual point in time.

These limits on connecting proxies to entities describe perhaps the most prominent challenges to disambiguation and entity resolution amongst very similar proxies: the concept of discreteness, relative to physical location, and durability, relative to proxies. Together these capture the limitations of the modern world that are passed through to the analytical process underpinning ABI.

References

[1] Ryan, S., "ABI Draft Lexicon Discussion (Approved for Public Release 13-437)," National Geospatial-Intelligence Agency, 15 August 2013, slide 3.

[2] Mayer, E., et al., "Prosopagnosia," in Godefroy, O., and J. Bogousslavsky, *The Behavioral and Cognitive Neurology of Stroke* (first ed.), New York: Cambridge University Press, 2007, pp. 315–334.

[3] "UUID Structure (Windows)," Microsoft Developer Network. [Online], available: http://msdn.microsoft.com/en-us/library/aa379358%28v=vs.85%29.aspx. [Accessed: 12-Oct-2014].

[4] "RFC 4122—A Universally Unique Identifier (UUID) URN Namespace," IETF Tools, July 2005.

[5] "RFC 760—DOD Standard Internet Protocol," *IETF Tools,* January 1980.

[6] Lowenthal, T., "IP Address Can Now Pin Down Your Location to Within a Half Mile," *Ars Technica,* April 22, 2011.

[7] Hill, K., "How Target Figured Out A Teen Girl Was Pregnant Before Her Father Did," *Forbes,* February 6, 2012.

[8] Getoor, L., and A. Machanavajjhala, "Entity Resolution: Theory, Practice & Open Challenges," workshop at the *Very Large Databases Conference 2012,* 2012, Vol. 5, pp. 2018–2019. [Online]. Available: http://vldb.org/pvldb/vol5/p2018_lisegetoor_vldb2012.pdf. [Accessed: 12 Oct 2014].

[9] Nelson, W., "Use of Circular Error Probability in Target Detection," MITRE Corporation, ESD-TR-88-109, May 1988.

[10] Talburt, J., "Entity and Identity Resolution," presented at the *MIT IQ Industry Symposium,* 14 Jul 2010. [Online]. Available: http://mitiq.mit.edu/IQIS/2010/Addenda/T2A%20-%20JohnTalburt.pdf. [Accessed: 12 Oct 2014].

7

Discreteness and Durability in the Analytical Process

Chapter 6 discussed the use of space-time to disambiguate among proxies and resolve entities. In the real world, however, there are many factors that limit these two processes. The two most important factors in ABI analysis are the discreteness of locations and durability of proxies. For shorthand, these two concepts are often referred to simply as discreteness and durability. Discreteness of locations deals with the different properties of physical locations, focusing on the use of particular locations by entities and groups of entities that can be expected to interact with given locations, taking into account factors like climate, time of day, and cultural norms. Durability of proxies addresses an entity's ability to change or alter given proxies and therefore, the analyst's need to periodically revalidate or reconfirm the validity of a given proxy for an entity of interest.

7.1 Real World Limits of Disambiguation and Entity Resolution

Discreteness and durability are designed as umbrella terms: They help express the real-world limits of an analyst's ability to disambiguate unique identifiers through space and time and ultimately, match proxies to entities and thereby perform entity resolution. They also present the two greatest challenges to attempts to automate the core precepts of ABI: Because the concepts are "fuzzy," and there are no agreed-upon standards or scales used to express discreteness and durability, automating the resolution process remains a monumental challenge. This section illustrates general concepts with an eye toward use by human analysts.

This chapter will not attempt to provide a definitive answer to providing mathematically sound constructs that can be used by software developers and programmers; it will be used to outline general principles and concepts as well as provide, where appropriate, high-level constructs that can be extended by technology. Chapters 10–20 deal with the technological challenges and as well as potential technology solutions, but it is, by no means, an exhaustive survey. Rather, the basic concepts expressed in Chapters 3–9 can serve as a foundation for further, more exhaustive research into applications that help human analysts solve complex problems.

7.2 Applying Discreteness to Space-Time

Understanding the application of discreteness (of location) to space-time begins with revisiting the concept of disambiguation. As discussed in Chapter 6, one of ABI's unique methodological contributions to the field of intelligence is the use of georeferenced information to disambiguate proxies from massive data sets and ultimately resolve those proxies to entities. Disambiguation, then, is one of the most important processes for both algorithms and human beings, and one of the major challenges involves assigning confidence values (either qualitative or quantitative) to the results of disambiguation, particularly with respect to the character of given locations, geographic regions, or even particular structures.

But why does the character of a location matter? The answer is simple, even intuitive: Not all people, or entities, can access all locations, regions, or buildings. Thus, when discussing the discreteness value of a given location, whether it is measured qualitatively or quantitatively, the measure is always relative to an entity or group/network of entities. This is typically expressed inverse to the traditional HUMINT collection approach, which focuses on assessing a particular entity's access to locations (and ultimately information) of interest [1]. In ABI, the analyst focuses first on the characteristics of a location, and then on the entities and entity networks that have access to the location, in order to evaluate discreteness.

Figure 7.1 provides a basic example of how characteristics of a location translate into specific access for entities and networks of entities sharing common attributes. Knowing which entities (or key attributes therefore defining formal or informal entity networks) possess natural access to given geographical locations is a powerful tool for disambiguation. Considering that the process of disambiguation begins with the full set of "all entities," the ability to subsequently narrow the potential pool of entities generating observable proxies in a given location based on the entities who would have natural access to a given location can be an extraordinarily powerful tool in the analysis process.

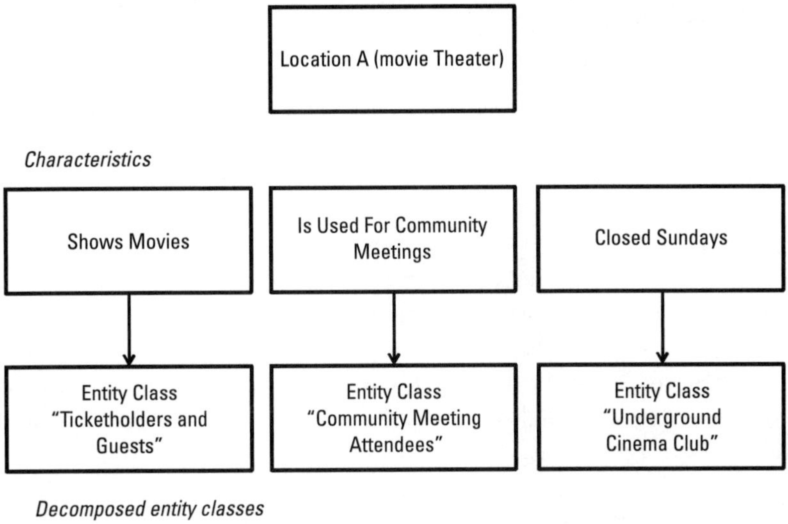

Figure 7.1 An example decomposition from characteristics of a location to relevant entity classes and/or networks.

There is enormous danger, however, in analysts unconsciously applying mirror imaging to assessing locations that operate under different sociocultural rules and norms. For example, some sections of homes in Afghanistan are limited to family members only, and visitors will never be allowed inside the family sections. Without detailed knowledge of these cultural norms, analysts might not only misunderstand the potential set of entities with access to a given location (or subset of a given location), they might also incorrectly substitute more familiar cultural knowledge resulting in a cascade of errors [2].

ABI's analytic process uses a simple spectrum to describe the general nature of given locations. This spectrum provides a starting point for more complex analyses, but the significant gap of a detailed quantitative framework to describe the complexity of locations remains. This is an open area for research and one of ABI's true hard problems.

7.3 A Spectrum for Describing Locational Discreteness

In keeping with ABI's development as a grassroots effort among intelligence analysts confronted with novel problems, a basic spectrum is used to divide locations into three categories of discreteness:

- Nondiscrete;
- Discrete;

- Semidiscrete.

The categories of discreteness are temporally sensitive, representing the dynamic and changing use of locations, facilities, and buildings on a daily, sometimes even hourly, basis. Culture, norms, and local customs all factor into the analytical "discreteness value" that aids ABI practitioners in evaluating the diagnosticity of a potential proxy-entity pair.

Diagnosticity of evidence is a critically important subconcept raised by Heuer with regard to information for intelligence analysts, and he correctly notes that it is a concept that is all too often ignored or misunderstood even by seasoned officers. The critical question raised by the concept of diagnosticity is, "How indicative is this single piece of data in answering the overall information need or gap?" Heuer writes,

> Evidence is diagnostic when it influences an analyst's judgment on the relative likelihood of the various hypotheses. If an item of evidence seems consistent with all hypotheses, it may have no diagnostic value at all. It is a common experience to discover that the most available evidence really is not very helpful, as it can be reconciled with all the hypotheses [2, p. 102].

This concept can be directly applied to disambiguation among proxies and resolving proxies to entities. Two critical questions are used to evaluate locational discreteness—the diagnosticity—of a given proxy observation. The first question is, "How many other proxies are present in this location and therefore may be resolved to entities through spatial co-occurrence?" This addresses the disambiguation function of ABI's methodology. The second question is, "What is the likelihood that the presence of a given proxy at this location represents a portion of unique entity behavior?" This question addresses the unique behaviors of entities over time, something that will be discussed in depth in Chapter 8 as "pattern of life." Public places present unique challenges to both of these questions and potential workflows, as seen in Figure 7.2.

In Figure 7.2, there are a number of four-digit identifiers, exemplars representing multiple proxies of entities that move freely throughout the public marketplace during the course of their everyday activities. Because of this movement of entities (and entity-proxy pairs) at most times of the day, this falls into the third category, or what ABI describes as a nondiscrete location. A nondiscrete location is not unique to any one entity or network of entities at the time in question (the concept of time's importance to discreteness is discussed further in Section 7.1). Although the single observation of an entity-proxy pairing at a nondiscrete location provides some information about that entity's pattern of life, it has less diagnosticity for the purpose of disambiguation.

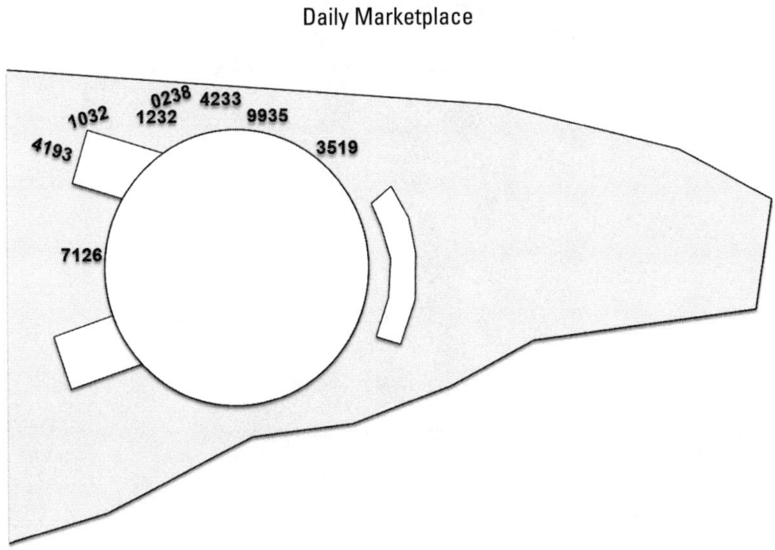

Figure 7.2 A nondiscrete location (public market) with a number of detected proxies.

Most public locations—including transportation hubs, public markets and shopping areas, and even religious institutions in major urban areas—fall into this category. As the density of proxies in a given area increases (and therefore the number of entities present also increases, though not at a 1:1 ratio), disambiguation becomes a more and more difficult task, especially with spatially coarse data. Fifty meters can become very relevant when that distance covers six different private residences and a public market.

Despite these difficulties, multiple proxy observations over space and time (even at nondiscrete locations) can be chained together to produce the same kind of entity resolution [1]. An analyst would likely need additional observations at nondiscrete locations to provide increased confidence in an entity's relationship to a location or to resolve an unresolved proxy to a given entity.

A *discrete* location is a location that is unique to an entity or network of entities at a given time. Observations of proxies at discrete locations, therefore, are far more diagnostic in nature because they are restricted to a highly limited entity network. The paramount example of a discrete location is a private residence. Metadata about private residences may also be correlated with relational or biographical information about one or more entities or networks. Figure 7.3 displays (in graph form) a notional entity network associated with a private residence. It shows "Pat," an entity directly connected to residence 1, as well as entities connected to "Pat" via other entities, increasing the degrees of separation between the residence and the entities. Observations of these entities at the discrete residence would potentially serve as evidence of the links between entities; for instance, if the "Tim" entity visited residence 1, it might provide an

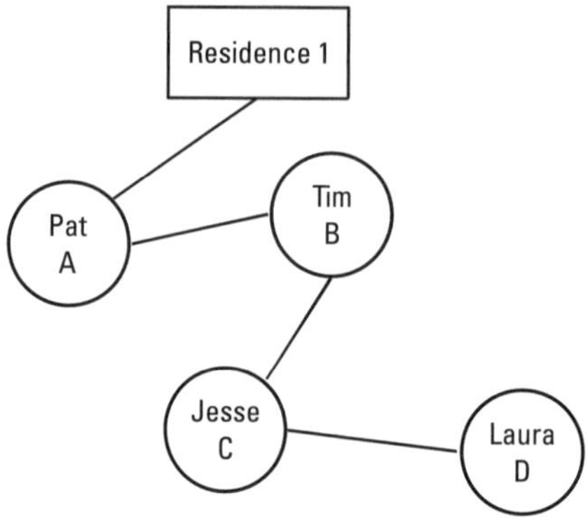

Figure 7.3 A notional entity network associated with a private residence. Note that Pat (entity A) is the only entity directly associated with the residence.

additional piece of evidence for the relationship between the "Pat" entity and the "Tim" entity. Thus, the observation of the "Tim" entity at residence 1 is highly diagnostic.

Revisiting the two principal questions from above, the following characteristics emerge regarding a discrete location:

- Proxies present at a private residence can be associated with a small network of entities, the majority of whom are connected through direct relationships to the entity or entities residing at the location;
- Entities present at this location can presumptively be associated with the group of entities for whom the location is a principal residence.

As discussed earlier, discrete locations can be far from perfect. In fact, the limitations of two-dimensional X/Y space for disambiguation become apparent when the problem of skyscrapers is examined. If a proxy is detected at a given location without a corresponding Z value (altitude, in three-dimensional space) and that proxy belongs to an entity in a skyscraper, the resulting mass of entities becomes very difficult to analyze. This is an illustrative example of how ABI's tradecraft drives a technological need with regard to certain kinds of dense urban environments, where Z values suddenly become very important for the analytical process.

Semidiscrete locations are those locations associated with several entities or multiple networks of entities. The distinction of a semidiscrete location is more subjective, but examples include fitness clubs (membership required), churches (entities are members of the same congregation), and private schools. Defining locations as semidiscrete highlights one of the challenges for proponents of automation; discreteness of a location is not absolute, but relative. A majority of

Table 7.1
Summary Description of the Scale and Characteristics of the Three Major Levels of Discreteness Along with Associated Examples of Each Level

Discreteness Value	Description	Example location
Discrete	Locations restricted to a highly limited entity network that provide highly diagnostic proxy observations	Private residence
Semi-discrete	Locations that maintain some degree of access control but still have many potential proxy-entity relationships	Restricted military installation with nonmilitary residents; sporting event restricted to ticketholders
Non-discrete	A location not unique to any one entity or network of entities at a given time; therefore, a location that has somewhat less value for the purpose of disambiguation.	Public market, square, or park

locations worldwide are semidiscrete, and whether a location is semidiscrete or nondiscrete is in many cases a judgment call.

Access controls play a major feature in semidiscrete locations; a classic example is a military base. While there are still public elements to a military base, and many entities may come from off base to work within the base's perimeter, the location is still restricted. It is not restricted enough to be considered discrete, but it is also not public enough to be nondiscrete. There are many more examples, but all follow this similar pattern of generally limited access to a group of entities sharing a common network or common attributes.

The complexity of discreteness is far from limited to X, Y, or even Z space. An important factor for considering the discreteness of a given location is temporality: From time of day to month of year, the networks and groups of entities and corresponding proxies expected at a given location actually change based on temporal factors. These will be discussed in depth in Section 7.4, along with concrete examples of temporal sensitivity across several time scales.

As ABI developed, the majority of analysts did not even consciously apply labels for discreteness to locations, and the factors surrounding the levels previously discussed were "factored in" as a part of the overall mental process. This information acted as a tacit model of the contextual, biographical, and relational information governing the analysis of events and transactions.

7.4 Discreteness and Temporal Sensitivity

Temporal sensitivity with respect to discreteness is a concept used to describe how the use of locations by entities (and therefore the associated discreteness values) changes over time; the change in function affects a change in the associated discreteness. While this may seem quite abstract, it is actually a concept many are comfortable with from an early age.

The classic example of a location where discreteness varies with time (and time scale) is a school. During school hours, the location is semidiscrete; a large population of entities (varying with school size) is present, which can make disambiguation among proxies difficult. The entities, however, mainly consist of those directly affiliated with the school (e.g., students, faculty, employees, and contractors) and then first- and second-degree affiliates of the aforementioned groups. After school hours, various extracurricular activities—such as band, art, and sports teams—might also be present in the school or its associated athletic fields. Parents and teachers might be present during parent-teacher conferences. All of these events could take place over the course of a single day or a week, offering different views of discreteness across multiple time scales.

In Figure 7.4, a school building is examined based on daily and weekly schedules. In this daily view (likely a weekday, given that classes are in session), band, classes, and sports practices are all incorporated. As each one takes place, the expectation of relevant entity networks would likely change over time. This emphasizes the absolute importance of regional and cultural expertise, both at the macro- (broad-brush) and micro- (specific to individual locations) levels.

This view of rhythmic, repeating activity is one that is likely familiar to those with a military background; indeed, the military refers to this daily view as a battle rhythm—synchronized activity and process among distributed warfighters [3]. In addition, when viewed at the macro level, the daily and subdaily variance in activity levels across multiple entities is referred to as a pattern of activity, which will be explored deeper in Chapter 8. This concept is equally familiar to high school students as well as educators. In the last several years,

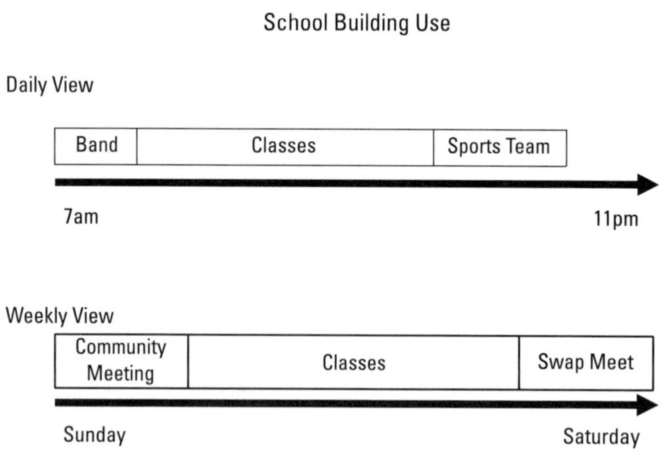

Figure 7.4 An example of varying use of a school building based on daily and weekly views of time.

substantial research and discussion has centered on block scheduling (in which classes meet for longer periods, but not every day) versus traditional scheduling (in which classes meet for shorter periods) [4]. Like battle rhythm, both offer a different view of daily activity that is time-dependent and, therefore, temporally sensitive. Relevant to ABI, the classrooms in question would have different associated entity classes and networks dependent upon time of day (in a traditional schedule) as well as day of week (in a block schedule format).

Understanding discreteness values—and how those values change over time, relative to the region and even specific location of interest—is absolutely critical for ABI. These factors make or break potential efforts at disambiguation and can, if not understood, cause analysts to make faulty associations between entities and proxies, which can cause problems much further downstream in both ABI and other analytic processes that rely on judgments derived from ABI's analytic methodology.

Discreteness is not alone, however, in its ability to affect the analytic process in ABI. The other major factor that must be considered by analysts is the durability of proxies relative to connected entities. With the goal of disambiguation and entity resolution, understanding the specific nature of proxies and associated durability is perhaps the most important external factor in the process of entity resolution.

7.5 Durability of Proxy-Entity Associations

The durability of proxies remains the other major factor contributing to the difficulty of disambiguation and entity resolution. Chapter 6 discusses the concept of iterative resolution, that is, the need to continuously "re-resolve" or "revalidate" proxy-entity associations. Though many proxies can (and are often) associated with single entities, these associations can range from nearly permanent to extraordinarily fleeting. The concept of durability represents the range of potential values for the duration of time of the proxy-entity association.

Like discreteness, however, discussing durability is not as simple as providing absolute "durability values" for various proxies and then routinely applying those values throughout the analytic process. While there are general categories of "more" and "less" durable, durability is as much determined by local customs, cultural beliefs and behaviors, and other factors affecting the daily life of individual entities as it is associated with the properties of various proxies.

A simple example of local practice and custom altering the durability of a class of proxy relates to cellular telephone use and retention. Currently, in the United States, two-year contracts are the norm for major cellular carriers like AT&T and Verizon, with a handful of smaller companies offering no-commitment services. This allows the larger carries to heavily subsidize the price of brand new telephones, a particularly significant practice in light of the high costs of new smartphone devices [5]. In many parts of Europe, by contrast, contracts are rare, exclusive deals linking handsets to carriers nearly nonexistent (or illegal), and use of prepaid plans far more common [6]. One potential consequence of this is categorizing a cell phone as a generally more durable proxy for an entity in the United States with a Verizon contract, versus a cell phone for a Spain- or Holland-based entity, who might swap numbers based on carrier location, movement, or other factors. This reinforces the need for regional expertise when assessing the behavior of entities or attempting to disambiguate in complex environments, just as was true when considering the variability of discreteness; durability, however, tends to lack the degree of temporal sensitivity that characterizes discreteness.

This is not to say that durability is temporally insensitive. Time plays an important factor in evaluating whether a discovered proxy-entity link is still valid, but rather than relative time (which is most relevant to discreteness, as in day of the week or hour of the day), the important temporal factor is time elapsed since proxy observation. Continuing with the example of cellular telephones as proxies, a resolution of a phone number to an entity using three-year old data might in fact provide a relevant proxy at the time of observation (e.g., which handset was the individual using in 2011?), but the current probative value of the proxy might be in question—how likely is the entity of interest to have retained a phone since 2011? Thus, absolute time elapsed from proxy observation is an important factor in considering the utility of a potential proxy-entity pairing.

Discreteness and durability are difficult but important concepts for analysts to consider as they move through the analytic process. Answering "who-where-who" and "where-who-where" workflow questions becomes exponentially more difficult when varying degrees of uncertainty in spatial-temporal correlation are introduced by the two major factors discussed in this chapter. Accordingly, structured approaches for analysts to consider the effects of discreteness and durability are highly recommended, particularly as supporting material to overall analytical judgments.

One continuous recommendation in all types of intelligence analysis is that assumptions made in the analytical process should be made explicit, so that intelligence consumers can understand what is being assumed, what is being assessed, and how assessments might change based on changes in the underlying assumptions presented by an analyst [2, pp. 9, 16]. One recommended technique is using a matrix during the analytic process to make explicit discreteness and durability factors in an effort to incorporate them into the overall judgments and conclusions. In addition, the use of a matrix can provide key values that can later be used to develop quantitative expressions of uncertainty, but these expressions are fundamentally meaningless without the underlying quantifications clearly expressed (in essence, creating a "values index" so that the overall quantified value can be properly contextualized).

In Table 7.2, the analyst has clearly identified some absolute factors (e.g., type of proxy—in this case cellular phone) as well as some factors unique to the area of interest based on domain knowledge (in this case, that users of prepaid phones on XYZ network tend to maintain numbers for 60–90 days on average). The analyst also assesses the discreteness of the location, noting it as a discrete private residence, but further notes that Thursdays have extended family gatherings that would therefore cause the location to be somewhat less discrete.

Above all, analysts must be continually encouraged by their leadership and intelligence consumers to clearly express uncertainty and "show their work." Revealing flaws and weaknesses in a logical assessment are unfortunately often perceived as weakness, and this tendency is reinforced by consumers that attack probabilistic assessments and express desires for stronger, "less ambiguous" results of analyses. The limitations of all analytic methodologies must be expressed, but in ABI this becomes a particularly important point.

One item not covered in Table 7.2 but discussed previously in Chapter 6 is the critical need for multiple spatiotemporally correlated observations in order to validate the relationship between a proxy and an entity. Thus, a proper expression of a proxy-entity correlation would document, in short, multiple instances of correlation as well as consider "weighting" the various observations. While quantifying "squishy" concepts such as this can be extraordinarily difficult and can even at times mask true uncertainty by assigning false precision to judgments through the use of numbers, this process might at least offer insight into the method used by an analyst to reach a conclusion and offer a means to compare effective correlations between analysts and provide assessments and feedback as a training mechanism to analysts.

One factor not discussed here that continues to stymie attempts at automation as well as proper attempts at quantitative expressions of uncertainty is the notion of "unknown" *n* values, where *n* is the full expression of a given data set. One of the challenges peculiar to intelligence analysis is the adversary's active role in denying the collection of data and providing false data to skew results [known collectively as denial and deception (D&D) in the intelligence community]. Lacking knowledge of the "full picture of collection" is what

Table 7.2
A Simple Single-Row Example of a Proxy-Entity Correlation and Surrounding Factors

Proxy	Proxy durability	Correlation location	Entity
Cellular telephone, XYZ pre-paid carrier	Prepaid telephone, minimally durable; local use suggests 60–90 days of use	Private residence, highly discrete, though extended family gatherings occur on Thursday evenings	Entity A or Entity B, brothers in family network

makes setting an absolute number of correlations needed to validate a proxy-entity pair nearly impossible, and it forces analysts to resort to relative measures and ultimately, analyst judgment regarding the validity of proxy-entity pairs. This problem will be examined in more detail in Chapter 9, where the concept of incidental collection is explained and contrasted with the classic model of targeted collection.

7.6 Summary

Chapter 6 introduces the core concepts of proxies and entities, while Chapter 7 discusses various limitations on matching proxies to entities through the process of entity resolution. This process, performed continuously for an entity, ultimately enables analysts to understand the particular pattern of life of a specific entity to enable specific actions. The details and uses for entity patterns of life will be discussed in detail in Chapter 8.

References

[1] "FM 2-22.3: Human Intelligence Collector Operations," Headquarters, Department of the Army, September 2006, p. 47, [online]. Available: http://fas.org/irp/doddir/army/fm2-22-3.pdf. [Accessed: 09 Nov 2014].

[2] Heuer, R. J., *The Psychology of Intelligence Analysis,* Center for the Study of Intelligence, 1999, p. 70.

[3] Duffy, L., et al., "A Model of Tactical Battle Rhythm," in *2004 Command and Control Research and Technology Symposium,* 2004. [Online]. Available: http://www.dtic.mil/get-tr-doc/pdf?AD=ADA466128. [Accessed: 09 Nov 2014].

[4] Aldermane, D., "Comparison Study of the Relationships of 4/4 Block Scheduled Schools and 7.Period Traditional Scheduled Schools on the Standards of Learning Tests for Virginia Public Secondary Schools," Virginia Polytechnic Institute and State University, Blacksburg, VA, 2000. [Online].

[5] Klein, D., "Will Sprint, T-Mobile, AT&T, and Verizon End Subsidized Phones?," *The Motley Fool,* September 14, 2014.

[6] Brustein, J., "Wireless, But Leashed," *The New York Times,* January 15, 2011.

8

Patterns of Life and Activity Patterns

Previous chapters briefly introduced the concept of pattern of life and its relevance to entities, the focus of analysis in ABI. The importance of this concept necessitates a fuller explanation of pattern of life to distinguish it from the concept of "activity patterns," also of relevance as contextual data in the ABI analytic process. This chapter discusses the basics of constructing patterns of life using indirect observations available to intelligence analysts and the importance of this process to ABI as a whole.

8.1 Entities and Patterns of Life

"Pattern(s) of life," like many concepts in and around ABI, suffers from an inadequate written literature and varying applications depending on the speaker or writer's context. The concepts underlying patterns of life are not new ones but have seen renewed interest due to the development of persistent surveillance technologies (discussed in greater detail in Chapter 12) that allowed intelligence officers to observe the movements of individuals over many consecutive hours. These concepts are familiar to law enforcement officers, who through direct surveillance techniques have constructed patterns of life on suspects for many years. One of the challenges in ABI explored in Section 8.2 is the use of indirect, sparse observations to construct entity patterns of life.

With discreteness, the varying uses of geographic locations over days, weeks, months, and even years is examined as part of the analytical process for ABI. Patterns of life are a related concept: A pattern of life is defined as *the specific set of behaviors and movements associated with a particular entity over a given period of time.* In simple terms, this is what people do everyday: wake up, eat breakfast, go to work or school, run errands, meet friends, attend meetings, and otherwise conduct specific activities. Among these activities are common threads: locations that entities go to and other entities with which they associate. These often repeat over the course of a day, week, or even month, depending on the location or type of activity. As an example, an entity's home and place of employment are two common locations for an entity. An entity might also go grocery shopping, attend a weekly community group meeting, attend religious services, obtain gasoline for a car or motorcycle, and mail letters at a post office. All of these are a part of an individual entity's pattern of life.

Important in this definition is the concept of "a particular entity." At times, the term "pattern of life" has been used to describe behaviors associated with a specific object (for instance, a ship) as well as to describe the behaviors and activity observed in a particular location or region. An example would be criminal investigators staking out a suspect's residence: They would learn the various comings and goings of many different entities, and see various activities taking place at the residence. In essence, they are observing small portions of individual patterns of life from many different entities, but the totality of this activity is sometimes also described in the same way.

A more accurate way to describe the various activities occurring at a given location is a "pattern of activity" or "activity pattern," defined as the larger unit of behaviors and activities not specific to an individual entity.

The surface harbor traffic in and around Los Angeles, for example, is a pattern of activity. The individual ships below also have activity patterns and have individuals on board who have patterns of life. An example of activity nesting is given in Figure 8.1. At the top level is the full activity pattern for an entire day of maritime traffic in Los Angeles, while a small part of that (at the

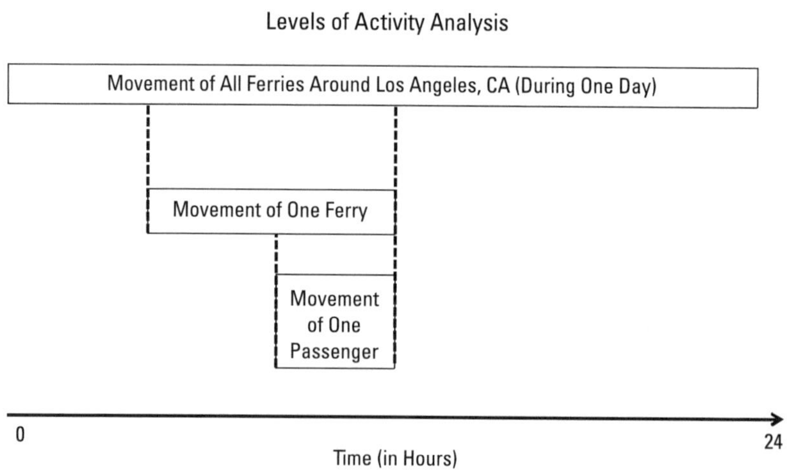

Figure 8.1 Different levels of activity analysis are always present in the environment: a large set, a specific physical object (a ship), and finally one entity.

next level down) is one specific movement of a specific ferry. On the ferry, numerous entities are on-board conducting a movement that is part of their larger pattern of life, visible in Figure 8.2.

One truth about patterns of life is that they cannot be observed or fully understood through periodic observations. Earlier, the relevance of persistent surveillance to patterns of life was briefly discussed. LTG Michael Flynn, former director of the Defense Intelligence Agency, hits on this importance while discussing differences between employment of intelligence assets between conventional and special operations forces: "The conventional force approach reveals a desire to service a large number of targets and units instead of developing the pattern of life of an enemy network. The tendency to think of persistence in terms of space rather than time results in sprinkling assets in multiple areas rather than focusing them on a limited number of locations." Flynn and his coauthors, in this analysis, explicitly recognize the importance of temporal duration in understanding patterns of life [1].

In sum, four important principles emerge regarding the formerly nebulous concept of "pattern of life":

- A pattern of life is specific to an individual entity;
- Longer observations provide better insight into an entity's overall pattern of life;
- Even the lengthiest surveillance cannot observe the totality of an individual's pattern of life;

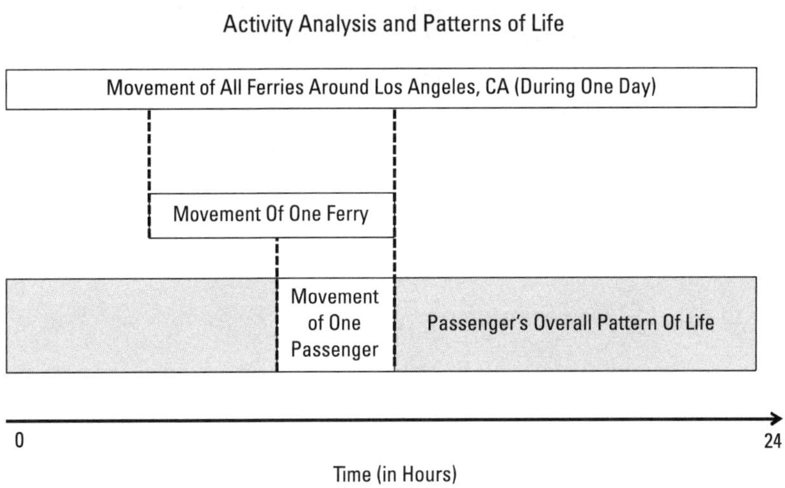

Figure 8.2 A specific entity's movement on a ferry in context of its overall pattern of life.

- Traditional means of information gathering and intelligence collection reveal only a snapshot of an entity's pattern of life.

At first, the second and third statements may appear contradictory. If longer observations provide better insight, it seems to follow that at some point, a long enough observation might reveal an entire pattern of life. In a theoretical construct, this is true: an unblinking eye, following an entity 24 hours a day, seven days a week, would provide a detailed and exact pattern of life. The reality, however, is that even the best surveillance is marred by gaps or breaks in observation. As a result, it is always important for the analyst to understand the parts of an entity's pattern of life that are not visible or collected. While it can be tempting to generalize or assume on the basis of what is observed, it is important to account for the possibilities during times in which an entity goes unobserved by technical or human collection mechanisms. In the context of law enforcement, the manpower cost of around-the-clock surveillance quickly emerges, and the need for officers to be reassigned to other tasks and investigate other crimes can quickly take precedence over the need to maintain surveillance on a given entity. Naturally, the advantage of technical collection versus human collection in terms of temporal persistence is evident [2].

Small pieces of a puzzle, however, are still useful. So too are different ways of measuring the day-to-day activities conducted by specific entities of interest (e.g., Internet usage, driving habits, and phone calls). Commercial marketers have long since taken advantage of this kind of data in order to more precisely target advertisements and directed sales pitches. However, these subaspects of an entity's pattern of life are important in their own right and are the building blocks from which an overall pattern of life can be constructed.

There is an underlying assumption present in the discussion of an entity's pattern of life; that assumption is a disambiguated or resolved entity. While proxy observations can be used to construct an element of an unresolved entity, it is ultimately impossible to construct a useful pattern of life without also resolving an entity. That said, however, there are ways in which the construction of patterns of life and the process of disambiguation and entity resolution go hand in hand, discussed in Section 8.5

8.2 Pattern-of-Life Elements

Pattern-of-life elements are the "building blocks" of a pattern of life. As discussed in Section 8.1, these elements can be measured in one or many different dimensions, each providing unique insight about entity behavior and ultimately contributing to a more complex overall picture of an entity. These elements can be broken down into two major categories:

- Partial observations, where the entity is observed for a fixed duration of time;
- Single-dimension measurements, where a particular aspect of behavior or activity is measured over time in order to provide insight into that specific dimension of entity behavior or activity.

Recall the previously discussed concept of levels of activity analysis (see Figures 8.1 and 8.2). On the third line of both Figures 8.1 and 8.2 is a perfect example of a partial observation, focused on the specific travel of one entity on one ferry in a given 24-hour period. To further extend the hypothetical ferry example, if an unmanned aerial system (UAS) surveils the ferry in an effort to maintain visual "custody" of a particular entity, the time period during which the aircraft maintained custody would be characterized as a partial observation. While in an ideal world the platform would be able to maintain continuous custody of the entity, the limitations of reality play a major part in causing breaks in observation and thereby introducing an element of uncertainty into the surveillance process. For example, the entity might walk inside a room with other entities, and then some time later, an entity might emerge. The limitations of the sensor platform (resolution, spectrum, field of view) all play a role in the operator's ability to assess whether the individual who emerged later was the same individual entity who entered the room, but even a high-confidence assessment is still an assessment, and there remains a nonzero chance that the entity of interest did not emerge from the room at all.

Single-dimension measurements, by contrast, focus on measuring the particular activity of a given entity in one dimension. As discussed in Section 8.1, Internet usage from a particular residence is one potential measure of activity and could constitute important information about an entity's pattern of life. Uncertainty, however, plays a major role here as well. For one, entity activity is not measured directly, but proxy activity is; thus, if the proxy in question is associated with multiple entities (a computer terminal used by four family members), or the same entity uses multiple proxies to conduct a particular type of activity (an individual who logs on primarily from Internet cafés because they lack home Internet access), the single-dimensional measurement may be skewed or nearly impossible to construct.

8.3 The Importance of Activity Patterns

Figures 8.1 and 8.2 demonstrate the idea of different levels of activity analysis and identify at the top level the concept of a large amount of activity, such as physical movements of ocean-going vessels, conducted by numerous entities in a given geographic location. This is an example of an activity pattern, dealing with aggregate behaviors and activities not specific to individual entities.

Understanding the concept and implications of data aggregation is important in assessing both the utility and limitations of activity patterns. The first and most important rule of data aggregation is that aggregated data represents a summary of the original data set. Regardless of aggregation technique, no summary of data can (by definition) be as precise or accurate as the original set of data [3]. Therefore, activity patterns constructed from data sets containing multiple entities will not be effective tools for disambiguation. Effective disambiguation requires precise data, and summarized activity patterns cannot provide this. If, as per Chapter 5, the question is primarily "who is conducting this activity," summarized levels of "what activity is being conducted" will not provide an effective answer.

If activity patterns—large sets of data containing summarized activity or movement from multiple entities—are not useful for disambiguation, why mention them at all in the context of ABI? There are two primary reasons. One is that on a fairly frequent basis, activity patterns are mistakenly characterized as patterns of life without properly distinguishing the boundary between specific behavior of an individual and the general behaviors of a group of individuals [4, 5]. The second reason is that despite this confusion, activity patterns can play an important role in the analytical process: They provide an understanding of the larger context in which a specific activity occurs. Thus, activity patterns of many types are considered context data, as discussed in Chapter 3. Understanding the relative "rhythm" of a specific class of activity, like vehicle traffic over time, allows a deeper understanding of the context in which a specific physical transaction occurs. An entity driving from point A to point B (and thus conducting a physical transaction) at an hour of the day when few cars are on the road (and thus, there are relatively fewer transactions being conducted) may warrant additional examination to determine if the transaction is significant.

The above example might constitute an "abnormal" occurrence in an analyst's mind—abnormal relative to the general activity observed at a particular time, or abnormal relative to a particular entity's known behaviors. The inherent difficulty in assessing normal versus abnormal behavior based on incomplete data could be the subject of a book on its own, but it is worth discussing in brief related to the concept of pattern of life.

8.4 Normalcy and Intelligence

"Normal" or "abnormal" are descriptors that appear often in discussions regarding ABI. Examining the descriptions at a deeper level, however, reveals that these descriptors are often applied to activity pattern analysis, an approach to analysis distinct from ABI. The basis in logic works as follows:

- Understand and "baseline" what is normal;
- Alert when a change is made (thus, when "abnormal" occurs).

An example of this is given in Figure 8.3, where an activity baseline is established for a 24-hour period and an abnormal "spike" in activity identified. The underlying assumption is that "abnormal" is what is most important. This assertion is not without merit, particularly in the subdiscipline of intelligence known as indications and warning (I&W), or warning intelligence. This area specifically focuses on identifying potentially relevant activities and alerting decision-makers or warfighters in advance of the anticipated activity, thereby providing "warning." Cynthia Grabo, a former senior analyst at the Defense Intelligence Agency, defines warning intelligence as dealing with "(a) direct action by hostile states against the U.S. or its allies, involving the commitment of their regular or irregular armed forces; (b) other developments, particularly conflicts, affecting U.S. security interests in which such hostile states are or might become involved; (c) significant military action between other nations not allied with

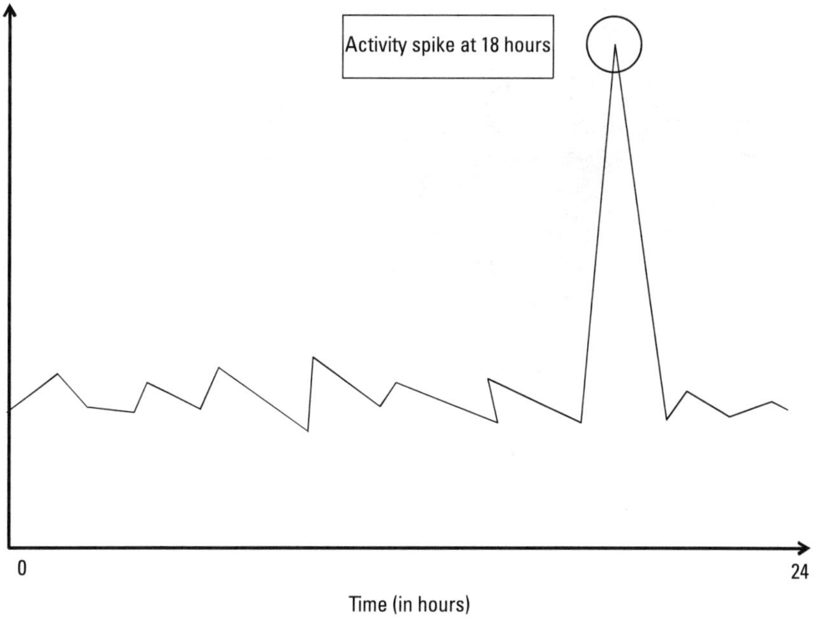

Figure 8.3 A notional graph of a 24-hour activity pattern. The spike at 18 hours might represent "abnormal" activity, but just as easily might not.

the U.S., and (d) the threat of terrorist action" [6]. Thus, warning is primarily concerned with what may happen in the future.

Pattern analysis supports this nicely, focused on identifying "abnormal" activity: potential indicators of events that require warning. ABI, with its forensic mindset (that can nonetheless be practiced in near-real time), focuses not on indicators of future events (activity itself) but on the identity of particular actors (entities).

Thus, while warning seeks to answer "what" and "when" questions, ABI remains dedicated to resolving "who" questions and focuses primarily on data correlations as opposed to establishing activity baselines. These data correlations are used to disambiguate entities and ultimately understand patterns of life and entity networks, but the process of constructing an entity's pattern of life and resolving that entity are intertwined rather than sequential. The complex interplay between two difficult concepts is described in Section 8.5.

8.5 Representing Patterns of Life While Resolving Entities

Until this point, disambiguation/entity resolution and patterns of life have been discussed as separate concepts. In reality, however, the two processes often occur simultaneously. As analysts disambiguate proxies and ultimately resolve them to entities, pieces of an entity's pattern of life are assembled. Once a proxy of interest is identified—even before entity resolution fully occurs—the process of monitoring a proxy creates observations: pattern-of-life elements. One major challenge, however, is representing and recording these observations in a useful manner that allows analysts and operators to preserve knowledge gained for future endeavors and enterprise learning.

8.5.1 Graph Representation

One of the most useful ways to preserve nonhierarchical information is in graph form. Rather than focus on specific technology, this section will describe briefly the concept of a graph representation and discuss benefits and drawbacks to the approach. Graphs have a number of advantages, but the single most relevant advantage is the ability to combine and represent relationships between data points from different sources. This is particularly useful when relating information from across information systems that already exist [7].

Graphs offer the ability to focus on objects—"nouns"—and relationships—"verbs" of even the most complex networks. Graphs also offer the ability to abstract data at multiple levels of analysis, allowing analysts to quickly move between views and analyze summarized graphs and then dive into specific details of interest in the graph.

In Figure 8.4, the basic units of graph representation are linked to their corresponding terms in ABI (NB graph objects can also be used to represent other kinds of data, but this section will focus on the application of graphs to representing patterns of life and pattern-of-life elements). The core of the graph is objects and relationships: objects, which are typically entities in ABI, and their relationships. Objects can also be locations, however, and graphs can be used to represent the relationships between entities and locations of interest. This is common in graph representations of patterns of life. As seen in Figure 8.4, graphs can also be used to relate proxy detections (observations) to appropriate entities, or even "unknown" entities used as placeholders for further information.

Clearly, the flexibility afforded by a pattern-of-life graph is beneficial when dealing with the heterogeneous data commonly encountered in ABI analysis. The graph view, however, also has disadvantages. Graphs constructed primarily in two dimensions (or three, depending on the visualization interface) do not necessarily have the underlying data to support measured distance relationships; the distance between nodes and length of relationships does not necessarily have meaning, confusing some users of graphs. In addition, strict graphs do not necessarily offer an effective means of expressing temporal bounds or quantitative data (though there are approaches that can remedy these general disadvantages).

8.5.2 Quantitative and Temporal Representation

With quantitative and temporal data, alternate views may be more appropriate. Here, traditional views of representing periodicity and aggregated activity pat-

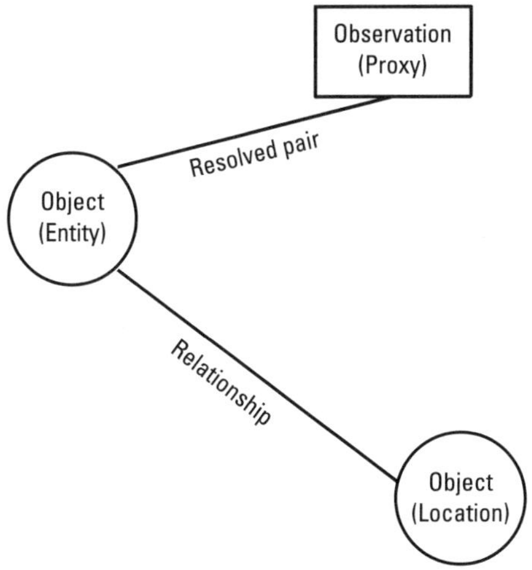

Figure 8.4 An index of graph attributes matched to their corresponding ABI terms.

terns are ideal; this allows appropriate generalization across various time scales. Example uses of quantitative representation for single-dimensional measurements (a pattern-of-life element) include the representation of periodic activity. One example of this is a bar graph indicating frequency of entity visits to a particular location of interest. This both shows useful data about one particular aspect of pattern of life and enhances the more general graph view.

Figure 8.5 shows an example of activity from a given entity-proxy pair at a location of interest based on day of the week. This allows analysts to discern any potential correlations between activity levels and day of the week and make recommendations accordingly. This view of data would be considered a single-dimensional measurement, and thus a pattern of life element. One important consideration when using single-dimensional measurement is the biases that are introduced by periodic collection; percent coverage must be evaluated, and the possibility that the portion collected cannot be properly generalized into a "complete" sample remains a serious concern of this method and particular view of data. Sampling in intelligence remains a less-than-preferred methodology in most circumstances (particularly with adversaries practicing operational security or D&D practices) but can provide useful insight when carefully applied.

8.6 Enabling Action Through Patterns of Life

One important element missing from most discussions of pattern of life is "Why construct patterns of life at all?" Having an entity's pattern of life,

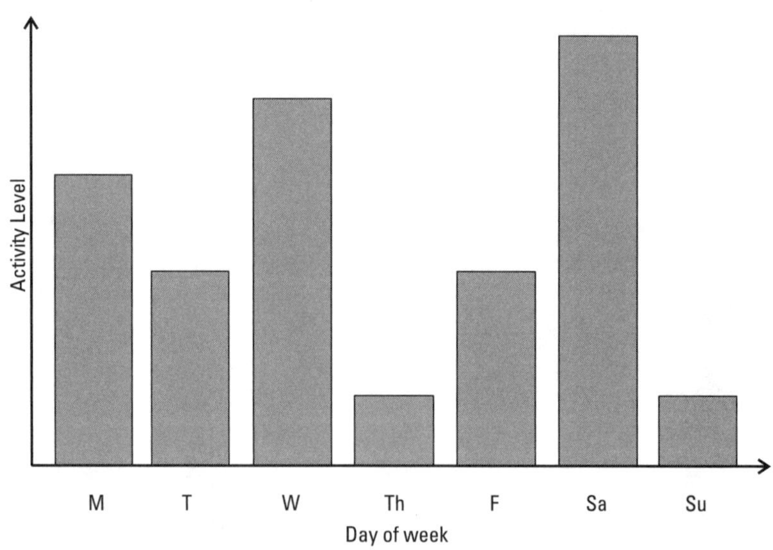

Figure 8.5 A single-dimensional measurement of activity at a particular location. The graph could also express the activity of particular entity at many different locations.

whether friendly, neutral, or hostile, is simply a means to an end, like all intelligence methodologies. The goal is not only to provide decision advantage at a strategic level but operational advantage at the tactical level. Examples of this include identifying key entities with potentially valuable information or influencers who can help turn the tide of opinion in a formerly hostile village, as well as more traditional kinetic operations against an irregular threat network. Understanding events, transactions, and activity patterns also allows analysis to drive collection and identifies areas of significance where further collection operations can help reveal more information about previously hidden networks of entities. Patterns of life and pattern-of-life elements are just one representation of knowledge gained through the analytic process, ultimately contributing to overall decision advantage.

References

[1] Flynn, M., et al,, "Employing ISR: SOF Best Practices," *Joint Forces Quarterly,* Vol. 3rd Quarter 2008, No. 50.

[2] Dees, T., "Surveillance Technology: An End to Stakeouts?," *Police Magazine,* December 14, 2010.

[3] Kabacoff, R., "Aggregation and restructuring data," *R-Statistics Blog,* September 9, 2012.

[4] "A Case Study: Patterns-of-Life and Activity-Based Intelligence Analysis." *AGI,* 13 March 2013. [Online]. Available: http://www.agi.com/downloads/support/productSupport/literature/pdfs/CaseStudies/031313_CaseStudy_Suritec.pdf [Accessed: 27-Nov-2014].

[5] "Data Fusion and Decision Making Systems," ESEN Sistem Entegrasyon. [Online]. Available: http://www.esensi.com.tr/data-fusion-and-decision-making-systems. [Accessed: 27-Nov-2014].

[6] Grabo, C., *Anticipating Surprise: Analysis for Strategic Warning,* Joint Military Intelligence College, 2002, p. 6.

[7] Sherman, M., "Advantages and Disadvantages of Document- and Graph-Based Databases," *Texas Enterprise,* 14 August 2014. [Online]. Available: http://www.texasenterprise.utexas.edu/2014/08/14/innovation/your-database-so-retro-old-data-new-databases. [Accessed: 27-Nov-2014].

9

Incidental Collection

This chapter explores the concept of incidental collection by contrasting the change in the problem space: from Cold War–era order of battle to dynamic targets and human networks on the 21st century physical and virtual battlefields. This demonstrates how collection concepts have evolved from the formation of the U.S. national security apparatus to the present day and explores new collection concepts that serve ABI's unique data needs.

9.1 A Legacy of Targets

The modern intelligence system—in particular, technical intelligence collection capabilities—was constructed around a single adversary, the Soviet Union. First the U-2 spy plane, and later the CORONA and subsequent reconnaissance satellites, were the crown jewels in America's intelligence arsenal: precise, highly specialized, and highly expensive. Because of the immense cost associated with collection, a process to ensure that collection focused on the highest priorities of the defense and intelligence enterprise had to be implemented. As understanding and observing Soviet military activity was one of the highest priorities, many of the original targets for early imaging systems were Soviet garrisons, missile facilities, and other military targets. The result was the creation of a collection list, or "deck" in intelligence parlance: a set of facilities to be imaged on a routine basis in order to provide advance warning of change and accurately estimate adversary capabilities [1, 2].

Imagery was also extremely valuable due to the predominant use of film-return systems on imaging satellites. Due in large part to technical challenges with digital transmission of images from space, film return technology would be the mainstay of the overhead imagery constellation until the 1970s [2]. One of the consequences of the employment of film is that film, in effect, dictated the operational lifespan of an imaging satellite. When the satellite was out of film, it was out of commission. Launching a new satellite was the only way to obtain additional images.

Even in the early days of the CORONA program, it was clear that the "take," or information coming back from the satellites, would require a large number of personnel for effective use. Just two months after the launch of the first CORONA platform in 1960, a formal proposal was made to the U.S. Intelligence Board to expand the number of personnel cleared into the KEYHOLE subcompartment, protecting overhead satellite imagery, by nearly 100 percent—from 1,431 personnel to 2,938 personnel. The overwhelming majority of these personnel were photo-interpreters and other intelligence analysts [3, p. 107].

Why did this relatively limited amount of data (by today's standards, measured in petabytes) require such a large number of personnel for exploitation? The answer is found in the TCPED process, discussed in Chapter 3. There remains a concerted effort to exploit every piece of imagery possible. The limitations of this process in the face of a rapidly increasing volume of information were apparent as early as 2000, when the NIMA Commission noted "that the future imagery architecture (FIA)-era increase in imagery of more than an order of magnitude does not, in and of itself, imply a need for a proportionate increase in exploitation capacity. Some increase may be needed, but an *N*-fold increase in imagery does not necessarily translate into an *N*-fold increase in information content, particularly when the additional imagery capacity is used to more frequently "sample the same target for activity analysis, or I&W" [4]. The commission anticipated the very problem the intelligence community is faced with today: The exploitation and operational paradigms are not suited to the current data environment.

9.2 Bonus Collection from Known Targets

Incidental collection is a relatively new term, but it is not the first expression of the underlying concept. In imagery parlance, "bonus" collection has always been present, from the very first days of "standing target decks." A simple example of this starts with a military garrison. The garrison might have several buildings for various purposes, including repair depots, vehicle storage, and barracks. In many cases, it might be located in the vicinity of a major population center, but with some separation depending on doctrine, geography, and other factors.

A satellite might periodically image this garrison, looking for vehicle movement, exercise starts, and other potentially significant activity. The garrison, however, only has an area of 5 km^2, whereas the imaging satellite produces images that span almost 50 km by 10 km. The result, as shown in Figure 9.1, is that other locations outside of the garrison—the "primary target"—are captured on the image. This additional image area could include other structures, military targets, or locations of potential interest, all of which constitute "bonus" collection. In Figure 9.1, the targeted collection is shown in white, while the bonus collection around the target appears in gray.

One fixture of "bonus" collection, however, is the original target deck: the reason the images were taken, in order to satisfy a request, or requirement, for an image of the target location. The same concept applies to HUMINT as well: A case officer might receive a requirement from an analyst to have a source identify the number of people who enter a building from 8 a.m. to 9 a.m. through physical surveillance. That source might also provide information on four specific people above and beyond the original task; the extra information is the "bonus" information in this context.

Neither of these examples fully captures the nuance of incidental collection, which represents a deliberate extension of "bonus" collection based on a very different paradigm of intelligence: data-driven rather than requirements-driven; multi-INT rather than stove-piped; inherently spatial and temporal rather than relational.

9.3 Defining Incidental Collection

In incidental collection, the first concept to be discarded is the notion of a deck of targets. Before even comparing incidental collection to other collection types based on the collection itself, the requirements driving the collection must be

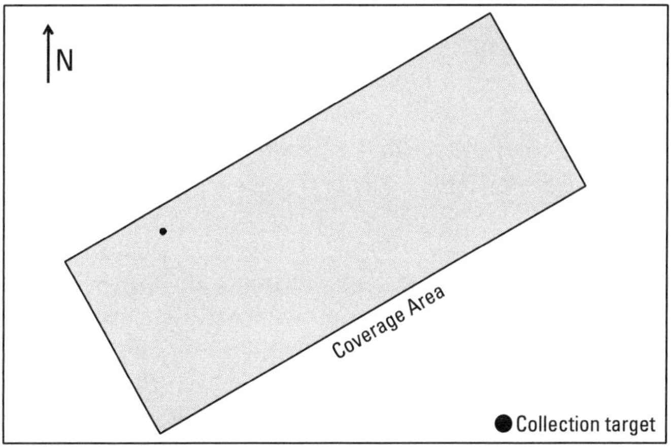

Figure 9.1 A simple illustration of targeted collection versus bonus collection.

examined. These play a key role in distinguishing the philosophy behind incidental collection.

With a target deck, as discussed in the Section 9.2, the goal is to collect just enough information to satisfy the requirement. Looking back to Chapter 5, this means that collection based on fixed requirements presumes a degree of a priori knowledge about what will be in the collection and whether the information collected has satisfied the relevant requirement, in many cases even before a human analyst has touched the data in question.

Incidental collection, rather than identifying a specific intelligence question as the requirement, focuses on the acquisition of large amounts data over a relevant spatial region or technical data type and sets volume of data obtained as a key metric of success. This helps address the looming problem of unknowns buried deep in activity data by maximizing the potential chances for spatiotemporal data correlations. Ultimately, this philosophy maximizes opportunities for proxy-entity pairing and entity resolution.

This can appear counterintuitive at first. Many senior officials in the intelligence and law enforcement communities have argued persuasively that one of the biggest challenges facing today's analytic corps is the overwhelming amount of data available to them, particularly when accounting for new data streams such as publicly available social media. The Congressional Research Services concluded in 2013, "While the intelligence community is not entirely without its legacy 'stovepipes,' the challenge more than a decade after 9/11 is largely one of information overload, not information sharing. Analysts now face the task of connecting disparate, minute data points buried within large volumes of intelligence traffic shared between different intelligence agencies [5]. Incidental collection appears to run against this, as it focuses on increasing data volume as a key goal. However, the sentiment that analysts are somehow "drowning" in data belies the truth: Analysts lack the tools and approaches to make use of the massive volumes of data available, particularly across highly heterogeneous data types. Armed with the proper tools, analysts can and will hungrily seek out voluminous streams of data. ABI's focus on spatiotemporal data correlation provides a framework to which tools and analytics can be attached to help analysts derive information from the data with which they work each day. The true issue is not volume but effective analytic frameworks and tools.

9.4 Dumpster Diving and Spatial Archive and Retrieval

In intelligence, collection is focused almost exclusively on the process of prioritizing and obtaining through technical means the data that should be available "next." In other words, the focus is on what the satellite will collect tomorrow, as opposed to what it has already collected, from 10 years ago to 10 minutes ago. But vast troves of data are already collected, many of which are quickly triaged and then discarded as lacking intelligence value. ABI's pillar of sequence neutrality emphasizes the importance of spatial correlations across breaks in time, so maintaining and maximizing utility from data already being collected for very different purposes is in effect a form of incidental collection.

Chapter 3 presents the case of a notional analyst attached to a military unit to illustrate how the pillars of ABI were derived from real operational experiences during the early 2000s. Mountains of data were being generated and collected by the hypothetical unit, and it is with these data that the analyst created a spatiotemporal analytic environment. This process is colloquially called "dumpster diving" by some analysts: repurposing of existing data through application of ABI's georeference to discover pillar.

Repurposing data through the process of data conditioning (extracting spatial, temporal, and other key metadata features and indexing based on those features) is a form of incidental collection and is critical to ABI. This is because the information in many cases was collected to service-specific collection requirements and/or information needs but is then used to fill different information needs and generate new knowledge. Thus, the use of this repurposed data is incidental to the original collection intent. This process can be applied across all types of targeted, exquisite forms of intelligence. Individual data points, when aggregated into complete data sets, become incidentally collected data.

This process was used by the first analysts practicing ABI in Iraq and Afghanistan. *Trajectory Magazine* wrote in its Winter 2012 issue, "A group of GEOINT analysts deployed to Iraq and Afghanistan began pulling intelligence disciplines together around the 2004–2006 timeframe…these analysts hit upon a concept called 'geospatial multi-INT fusion.'" The article goes on to write that those analysts recognized that the one field that all data had in common was location [6].

This approach is made possible by focusing on geographic location as a primarily filtering and retrieval mechanism for data. By approaching the "where" in ABI's two predominant workflows (who-where-who and where-who-where) as a spatial environment with temporal attribution, the analyst has both a valid methodology and an effective mechanism to correlate information based on the additional metadata. Figure 9.2 shows an example process flow for turning targeted collection from single intelligence disciplines into incidentally collected data through the process of georeferencing. Figure 9.2 also shows how this process can be added to existing processing chains without disruption, enabling present exploitation methods to continue while enabling ABI through dedicated application of the georeference-to-discover pillar.

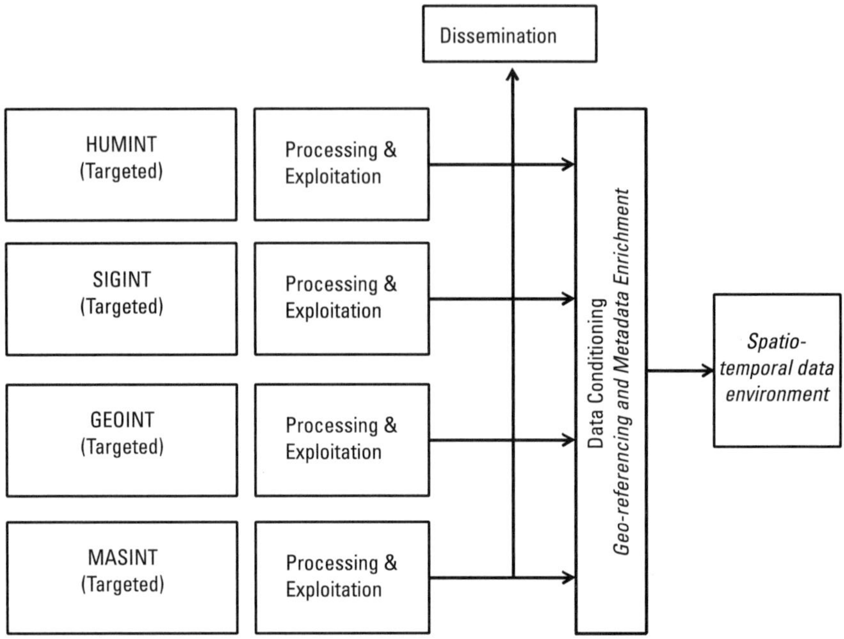

Figure 9.2 A process flow for transforming targeted collection into incidental collection.

9.5 Rethinking the Balance Between Tasking and Exploitation

Incidental collection has direct and disruptive implications for several pieces of the traditional TCPED cycle. The first and perhaps most significant is drastically re-examining the nature of the requirements and tasking process traditionally employed in most intelligence disciplines. The current formal process for developing intelligence requirements was established after the Second World War, and remains largely in use today. It replaced an ad hoc, informal process of gathering intelligence and professionalized the business of developing requirements [7].

Like most formal intelligence processes, the legacy of Cold War intelligence requirements was tuned to the unique situation between 1945 and 1991, a bipolar contest between two major state adversaries: the United States and the Soviet Union. Thus the process was created with assumptions that, while true at the time, have become increasingly questionable in the unipolar world with numerous near-peer state competitors and increasingly powerful nonstate actors and organizations.

These processes were adapted when IMINT stepped onto the intelligence scene with the development of high-altitude photoreconnaissance platforms and imaging satellites in the 1950s and 1960s. With point collectors and valuable sortie time for airplanes and satellites, requirements were defined specifically, with an effort to ensure that just enough information was collected to satisfy requirements and that excess system capacity was not wasted. This process of "satisficing"—collecting just enough that the requirement was fulfilled—required a clear understanding of the goals from the development of the requirement and management of the collection process. This, of course, meant that the information needs driving requirement generation, by definition, had to be clearly known, such that technical collection systems could be precisely tasked.

The result of this was an emphasis on tasking in the TCPED cycle, focusing on solving the problem of allocating intelligence resources. In an era that privileged clandestine information and technical sensors, this approach was both sensible and effective. Precise amounts of information were collected in service of narrow requirements, and privileged information was exploited in service of the overall intelligence cycle.

The shift of the modern era from clandestine and technical sensors to new, high-volume approaches to technical collection; wide-area and persistent sensors with long dwell times; and increasing use of massive volumes of information derived from open and commercial sources demands a parallel shift in emphasis of the tasking process. Because of the massive volumes of information gained from incidentally collected—or constructed—data sets, tasking is no longer the most important function. Rather, focusing increasingly taxed exploitation resources becomes critical; in addition, the careful application of automation to prepare data in an integrated fashion (performing functions like feature extraction, georeferencing, and semantic understanding) is necessary. "We must transition from a target-based, inductive approach to ISR that is centered on processing, exploitation, and dissemination to a problem-based, deductive, active, and anticipatory approach that focuses on end-to-end ISR operations," according to Maj. Gen. John Shanahan, commander of the 25th Air Force who adds that automation is "a must have" [8].

Focusing on exploiting specific pieces of data is only one part of the puzzle. A new paradigm for collection must be coupled to the shift from tasking collection to tasking exploitation. Rather than seeking answers to predefined intelligence needs, collection attuned to ABI's methodology demands seeking data, in order to enable correlations and entity resolution.

9.6 Collecting to Maximize Incidental Gain

The concept of broad collection requirements is not new. ABI, however, is fed by broad requirements for specific data, a new dichotomy not yet confronted by the intelligence and law enforcement communities. This necessitates a change in the tasking and collection paradigms employed in support of ABI, dubbed coarse tasking for discovery.

Decomposing this concept identifies two important parts: the first is the concept of coarse tasking, and the second is the concept of discovery. Coarse tasking first moves collection away from the historical use of collection decks consisting of point targets: specific locations on the Earth. These decks have been used for both airborne and space assets, providing a checklist of targets to service. Coverage of the target in a deck-based system constitutes task fulfillment, and the field of view for a sensor can in many cases cover multiple targets at once.

Figures 9.3 and 9.4 show the geographic and tabular view of traditional deck-based collection approaches. The focus is on obtaining coverage of fixed targets. As seen in Figure 9.4, this ultimately constitutes a checklist, with a binary "yes-no" evaluation in terms of whether a target was collected (or not). In many cases, these plans are created on a periodic basis, often every day, with modifications occasionally coming in the form of "ad hoc" requests, inserted through processes outside the typical channels used to validate collection needs, or reflecting request submitted on a "not-to-interfere" basis. Prioritization of targets plays an important role in this construct, instructing the collector to focus energy on certain targets, even at the expense of others. The tasking model used in collection decks is specific, not coarse, providing the most relevant point of contrast with collection specifically designed for supporting ABI analysis.

Collection using a coarse tasking model operates very differently. While geographic bounds still play an important role in the collection process, the specific goal of collection and metrics used for evaluation change dramatically. Rather than measuring fulfillment via a checklist model, coarse tasking's goal

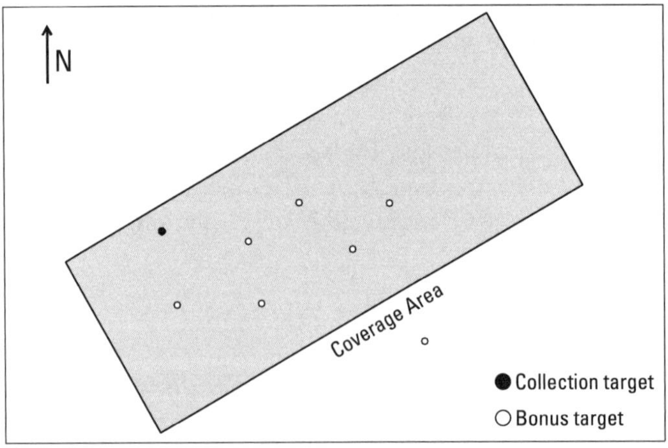

Figure 9.3 A notional series of targets for an assigned collector, located in geographic proximity to each other.

Collection Plan – 12 December

Target ID	Name	Location	Priority Level
140X02935-1	JAMES BARRACKS	34-20-55N 071-22-41E	1
123Z71238-4	JAMES SUPPLY DEPOT	33-12-48N 070-10-31E	1
140A10002-0	JAMES AMMO DUMP	33-51-55N 070-57-14E	3
102P25267-4	WENDELL TRAINING AREA	34-40-41N 072-42-50E	2

Figure 9.4 A sample collection plan for a static deck of four targets. These plans are created in advance and, while subject to change, most often collected as produced.

is to maximize the amount of data (and as a corollary, the amount of potential correlations) in a given collection window. This is made possible because the analytic process of spatiotemporal correlation is what provides information and ultimately meaning to analysts, and the pillar of data neutrality does not force analysts to draw conclusions from any one source, instead relying on the correlations between sources to provide value. Thus, collection for ABI can be best measured through volume, identification and conditioning of relevant metadata features, and spatiotemporal referencing.

This leads directly into enabling discovery as a consequence of coarse tasking. With collection volume maximized, potential discoveries via spatiotemporal correlations as well as metadata-based correlations are made possible. As the volume and number of specific data points increases, so too does the potential for new discoveries buried in the data. Unlike with the target deck model of collection, key intelligence questions are often still unknown at this point in the process, meaning that increasing collection in order to spur thinking about the right questions to ask is as important (and perhaps more important) than answering known questions using available collection resources.

9.7 Incidental Collection and Privacy

This approach can raise serious concerns regarding privacy. "Collect it all, sort it out later" is an approach that, when applied to signals intelligence, raised grave concern about the potential for collection against U.S. citizens. This is a problem that every new intelligence methodology, particularly those focused around new kinds of large-volume sensors, will have to confront.

Incidental collection has been portrayed in a negative light with respect to the Section 215 metadata collection program [9]. Part of this, however, is a direct result of the failure of intelligence policy and social norms to keep up with the rapid pace of technological development. The same is true of norms regarding privacy as well as the increasing accessibility of areas across the world to U.S. citizens, increasing the potential for the accidental collection of information.

U.S. intelligence agencies, by law, cannot operate domestically, with narrow exceptions carved out for disaster relief functions in supporting roles to lead federal agencies [10]. This restriction to overseas operations combined with very narrow, targeted approaches to gathering intelligence made the overall likelihood of accidental collection against U.S. citizens fairly minimal. However, the proliferation of open source information, in particular social networking information, raises new and interesting questions regarding the ability of intelligence agencies to leverage this information. Can intelligence agencies collect and mine publicly available information?

In most cases, this remains an unsettled issue. While this book will not delve into textual analysis of existing law and policy, one issue that agencies will be forced to confront is the ability of commercial "big data" companies like Google and Amazon to conduct the kind of precision analysis formerly possible only in a government security context. The ability to do this is made possible by the reams of data now available through location- and context-aware devices; datafication of everyday behavior through devices like FitBits; and publicly-accessible social media information from sites like LinkedIn, Twitter, and Facebook. Chapter 12 offers an in-depth examination of persistent surveillance technology.

These sources offer the potential for large, persistent, incidentally collected data sets that can be both exploited and combined with other data sources. In the context of ABI, while only a relatively percentage of Tweets are geotagged, the resulting subset of tweets still numbers in the millions. As of 2013, the University of Southern California assessed the number to be approximately 20 percent (including location divulged from geotagging or metadata); restricting the size to geotagging alone would drop the overall number further [11].

9.8 Summary

With a thorough understanding of ABI's methodology, there is a need to understand the technological developments that enable the methodology. Chapters 10–17 explore technology concepts like "big data" that are central to ABI's processes and that must be understood by analysts and technologists alike.

References

[1] Clark, R., "Perspectives on Intelligence Collection," *The Intelligencer,* Vol. 20, No. 2, Autumn–Winter 2013.

[2] Waltrop, D., "Recovery of the Last GAMBIT and HEXAGON Film Buckets from Space, August–October 1984," *Studies In Intelligence,* Vol. 58, No. 2, pp. 19–34.

[3] Ruffner, K., ed., *CORONA: America's First Satellite Program,* Washington, DC: Center for the Study of Intelligence, 1995.

[4] "NIMA Commission Report—NIMA in Context," Federation of American Scientists. [Online]. Available: http://fas.org/irp/agency/nima/commission/article05.htm [Accessed: 12- Dec-2014].

[5] Erwin, M., "Intelligence Issues for Congress," Congressional Research Service, Washington, D.C., RL33539, 2013.

[6] Quinn, K., "A Better Toolbox," *Trajectory Magazine,* Winter 2012.

[7] Heffter, C., "A Fresh Look at Collection Requirements," Central Intelligence Agency, 09–18-1995.

[8] "Q&A: Major General John N.T. 'Jack' Shanahan," KMI Media Group, 05–08-2014.

[9] Stout, M., "Incidental Collection," War on the Rocks, 07–11-2014. [Online]. Available: http://warontherocks.com/2014/07/warchives-incidental-collection/#_. [Accessed: 12-Dec-2014].

[10] "About NGA," National Geospatial-Intelligence Agency. [Online]. Available: https://www1.nga.mil/about/Pages/default.aspx. [Accessed: 12-Dec-2014].

[11] "Twitter and Privacy: Nearly one-in-five Tweets divulge user location through geotagging or metadata," University of Southern California, 09–03-2013. [Online]. Available: https://pressroom.usc.edu/twitter-and-privacy-nearly-one-in-five-tweets-divulge-user-location-through-geotagging-or-metadata/. [Accessed: 12-Dec-2014].

10

Data, Big Data, and Datafication

The principle of data neutrality espouses the use of new types of data in new ways. ABI represented a revolution in how intelligence analysts worked with a volume, velocity, and variety of data never before experienced. This chapter describes the basic principles of information management, introduces the "big data" revolution, and describes emerging efforts to integrate big data and intelligence.

10.1 Data

The amount of digital data is increasing exponentially. The U.S. Library of Congress, the largest library in the world, comprises about 158 million assets: 36 million books and print materials, 14 million photographs, 3.5 million recordings, 69 million manuscripts, and 5.5 million maps. Its digitized holdings are about 200 terabytes [1]. By April of 2014, social networking service Facebook was archiving over three times its total holdings per day [2]. This explosion of digital data is driven by the proliferation of Internet-connected mobile devices and increasingly pervasive sensors. Deriving value from large volumes of disparate data is the primary objective of an intelligence analyst.

Data is comprised of the atomic facts, statistics, observations, measurements, and pieces of information that are the core commodity for knowledge workers like intelligence analysts. Data represents the things we know.

The discipline of intelligence used to be data-poor. The things we did not know, and the data we could not obtain far outnumbered the things we knew and the data we had. Today, the digital explosion complicates the work environment because there is so much data that it is simply not possible to gather, process, visualize, and understand it all. Historical intelligence textbooks describe techniques for reasoning through limited data sets and making informed judgments, but analysts today have the possibility to obtain exceedingly large quantities of data. The key skill now is the ability to triage, prioritize, and correlate information from a giant volume of data.

10.1.1 Classifying Data: Structured, Unstructured, and Semistructured

The first distinction in data management relies on classification of data into one of three categories: structured data, unstructured data, or semistructured data. The former comprised the predominant information type in the 20th century, while unstructured and semistructured data have exploded dramatically in volume and popularity, introducing new challenges for managing and understanding information. Each class of data requires fundamentally different approaches to storage, ingestion, management, and analysis. Analysis to support ABI requires integration of multi-INT data from all three classes to resolve entities and understand patterns.

Structured Data

Structured data is "data that resides in fixed fields within a record or file" [3]. Examples include tables, spreadsheets, or databases. The field of information management evolved since the 1950s to store structured data. The most popular database format for structured data is called a relational database management system (RDBMS) based on the relational model developed by IBM's E. F. Codd in 1969 [4].

Relational databases are ideal for storage of transactional information like financial transactions, logistics records, sales records, call histories, and other information that fits an easily standardized, well-defined approach. The structure of the database is called the data model or schema. In a relational database, the schema must be defined a priori to match a desired business process. For example, consider: "Big bank requires the ability to store financial transactions from credit card purchases made worldwide." From this description, the schema must contain the credit card number, the amount of the purchase, and the country of the purchase. We also determine that the "amount" field must contain units for multiple currencies or that a second field is required for the type of currency.

Different tables link together using a key, a common attribute between two tables. If table A contains a list of transactions for multiple bank accounts and table B contains a list of account owners for each account number, the account number is the common key that links the two tables as shown in Figure 10.1. User Al Dee with account number 257417014 runs a simple calculation upon login to calculate the balance of his account by summing all the transac-

Table A

Trans #	Account #	Amount ($)
1	373641324	$57.99
2	257417014	$61.50
3	257417014	$1.21

Table B

Account #	First	Last
239872452	Steve	Erino
234987897	Doc	Brown
257417014	Al	Dee

Account # is a Key from A to B

Figure 10.1 Example of a key joining two database tables in a relational database.

tions for his account from table A and subtracting them from his previously monthly balance. This is how analytics are typically applied in RDBMSs.

Transactional data, which may also be thought of as short simple records in well-defined fields, fits well into the relational model. When digital multimedia proliferated in the 1990s, the relational model adapted to store the digital data—also called the package—either within a database field or as a link to a directory of files. This type of database table is called a catalog. One example is a catalog of imagery that contains fields like the file name, time/date taken, camera type, latitude/longitude, image size, bit depth, a free text field for descriptive text or tags, and a link to the digital package. The Department of Defense stores national imagery using a relational database model with similar fields while the image package is stored in the National Imagery Transmission Format—itself a highly structured database wrapper around a digital image [5]. Using such a relational database and a table joined with a database of facilities, an analyst can query for all the imagery that contains a target facility over a period of time.

Queries to a relational database use a special-purpose programming language called the structured query language (SQL), pronounced "es-cue-el" or "see-kwell." SQL uses Boolean operations and conditional expressions (where, then, else, if) to retrieve data from an RDBMS. It can also be used to insert and manipulate data in tables. SQL works well with relational databases, but critics highlight the lack of portability of SQL queries across RDBMSs from different vendors due to implementation nuances of relational principles and query languages.

As data tables grow in size (number of rows), performance is limited, because many calculations must search the entire table. Join operations that use relational keys to combine values from multiple tables become exponentially more complicated as the size and complexity of data grows. When the attributes of the data and the desired business processes cannot be stated a priori, modifications and alterations to database structured and schema are costly and error-prone. New models have been developed to address these challenges and those introduced by a widely proliferating class of data: unstructured data.

Unstructured Data

The introduction of the World Wide Web in 1996 and the rapid proliferation of e-commerce and Internet-connected mobile devices complicated the field of information management. For the first time, a majority of the data being produced was unstructured and did not fit well into the popular RDBMS model for information storage. Unstructured data does not follow a preformatted schema. This includes information in free-flowing text documents, Microsoft-Word files, web sites, blogs, patents, manuscripts, and tweets.

"Not only SQL" (NoSQL) is a database concept for modeling data that does not fit well into the tabular model in relational databases. There are two classifications of NoSQL databases, key-value and graph. Column family or document stores are subtypes of key-value databases. The types of NoSQL databases are summarized in Table 10.1.

NoSQL databases are referred to as "schema-less" databases, but more accurately, their schema does not need to be defined a priori before the database is instantiated. This means that a user can add new keys or columns without recreating the entire database and migrating from one table to another. This property is also useful for evolving problems (and intelligence problems) where the attributes of the problem and desired business processes are difficult to define.

One of the advantages of NoSQL databases is the property of *horizontal scalability,* which is also called *sharding.* Sharding partitions the database into

Table 10.1
Four Types of NoSQL Databases

Type	Description	Examples
Key-Value	The simplest type of NoSQL database to implement; an associative array of (key, value) pairs where each key maps uniquely once to a value in the table.	Dynamo, FoundationDB, MemcacheDB, Redis, Riak
Column Family	Subset of a key-value database, except that key may point to multiple values differentiated by a timestamp. Instantiates multiple records across column space which may improve performance and disk access	Accumulo, Cassandra, HBase
Document	Subset of a key-value database, except that the value contains a data package that is a text document. The document is not stored as a binary object, but as a series of words that can be queried and processed in addition to the primary key.	Couchbase, MarkLogic, MongoDB, LotusNotes
Graph	Uses a graph structured with nodes, properties and edges to store data. Highly flexible to represent complex data dominated by many relationships. (This type of data is extremely difficult to model at scale using RDBMS).	Allegro, Neo4J, Virtuoso, Brightstar DB, DEX, Horton, Oracle Spatial and Graph

smaller elements based on the value of a field and distributes this to multiple nodes for storage and processing. This improves the performance of calculations and queries that can be processed as subelements of a larger problem using a model called "scatter-gather" where individual processing tasks are farmed out to distributed data storage locations and the resulting calculations are reaggregated and sent to a central location.

Semistructured Data

The term semistructured data is technically a subset of unstructured data and refers to tagged or taggable data that does not strictly follow a tabular or database record format. Examples include markup languages like XML and HTML where the data inside a file may be queried and analyzed with automated processes, but there is no simple query language that is universally applicable. Semistructured data does not require a data model, so it may be "easier" or "quicker" to set up an information archive; however, as these data sets proliferate and are used for different purposes, it becomes increasingly costly and difficult to maintain them at scale. This is because different people use different tags to mean the same thing or the same tag to mean different things. Semistructured databases do not require formal governance, but operating a large data enterprise without a governance model makes it difficult to find data and maintain interoperability across data sets.

10.1.2 Metadata

Metadata is usually defined glibly as "data about data." The purpose of metadata is to organize, describe, and identify data. The schema of a database is one type of metadata. The categories of tags used for unstructured or semistructured data sets are also a type of metadata.

The metadata in a library card catalog contains the author, title, subject, call number, and category. Digital photographs may contain metadata about the date/time of the image, location (if using a GPS-enabled camera), image size, camera model, settings, and dozens of other fields. (The image itself is not metadata. It is the package).

Metadata may include extracted or processed information from the actual content of the data. Chapter 12 details techniques for automatically detecting moving objects in video. The number of movers in each time slice of the video may be tagged as a metadata stream that is associated with the raw data. In this example, an analyst could query the video for "the frame with the largest number of movers" or "the frames (and timestamps) where the mover count is greater than 50." Advanced algorithms that identify activities in video like digging, dancing, running, or fighting can also tag activities as metadata.

Clip marks—analyst-annotated explanations of the content of the video—are considered metadata attached to the raw video stream.

Sometimes, the only common metadata between data sets is time and location. We consider these the central metadata values for ABI. The third primary metadata field is a unique identifier. This may include the ID of the individual piece of data or may be associated with a specific object or entity that has a unique identifier. Because one of the primary purposes of ABI is to disambiguate entities and because analytic judgments must be traced to the data used to create it, identifying data with unique identifiers (even across multiple databases) is key to enabling analysis techniques.

10.1.3 Taxonomies, Ontologies, and Folksonomies

A taxonomy is the systematic classification of information, usually into a hierarchical structure of entities of interest. A commonly recognized taxonomy is the biological classification scheme developed by Carolus Linnaeus, who grouped species according to shared or similar physical characteristics. The seven-level taxonomy (kingdom, phylum, class, order, family, genus, species) is standardized by international bodies to facilitate the ease of information sharing on the discovery and classification of new species.

Because many military organizations and nation-state governments are hierarchical, they are easily modeled in a taxonomy. Also, because the type and classification of military forces (e.g., aircraft, armored infantry, and battleships.) are generally universal across different countries, the relative strength of two different countries is easily compared. Large businesses can be described using this type of information model. Taxonomies consist of classes but only one type of relationship: "is child/subordinate of."

An *ontology* "provides a *shared vocabulary* that can be used to model a domain, that is, the type of objects and or concepts that exist and their properties and relations" (emphasis added) [6, p. 5]. Ontologies are formal and explicit, but unlike taxonomies, they need not be hierarchical. Also, the number of relationships allowed between objects and concepts is greater than one. This type of knowledge model is more readily applied to networks or unconventional organizations (e.g., criminal organizations and social networks) that have many complex, interrelated relationships and rapidly evolving properties. Most modern problems have evolved from taxonomic classification to ontological classification to include the shared vocabulary for both objects and relationships. Ontologies pair well with the graph-based NoSQL database method described in Chapter 15. It is important to note that ontologies are formalized, which requires an existing body of knowledge about the problem and data elements.

With the proliferation of unstructured data, user-generated content, and democratized access to information management resources, the term *folksonomy* evolved to describe the method for collaboratively creating and translating tags to categorize information [7]. Unlike taxonomies and ontologies that are formalized, folksonomies evolve as user-generated tags are added to published content. Also, there is no hierarchical (parent-child) relationship between tags. This technique is useful for highly emergent or little understood problems where an analyst describes attributes of a problem, observations, detections, issues, or objects but the data does not fit an existing model. Over time, as standard practices and common terms are developed, a folksonomy may be evolved into an ontology that is formally governed. The key aspects of these three organizational models are summarized in Table 10.2.

10.2 Big Data

Big data is an overarching term that refers to data sets so large and complex they cannot be stored, processed, or used with traditional information management techniques. Altamira's John Eberhardt defines it as "any data collection that cannot be managed as a single instance" [8]. Joe Hellerstein, a computer scientist at the University of California in Berkeley, calls it "the industrial revolution of data" [9]. The data revolution is impacting government, commercial enterprise, and individuals. Some examples include the following:

- Astronomy: Collecting over 200 gigabytes per night, the Sloan Digital Sky Survey collected more information in its first few weeks than all the data collected in the history of astronomy [9].
- Physics: The Large Hadron Collider gathers the information from over 600 million collisions per second in a 1,450-m^2, 11,000-server data center that shares data in real time with over 8,000 physicists around the world. The grid runs more than two million processing jobs per day [10].
- e-Commerce: On "cyber Monday" in 2013, Amazon.com processed more than 36.8 million transactions—426 purchases per second [11].

Table 10.2
Summary of Taxonomies, Ontologies, and Folksonomies

Term	Primary Attribute	Structure	Flexibility	Complex Relationships?
Taxonomy	Hierarchical Structure	High	Low	No
Ontology	Formal, shared vocabulary	Medium	Medium	Yes
Folksonomy	Groups of tags; collaboratively created	Very Low	High	No

- Social media: Tumblr blog authors publish 27,000 posts every minute. Facebook has over 50 billion photos. Twitter users tweet more than 340 million times per day [12].

Hundreds of eye-popping statistics and infographics proliferate across the Internet, but the fundamental aspect of this information revolution is the change in focus from the way large enterprises store, process, and use information to derive actionable intelligence and provide strategic advantage.

10.2.1 Volume, Velocity, and Variety...

In 2001, Gartner analyst Doug Laney introduced the now ubiquitous three-dimensional characterization of "big data" as increasing in volume, velocity, and variety [13]:

- Volume: The increase in the sheer number and size of records that must be indexed, managed, archived, and transmitted across information systems.
- Velocity: The dramatic speed at which new data is being created and the speed at which processing and exploitation algorithms must execute to keep up with and extract value from data in real time. In the big data paradigm, "batch" processing of large data files is insufficient.
- Variety: While traditional data was highly structured, organized, and seldom disseminated outside an organization, today's data sets are mostly unstructured, schema-less, and evolutionary. The number and type of data sets considered for any analytic task is growing rapidly.

Since Laney's original description of "the three V's," a number of additional "V's" have been proposed to characterize big data problems. Some of these are described as follows:

- Veracity: The truth and validity of the data in question. This includes confidence, pedigree, and the ability to validate the results of processing algorithms applied across multiple data sets. Data is meaningless if it is wrong. Incorrect data leads to incorrect conclusions with serious consequences.
- Vulnerability: The need to secure data from theft at rest and corruption in motion. Data analysis is meaningless if the integrity and security of the data cannot be guaranteed.

- Visualization: Including techniques for making sense of "big data." (See Chapter 13.)
- Variability: The variations in meaning across multiple data sets. Different sources may use the same term to mean different things, or different terms may have the same semantic meaning.
- Value: The end result of data analysis. The ability to extract meaningful and actionable conclusions with sufficient confidence to drive strategic actions. Ultimately, value drives the consequence of data and its usefulness to support decision making.

While numerous other "V's" have been proposed, the above definitions highlight the most significant attributes of "big data" for intelligence applications. Because intelligence professionals are called on to make judgments, and because these judgments rely on the underlying data, any failure to discover, correlate, trust, understand, or interpret data or processed and derived data and metadata diminishes the value of the entire intelligence process. In the traditional data-poor model of intelligence, validation of each piece of data was simple because of the sparsity and consistency of known sources on known issues. Critics of ABI and the analysis of "big data" warn of "Deus ex machina" (God from the machine) or "garbage-in-garbage-out" that could result from ignorance of the aforementioned "V's."

10.2.2 Big Data Architecture

"Big data" definitions say that a fundamentally different approach to storage, management, and processing of data is required under this new paradigm, but what are some of the technology advances and system architectural distinctions to enable "big data?"

In 2004, Google began developing and implementing a large-scale column-family key-value store called BigTable. In one of the first public discussions of the project, Google's Jeff Dean described the table where rows were URLs and the columns stored file data or other contents [14]. This arrangement allows for efficient search to locate a URL where the file contents (HTML descriptor of the page content) are used for a keyword search. The table is a sparse, multidimensional sorted map that is easily distributed across multiple processing nodes, allowing Google to affordably scale to petabyte-scale databases across thousands of commodity computers and easily add more hardware without reconfiguration. Because hardware failures are inevitable, this scaling property also allowed Google to efficiently replicate data across multiple nodes and remove the potential for data loss due to hardware failure. A traditional approach to this problem would have required exponentially more expensive "tier-1" low-failure rate, redundant equipment.

Most "big data" storage architectures use a key-value store based on Google's BigTable. Accumulo is a variant of BigTable that was developed by the National Security Agency (NSA) beginning in 2008. Accumulo augments the BigTable data model to add cell-level security, which means that a user or algorithm seeking data from any cell in the database must satisfy a "column visibility" attribute of the primary key [15]. In a surprise move, NSA released 200,000 lines of mostly Java code and hundreds of pages of documentation to the Apache Foundation, making the project open-source.

Google further innovated by implementing a framework for distributed processing now referred to as MapReduce, the processing analogue to BigTable's storage paradigm. The Map() procedure distributes a processing request to multiple nodes. Because BigTable is a sparse, column-family matrix, it is easily partitioned into subproblems. A Map() function could include summarizing the number of data elements that include a key word or locating a file based on user-defined criteria. The Reduce() operator then gathers the output of multiple Map() commands into a single aggregated result. These operations are shown in Figure 10.2.

One of the most popular open-source instances of the MapReduce model is Apache Hadoop. Hadoop relies on a distributed, scalable Java file system, the Hadoop distributed file system (HDFS), which stores large files (gigabytes to terabytes) across multiple nodes with replication to prevent data loss. Typically, the data is replicated three times, with two local copies and one copy at a geographically remote location. According to Hadoop vendors, more than half

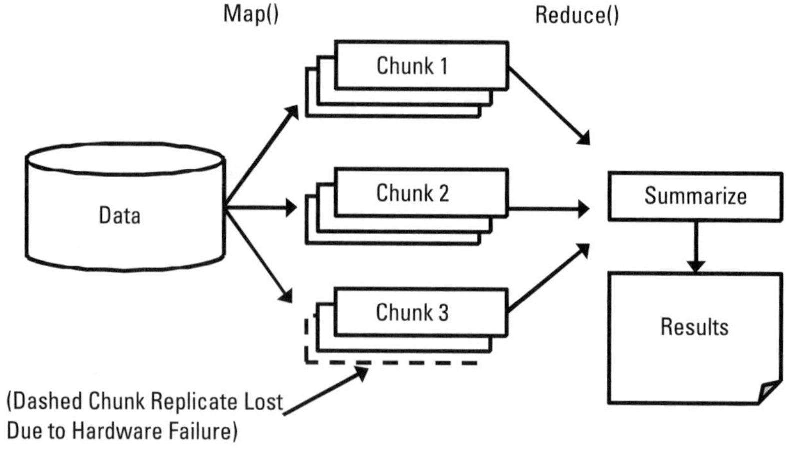

Figure 10.2 The MapReduce model.

of Fortune 50 companies use the technology [16]. Data management heavyweights IBM, Microsoft, and Oracle have adopted Hadoop within their architectures and offerings [17].

Recognizing that information is increasingly produced by a number of high-volume, real-time devices and must be integrated and processed rapidly to derive value, IBM began the System S research project as "a programming model and an execution platform for user-developed applications that ingest, filter, analyze, and correlate potentially massive volumes of continuous data streams" [18]. Commercialized as a product called InfoSphere® Streams, the software manages up to millions of events or messages per second with submillisecond response times. IBM refers to this model of processing "data in motion" as "stream computing." InfoSphere® Streams enables implementation of analytics on real-time data. One type of analytic identifies tweets with a given hashtag, calculates the sentiment associated with the tweet, and aggregates the results into a real-time display. Commercial companies use this capability to get real-time customer sentiment about their brand. They use stream computing to derive real-time feedback on new advertisements, stock movements, public statements, promotions, or viral videos and comments [19].

Figure 10.3 shows a generic architecture stack to support big data analysis. The architecture is divided into service tiers: infrastructure services, data services, platform services, and software services. Shaded boxes highlight architectural elements that are new or significantly changed under the "big data" model. Big data architectures are increasingly being implemented to support ABI data management and analytics.

10.2.3 Big Data in the Intelligence Community

Recognizing that information technology spending across the 17 intelligence agencies accounts for nearly 20% of National Intelligence Program funding, the intelligence community embarked on an ambitious consolidation program called the intelligence community information technology environment (IC-ITE), pronounced "eye-sight" [20]. IC-ITE includes a single IC desktop, an "apps mall" for community applications, and a common platform and infrastructure that implements cloud-computing technologies

IC-ITE includes the deployment of two large-scale cloud computing technologies as shown in Figure 10.4. The first, IC GovCloud, is deployed by NSA and is based on the Google cloud paradigm. The second cloud was acquired by the CIA in a $600-million deal with Amazon.com [22]. Based on Amazon Web Services (AWS), the Commercial Cloud Services (C2S) instance provides pay-as-you-go on demand computing resources across intelligence community implementations [20].

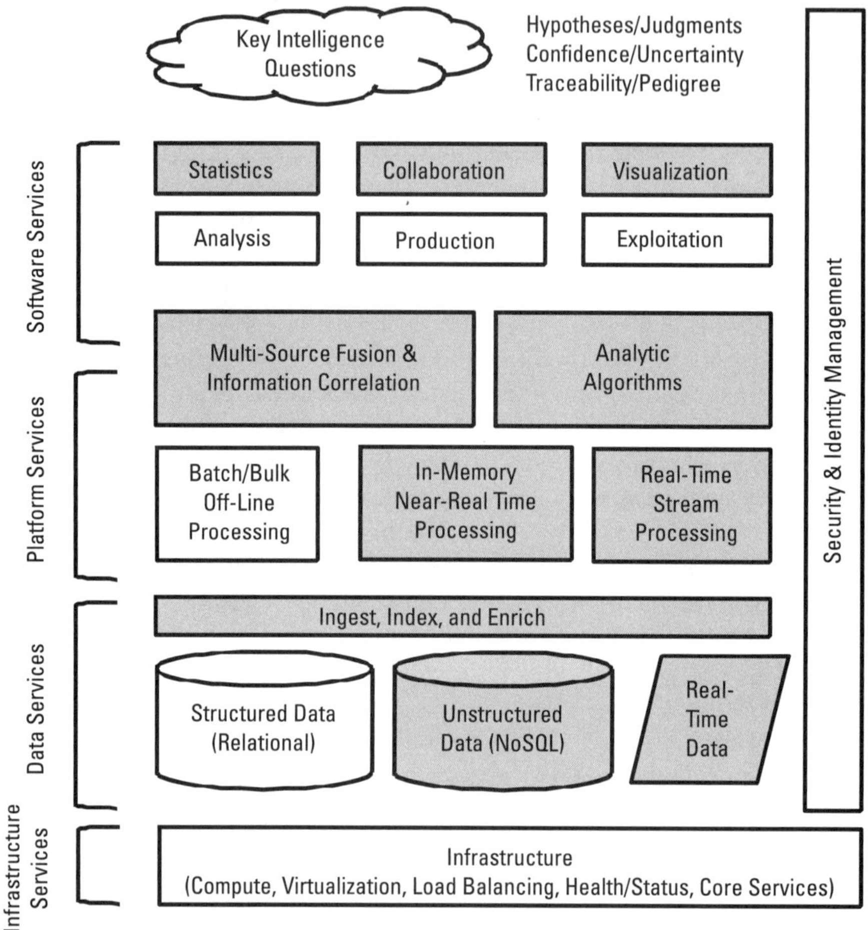

Figure 10.3 Generic architecture and key components of a "big data" technology stack.

IC GovCloud provides high performance analytics and large-scale computing on primarily document-based information. This type of architecture excels at keyword searches and other large-scale queries. The Google-based technology stack requires specialized programming techniques to take advantage of parallelization features. C2S on the other hand is not optimized for large-scale, high-performance computing. It is optimized around low-intensity, efficient utilization of resources. C2S deployments are ideal for web servers and some infrequently used analytic processes while IC GovCloud is more appropriate for repetitive "big data" analytics across massive databases. Ongoing efforts seek to deploy IT capabilities onto the community-provided infrastructure to enhance information sharing and lower IT costs.

Figure 10.4 Major components of IC-ITE [21].

10.3 The Datafication of Intelligence

In 2013, Kenneth Neil Cukier and Victor Mayer-Schöenberger introduced the term "datafication" to describe the emergent transformation of everything to data. "Once we datafy things, we can transform their purpose and turn the information into new forms of value," they said [23]. A revolution is under way within the intelligence community to datafy all information into forms that are more easily discoverable, shareable, and correlatable across the spectrum of multi-INT.

The impetus for datafication is partially driven by the recommendations of the 9/11 Commission, which noted that a decentralized network model allows agencies to search across agency lines so that the maximum number of users can access some form of all information [24, p. 418]. The report also recommends that the intelligence community establish a formalized policy for information sharing. Intelligence Community Directive 501, Discovery and Dissemination or Retrieval of Information Within the Intelligence Community recognizes the need to "foster an enduring culture of responsible sharing and collaboration within an integrated [intelligence community]" [25].

Much has been made about the revolution in big data analytics and what it means to the everyday lives of ordinary citizens. Over the last 10 years, direct application of commercial "big data" analytic techniques to the intelligence community has thus far missed the mark. There are a number of reasons for this, but first and foremost among them is the fact that a majority of commercial "big data" is exquisitely structured and represents near complete data sets. For example, the record of credit card transactions at a major department store includes only credit card transactions at that department store, and not random string of numbers that might be UPC codes for fruits and vegetables at a cross-town grocery store. In contrast, intelligence data is either typically unstructured text captured in narrative form or arrives as a mixture of widely differing structures.

Furthermore, the nature of intelligence collection—the quest to obtain information on an adversary's plans and intentions through a number of collection disciplines—all but ensures that the resulting data sets are "sparse," representing only a small portion or sample of the larger picture from which they are drawn. The difficulty is that unlike the algorithm-based methods applied to commercial big data, it is impossible to know the bounds of the larger data set. Reliable and consistent inference of larger trends and patterns from a limited and unbounded data set is impossible.

This does not mean intelligence professionals cannot learn from and benefit from the commercial sector's experiences with big data. Indeed, industry has a great deal to offer with respect to data conditioning and system architecture. These aspects of commercial systems designed to enable "big data" analysis will be critical to designing the specialized systems needed to deal with the more complex and sparse types of data used by intelligence analysts.

10.3.1 Collecting It "All"

While commercial entities with consistent data sets may have success using algorithmic prediction of patterns based on dense data sets, the key common methodology between "big data" and ABI is the shift away from sampling information at periodic intervals toward examining massive amounts of information abductively and deductively to identify correlations. Cukier and Mayer-Shoeberger write:

> When collecting data was costly and processing it was difficult and time consuming, the sample was a savior. Modern sampling is based on the idea that, within a certain margin of error, one can infer something about the total population from a small subset, as long the sample is chosen at random. Hence, exit polls on election night query a randomly selected group of several hundred people to predict the voting behavior of an entire state. For straightforward questions, this process works well. But it falls apart when we want to drill down into subgroups within the sample. What if a pollster wants to know which candidate single women under 30 are most likely to vote for? How about university-educated, single Asian American women under 30? Suddenly, the random sample is largely useless, since there may be only a couple of people with those characteristics in the sample, too few to make a meaningful assessment of how the entire subpopulation will vote. But if we collect all the data—'n = all,' to use the terminology of statistics—the problem disappears [23].

Cukier and Mayer-Schoenberger, in their assessment of the advantages of "n = all," effectively argue for a move to a more deductive workflow based on data correlations, rather than causation based on sparse data. "n = all" and georeference to discover share the common intellectual heritage predicated on collecting all data in order to focus on correlations in a small portion of the dataset: Collect broadly, condition data, and enable the analyst to both explore and ask intelligence questions of the data.

The approach of "n = all" is the centerpiece of former NSA director general Keith Alexander's philosophy of "collect it all." According to a former senior U.S. intelligence official, "rather than look for a single needle in the haystack, his approach was, 'Let's collect the whole haystack. Collect it all, tag it, store it… and whatever it is you want, you go searching for it'" [26]. In the past, when collection was limited, collection managers had to decide what to collect and how it would be used. One of the key advantages of collecting and indexing all (of the available and allowable data) is that you do not need to decide beforehand how the data will be used, the premise of incidental collection as described in Chapter 9.

This is also an eloquent argument for sequence neutrality as a function of one's approach to data collection. By explicitly stipulating that the goal is to collect as much information as possible without predefining an intended purpose, the manner in which intelligence information is collected changes as well. The same advances that have already begun to pay dividends in the commercial sector can revolutionize how the U.S. government looks at intelligence information and how it collects such information. In the "big data" world of the early 21st century, both commercial businesses and intelligence enterprises embraced the four pillars of ABI without initially realizing they were doing so. Sections 10.3.2 and 10.3.3 provide high-level unclassified overviews of some IC programs for the datafication of intelligence.

10.3.2 Object-Based Production (OBP)

In 2013, Catherine Johnston, director of analysis at the Defense Intelligence Agency (DIA), introduced object-based production (OBP), a new way of organizing information in the datafication paradigm. Recognizing the need to adapt to growing complexity with diminishing resources, OBP implements data tagging, knowledge capture, and reporting by "organizing intelligence around objects of interest" [27, p. 3]. OBP addresses several shortfalls. Studies have found that known information was poorly organized, partially because information was organized and compartmented by the owner. Reporting was within INT-specific stovepipes. Further compounding the problem, target-based intelligence aligned around known facilities. Analysts spend most of their time assembling known data about these facilities, described in Figure 1.4 as monitoring. This stove-poped, owner-centric model is shown on the left of Figure 10.5.

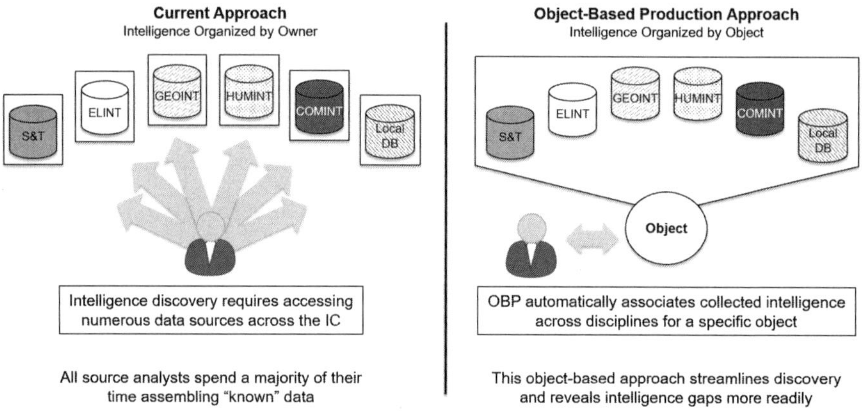

Figure 10.5 Implementation of the OBP approach. (Source: Defense Intelligence Agency [27, p. 4].)

An object- and activity-based paradigm is more dynamic. It includes objects that move, vehicles and people, for which known information must be updated in real time. This complicates timely reporting on the status and location of these objects and creates a confusing situational awareness picture when conflicting information is reported from multiple information owners. By organizing information around the object instead of the INT-specific stovepipe, analysts reduce the time spent searching for data, streamline discovery, and reveal intelligence gaps more readily as described on the right of Figure 10.5.

The goal of OBP is to provide current status on objects of interest by associating information with a shared set of real-world objects, providing a consolidated community perspective about the behavior of these objects and how they interact with one another. One way of capturing this information is through the means of a "baseball card." The baseball card displays the durable attributes of an object (e.g., length, speed, and number of missiles) and the activity attributes of the object, namely its current location and type of activity. An example of a baseball card for the pirate ship *Queen Anne's Revenge* is shown in Figure 10.6.

According to Johnston, QUELLFIRE is the intelligence community's program to deliver OBP as an enterprise service where "all producers publish to a unifying object model" (UOM) [27, p. 6]. Under QUELLFIRE, OBP objects are incorporated into the overall common intelligence picture (CIP)/common operating picture (COP) to provide situational awareness [29]. This focus means that the pedigree of the information is time-dominant and must be continually updated. Additional work on standards and tradecraft must be developed to establish a persistent, long-term repository of worldwide intelligence objects and their behaviors.

Vessel Name: *Queen Anne's Revenge*

Captain: Edward Thatch (aka: *Blackbeard*)

Current Location: 35.706119, -76.652025
Named Location: Beauford Inlet, North Carolina, USA

Year_Launched: 1710
Displacement: 300-tons
Length: 31.4 m
Beam 7.1 m
Max_Speed: 15 knots
Complement: 125
Num_Guns: 40
Num_Torpedoes: 0
Num_Missiles: 0
Radar: None

Current_Activity: Plundering
Previous_Activity: Rum_Running
Issue: Piracy, Commerce
Home_Port: East Indies Sea

Figure 10.6 Example of a "baseball card" for the pirate vessel Queen Anne's Revenge. (Vessel information source: Wikipedia [28].)

The National Geospatial-Intelligence Agency (NGA) adopted the OBP organizing model (Figure 10.7) to integrate multiple classes of data. Objects, relationships, and behaviors are integrated with multiple sources of data on a rich foundation of contextual GEOINT. Gauthier augments OBP with the addition of models that are "exposed and discovered by building them from simple and universal data building blocks" and proposes that collection strategies be designed around these models [30, p. 8].

10.3.3 Relationship Between OBP and ABI

There has been a general confusion about the differences between OBP and ABI, stemming from the fact that both methods focus on similar data types and are recognized as evolutions in tradecraft. OBP, which is primarily espoused by DIA, the nation's all-source military intelligence organization, is focused on order-of-battle analysis, technical intelligence on military equipment, the status of military forces, and battle plans and intentions (essentially organizing the known entities). ABI, led by NGA, began with a focus on integrating multiple sources of geospatial information in a geographic region of interest—evolving with the tradecraft of georeference to discover—to the discovery and resolution of previously unknown entities based on their patterns of life. This tradecraft produces new objects for OBP to organize, monitor, warn against, and report as shown in Figure 10.8. OBP, in turn, identifies knowledge gaps, the things that are unknown that become the focus of the ABI deductive, discovery-based pro-

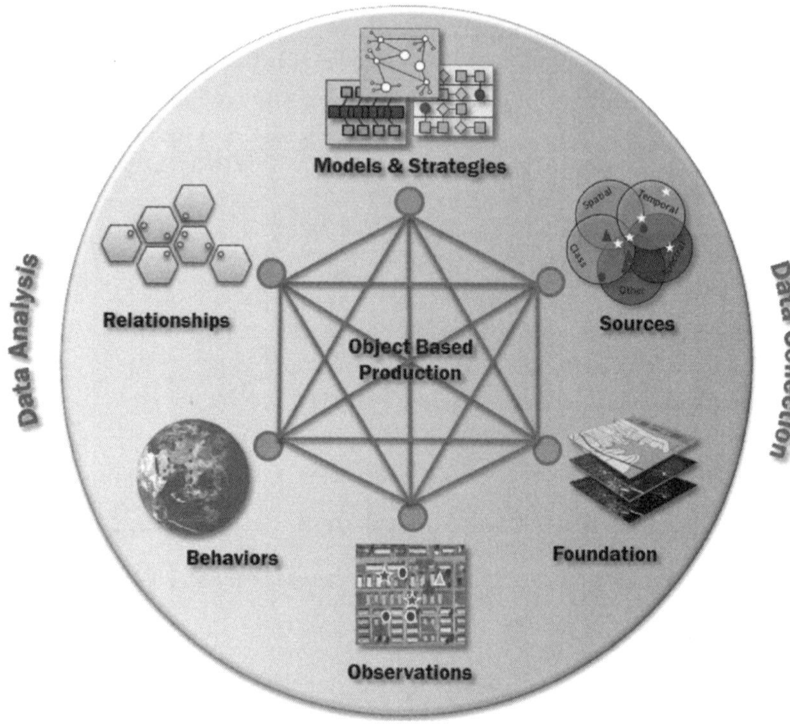

Figure 10.7 OBP as an organizing discipline for observations and supporting data. (Source: NGA. Approved for Public Release, 14-233 [30].)

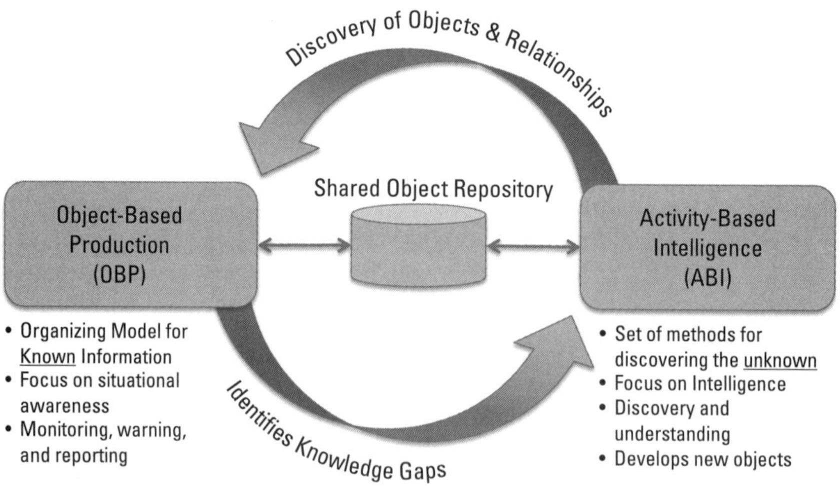

Figure 10.8 Relationship between ABI and OBP.

cess. Efforts to meld the two techniques are aided by the IC-ITE cloud initiative, which colocates data and improves discoverability of information through common metadata standards. This also includes a shared object repository that can be populated and analyzed using both approaches. This is the domain of knowledge management, described in Chapter 15.

10.4 The Future of Data and Big Data

Kevin Ashton, founder of MIT's Auto-ID center coined the phrase "Internet of Things" in 1999 to highlight a paradigm shift in the information landscape when a majority of digital information is created by machines rather than people. Ashton says, "nearly all of the roughly 50 petabytes of data available on the Internet were first captured and created by human beings—by typing, pressing a record button, taking a digital picture, or scanning a bar code" [31]. Research firm Gartner believes that by 2020, there will be over 26 million machines pumping real-time data onto the Internet [32]. Ashton continues:

> If we had computers that knew everything there was to know about things—using data they gathered without any help from us—we would be able to track and count everything and greatly reduce waste, loss, and cost. We would know when things needed replacing, repairing or recalling, and whether they were fresh or past their best. The Internet of Things has the potential to change the world, just as the Internet did. Maybe even more so. [31]

Former CIA director David Petraeus highlighted the challenges and opportunities of the Internet of Things in a 2012 speech at In-Q-Tel, the agency's venture capital research group: "As you know, whereas machines in the 19th century learned to do, and those in the 20th century learned to think at a rudimentary level, in the 21st century, they are learning to perceive—to actually sense and respond" [33]. He further highlighted some of the enabling technologies developed by In-Q-Tel investment companies, listed as follows:

- Item identification, or devices engaged in tagging;
- Sensors and wireless sensor networks—devices that indeed sense and respond;
- Embedded systems—those that think and evaluate;
- Nanotechnology, allowing these devices to be small enough to function virtually anywhere.

In-Q-Tel's portfolio includes (or has included) Digital Reasoning, Endeca, FireEye, geoIQ, MetaCarta, Narrative Science, Nervve Technologies, Palantir, Recorded Future, Spotfire, Stratify, and SRD. Many of the early partner firms have been acquired by large companies including Google, IBM, Microsoft, Oracle, and Raytheon [34].

The potential to support OBP and ABI are tremendous, as physical objects and entities may self-report their location and status information to the Internet in real time. Analysts will be challenged to sort the wheat from the chaff and identify those entities that are behaving abnormally. Increasingly, it will be impossible for entities to "live off the grid" as humans will depend on information access as a basic resource.

In his remarks at the GigaOM Structure:Data conference in New York in 2013, CIA chief technology officer (CTO) Hunt said, "It is nearly within our grasp to compute on all human generated information" [35]. This presents new challenges but also new opportunities for intelligence analysis.

10.5 Summary

This chapter introduces the basic building blocks of data and big data, which are the underpinning of ABI analysis. Chapters 11–17 describe new collection, analysis, analytics, knowledge management, and information sharing technologies and concepts that will reshape the way analysts use data to gather insight, discover unknowns, and drive strategic advantage.

References

[1] "Fascinating Facts" Library of Congress. Available: http://www.loc.gov/about/fascinating-facts/. Accessed: 16 Jul 2014.

[2] Cohen, D., "How Facebook Manages a 300-Petabyte Data Warehouse, 600 Terabytes Per Day," April 11, 2014, http://www.adweek.com/socialtimes/orcfile/434041.

[3] "Definition: Structured Data." *PC Magazine Encyclopedia,* http://www.pcmag.com/encyclopedia.

[4] Codd, E. F., "A Relational Model of Data for Large Shared Data Banks," *Commun. ACM,* Vol. 13, No. 6, June 1970, pp. 377–387.

[5] "Interface Standard, National Imagery Transmission Format Version 2.1 (MIL-STD-2500C)," Department of Defense. May 1, 2006.

[6] Arvidsson, F., and A. Flycht-Eriksson, "Ontologies I," Powerpoint presentation, available: http://www.ida.liu.se/~janma/SemWeb/Slides/ontologies1.pdf.

[7] Peters, I., *Folksonomies: Indexing and Retrieval in Web 2.0,* Berlin, Germany: Walter de Gruyter GmbH and Co., 2009.

[8] Law, D., and J. Eberhardt, "Do You Know Big Data?," June 9, 2014, web, http://www.ctovision.com/download/know-big-data/.

[9] "Data, Data Everywhere. A Special Report on Managing Information," *The Economist,* February 2010.

[10] "Computing," CERN, web, http://home.web.cern.ch/about/computing.

[11] Palladino, V., "Amazon Sold 426 Items per Second in Run-up to Christmas," *The Verge,* December 26, 2013.

[12] Connor, M., "Data on Big Data," July 18, 2014, web, http://marciaconner.com/blog/data-on-big-data/. Accessed July 26, 2014.

[13] Laney, D., "3D Data Management: Controlling Data Volume, Velocity, and Variety," META Group, Research Note, Application Delivery Strategies, February 2001.

[14] Hitchcock, A., "Google's BigTable," web, http://andrewhitchcock.org/?post=214.

[15] Hoover, J. N., "NSA Submits Open Source, Secure Database to Apache." *InformationWeek,* September 6, 2011.

[16] "Altior's AltraSTAR - Hadoop Storage Accelerator and Optimizer Now Certified on CDH4," December 19, 2012.

[17] Noyes, K., "Hadoop: How a little open source project took over big data," *Fortune,* June 30, 2014.

[18] "Stream Computing Platforms, Applications, and Analytics—System S: Application Areas, System Components, and Programming Model," IBM, http://researcher.watson.ibm.com/researcher/view_group.php?id=2531.

[19] Brownlee, J.,"IBM's Next Big Thing: Psychic Twitter Bots," Co. Design, March 3, 2014.

[20] Slabodkin, G., "How Cloud Is Changing the Spy Game," *Defense Systems,* August 22, 2014.

[21] Anderson, S., "Navy's Journey to the JIE and IC ITE, A Process Not a Destination," *CHIPS,* The Department of the Navy's Information Technology Magazine, September 15, 2014, web.

[22] Babcock, C.,"Amazon Wins Best Cloud in CIA Bake-Off," *Information Week,* June 25, 2013.

[23] Cukier, K. N., and V. Mayer-Schoenberger, "The Rise of Big Data," *Foreign Affairs,* May/June 2013.

[24] *The 9/11 Commission Report: Final Report of the National Commission on Terrorist Attacks upon the United States,* Washington, D.C.: U.S. Government Printing Office, 2004.

[25] Director of National Intelligence, "Intelligence Community Directive (ICD) 501, Discovery and Dissemination or Retrieval of Information Within the Intelligence Community." 21 Jan 2009. Web: http://www.ncix.gov/publications/policy/docs/ICD_501-Discovery_and_Dissemination_or_Retrieval_of_Information_within_the_IC.pdf

[26] Nakashima, E., and J. Warrick, "For NSA Chief, Terrorist Threat Drives Passion to 'Collect It All,'" *The Washington Post,* July 14, 2013.

[27] Johnston, C., "(U) Modernizing Defense Intelligence: Object Based Production and Activity Based Intelligence," *Defense Intelligence Agency (DIA) Innovation Day 2013,* 27 Jun 2013. Web: https://www.ncsi.com/diaid/2013/presentations/johnston.pdf

[28] "Queen Anne's Revenge," Wikipedia.

[29] "Quellfire (QF) Knowledge Manager (KM) at SAIC," web, http://www.simplyhired.com/job/qzdje7h4qs. [Accessed: 17 Jul 2014.]

[30] Gauthier, D., "Activity-Based Intelligence: Finding Things That Don't Want to be Found," presented at the *2013* GEOINT Symposium,* Tampa, FL, April 16, 2014. Approved for Public Release NGA Case #14-233. Web. http://geointv.com/archive/geoint-2013-gov-pavillion-nga-abi/.

[31] Ashton, K., "That 'Internet of Things' Thing," *RFID Journal,* June 2009.

[32] "Gartner Says the Internet of Things Installed Base Will Grow to 26 Billion Units By 2020," Gartner, December 12, 2013, web.

[33] "Remarks by Director David H. Petraeus at In-Q-Tel CEO Summit," Central Intelligence Agency, 2012, web. https://www.cia.gov/news-information/speeches-testimony/2012-speeches-testimony/in-q-tel-summit-remarks.html. [Accessed: 27 Jul 2014.]

[34] "In-Q-Tel," web, https://www.iqt.org/historical-snapshot/. [Accessed: 26 Jul 2014.]

[35] Sledge, M., "CIA's Gus Hunt on Big Data: We 'Try to Collect Everything and Hang on to It Forever,'" *The Huffington Post,* March 20, 2013.

11

Collection

Collection is about gathering data to answer questions. This chapter summarizes the key domains of intelligence collection and introduces new concepts and technologies that have codeveloped with ABI methods. It provides a high-level overview of several key concepts, describes several types of collection important to ABI, and summarizes the value of persistent surveillance in ABI analysis.

11.1 Introduction to Collection

Collection is the process of defining information needs and gathering data to address those needs. The collection discipline has evolved from traditional "INT" specific distinctions, but some of the major collection disciplines and their suitability for ABI are shown in Figure 11.1. The overarching term for remotely collected information is ISR (intelligence, surveillance, and reconnaissance).

In the U.S. intelligence community, intelligence agencies are aligned along INT-specific "stovepipes." In recent years, this division has led to criticism about a failure to share information and connect the dots, but the distinction was originally by design. The traditional INT distinctions are described as follows:

- Human intelligence (HUMINT): The most traditional "spy" discipline, HUMINT is "a category of intelligence derived from information collected and provided by human sources" [1]. This information is gathered through interpersonal contact; conversations, interrogations, or other like means.

- Signals intelligence (SIGINT): Intelligence gathering by means of intercepting of signals. In modern times, this refers primarily to electronic signals.
    - Communications intelligence (COMINT): A subdiscipline of SIGINT, COMINT refers to the collection of signals that involve the communication between people, defined by the Department of Defense (DoD) as "technical information and intelligence derived from foreign communications by other than the intended recipients" [2]. COMINT exploitation includes language translation.
    - Electronic intelligence (ELINT): A subdiscipline of SIGINT, ELINT refers to SIGINT that is not directly involved in communications. An example includes the detection of an early-warning radar installation by means of sensing emitted radio frequency (RF) energy. (This is not COMINT, because the radar isn't carrying a communications channel).
- Imagery intelligence (IMINT): Information derived from imagery to include aerial and satellite-based photography. The term "IMINT" has generally been superseded by "GEOINT."
- Geospatial intelligence (GEOINT): A term coined in 2004 to include "imagery, IMINT, and geospatial information" [3], the term GEOINT reflects the concepts of fusion, integration, and layering of information about the Earth.
- Measurement and signals intelligence (MASINT): Technical intelligence gathering based on unique collection phenomena that focus on specialized signatures of targets or classes of targets.
- Open-source intelligence (OSINT): Intelligence derived from public, open information sources. This includes but is not limited to newspapers, magazines, speeches, radio stations, blogs, video-sharing sites, social-networking sites, and government reports.

Each agency was to produce INT-specific expert assessments of collected information that was then forwarded to the CIA for integrative analysis called all-source intelligence. The ABI principle of data neutrality posits that all sources of information should be considered equally as a sources of intelligence.

There are a number of additional subdisciplines under these headings including technical intelligence (TECHINT), acoustic intelligence (ACINT), financial intelligence (FININT), cyber intelligence (CYBINT), and foreign instrumentation intelligence (FISINT) [4].

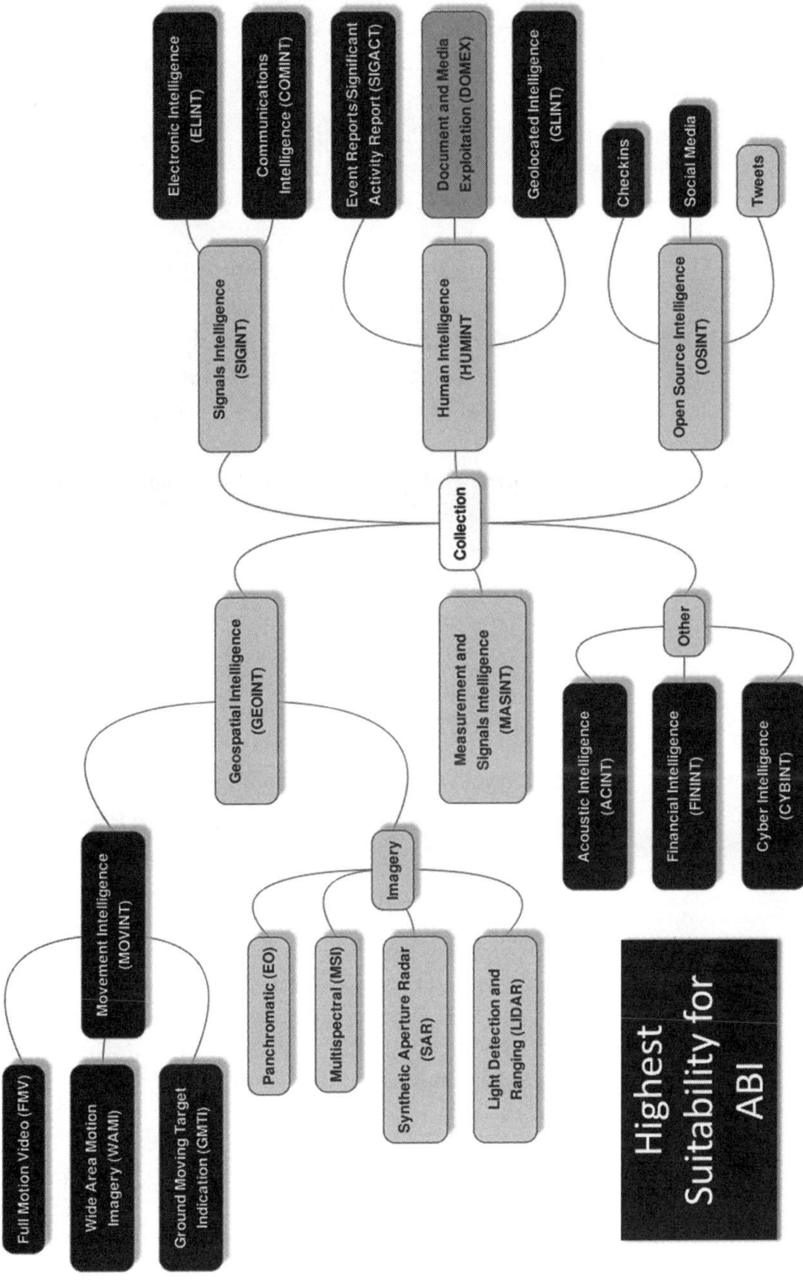

Figure 11.1 Taxonomy of some ABI-related collection disciplines.

In the first and second ages of intelligence, collection philosophy was based on *reconnaissance* of fixed targets to provide *indication and warning* of large-scale movements of conventional military forces. IMINT, for example, was limited by the ability of photoreconnaissance to take a snapshot in time. Understanding of activity was based upon this scheduled, periodic slice of the day. Imagery analysts had to infer the activities between low-frequency observations.

As mentioned in Chapter 1, traditional intelligence methods were ill suited to address mobile, fleeting, relocatable, and hiding targets prevalent in modern conflict. Despite thousands of airborne surveillance sorties during 1991's Operation Desert Storm, efforts to reliably locate Iraq's mobile SCUD missiles were unsuccessful [5]. The problem was further compounded during counterterrorism and counterinsurgency operations in Iraq and Afghanistan where the targets of intelligence collection are characterized by weak signals, ambiguous signatures and dynamic movement. The ability to capture *movement intelligence* (MOVINT) is one collection modality that contributes to ABI, because it allows direct observation of events and collection of complete transactions. Other advances like multisensor platforms and a shift from reconnaissance to surveillance collect the "big data" take that is required to enable ABI analysis over a wide area. This chapter addresses three subclasses of MOVINT collection: full motion video (FMV), wide-area motion imagery (WAMI)—both types of motion imagery—and ground moving target indication (GMTI) from radar.

11.2 MOVINT with Motion Imagery

The first, most basic type of dynamic MOVINT collection is the use of motion imagery as opposed to traditional still imagery. Motion imagery is a sequence of consecutive images that when viewed by humans tends to create the "illusion" of motion:

> Motion imagery is defined as a likeness or representation of any natural or man-made feature or related object or activity utilizing sequential or continuous streams of images that enable observation of the dynamic (temporal) behavior of objects within the scene. Motion Imagery temporal rates—nominally expressed in frames per second—must be sufficient to characterize the desired dynamic phenomena. [6]

The different categories of motion imagery are summarized in Figure 11.2.

Motion imagery also includes metadata related to the data stream, sensor, or collecting platform. The minimum frame rate for motion imagery is 1 Hz (one frame per second). The term "video" is used to describe a frame rate of between 6 and 120 Hz. Humans can comfortably process the "illusion of

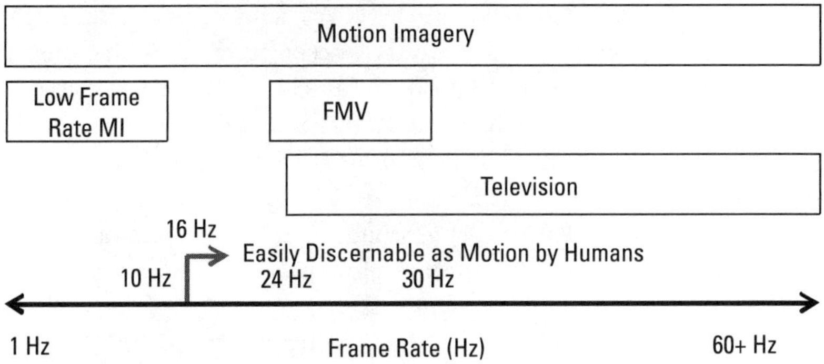

Figure 11.2 Categories of motion imagery.

motion" at about 16 frames per second. Commercial video is typically recorded at between 24 and 30 Hz. The Department of Defense uses the term FMV to refer to video that is typically captured at 24–30 Hz or greater.

Multiple phenomenologies can be used to capture motion imagery. Electro-optical (EO) imaging systems are passive sensors that operate in the visible light spectrum by sensing energy emitted or reflected off a target and converting the sensed light or change in light to an electronic signal. Commercial applications including digital cameras and video cameras are examples of electro-optical systems. Because EO systems require a reflected light source, they are typically used for daytime imaging.

Infrared cameras are passive sensors that operate in the infrared spectrum by sensing thermal radiation (heat) emitted by a target and producing an electronic signal that is processed into video. While EO systems require a reflected light source, infrared (IR) systems detect emitted radiation and therefore can be used to produce daytime or nighttime imagery by measuring the thermal contrast, the difference in signal between the target and its surroundings.

The midwave IR band, between 3 and 5 μm, is the portion of the electromagnetic spectrum where terrestrial targets have high thermal contrast due to black-body physics. Unfortunately, in this waveband, the thermally induced noise of the sensor may be greater than the collected signal, requiring that the detector be cryogenically cooled. This adds weight, cost, and complexity. A long-wave IR (LWIR) camera operates between 8 and 12 μm and can also detect thermal targets, but because infrared energy in this waveband is absorbed, scattered, and refracted by the atmosphere, long-range collection is more difficult.

11.2.1 FMV

FMV refers to motion imagery captured at 24 Hz or greater. The most widely recognized FMV collector is the U.S. Air Force MQ-1 Predator and the follow-on MQ-9 Reaper unmanned aerial vehicle (UAV). The medium-altitude MQ-1 was first developed under an advanced concept technology demonstration (ACTD) contract from 1994 to 1996. Beginning in 2001, the drone was widely used to provide airborne surveillance in Afghanistan. The aircraft is equipped with the Raytheon AN/AAS-52 EO/IR multispectral targeting system (MTS) featuring multiple wavelength sensors, near-IR, and color TV cameras, illuminators, eye-safe range finders, image merging, spot trackers, and other avionics [7]. Because of the limited field of regard of the so-called soda straw, collection must be highly targeted. Sensor operators slew the camera to follow the target. The system can therefore only track a single entity at one time [8].

According to a RAND report on motion imagery exploitation, there are similarities between the production workflow between an air force motion imagery exploitation team and reality television producers. The ISR mission commander is like the showrunner or executive producer, allocating resources and developing a seamless thread through the event. A three-person crew gathers the "footage" and assembles the results. The imagery analyst maintains "eyes on" the real-time video stream. The imagery analyst's job is to watch streaming video in real time and translate the observed activities into text write-ups called clip marks. The clip marks are time-tagged and searchable and provide a human-exploited summary of the video stream. A *correlation analyst* (CAN) looks for cross-cueing opportunities, making dynamic adjustments as he or she observes events. The *imagery report editor* (IRE) is like the story editor, reviewing the results of the screening and publishing products that include annotated JPEG snapshots, storyboards, and highlight videos [8, 9].

Analysts discern patterns of life against targeted entities using real-time FMV and exploited video streams. By persistently following a target, analysts identify the geospatial nodes of interest, which include places of residence, meeting places, frequented businesses, and the motion paths the entity typically takes between these places. This information can be aggregated to understand the motion of the entity and anticipate where he or she may go in the future.

Following heightened demand for airborne ISR from battlefield commanders, the Pentagon set a goal of increasing the number of Predator/Reaper combat air patrols (CAPs) or "orbits" to 65 by the end of FY2013 [10]. Because each UAV carried only a single MTS sensor ball, at least one aircraft was required to maintain a track on a target. By 2009, then head of the Air Staff's ISR Directorate, Lt. General David Deptula, began pushing for a reimagining of ISR capabilities and a shift from orbits and CAPs to a measure of the analytic output that could be produced per platform. This video spawned a revolutionary new capability, WAMI.

11.2.2 WAMI

Introducing the Air Force's WAAS program, called Gorgon Stare, Deptula defined the way ahead for airborne ISR that extends the single-ball FMV of the MQ-1 to a new model that can transmit live-video images of a wide area. The Gorgon Stare system, shown in Figure 11.3, can transmit up to 65 FMV-like video "chip outs" to troops on the ground and can track enemy movements [11].

The Gorgon Stare sensor is one of a new generation of collection capabilities called WAMI. Typically, this imagery is collected by multiple digital cameras that produce high-resolution composite imagery at frame rates of 1 Hz or greater over an area of 10 km^2 or more. WAMI systems may also be referred to as large-volume streaming data (LVSD) imagery, wide area persistent surveillance (WAPS), or wide area large format (WALF) systems [6]. In contrast to traditional FMV cameras that typically rely on a single charge coupled device (CCD) focal plane array, WAMI systems are made up of several complete cameras or a composite focal plane array of CCDs. The cameras are arranged to produce an overlap, which is digitally removed through image fusion and processing. This technique allows a smaller camera to capture imagery over a wide area without an excessively large focal plane array and telescope.

The FMV-like "video windows" or "chip outs" depicted in Figure 11.3 provide a unique capability to enable ABI because a single platform in orbit over a large area can map multiple long-duration motion transactions simultaneously. Typically, these are processed and streamed in real time to provide a situational awareness–quality view of targeted entities during the mission. The FMV-like chip outs are only a small fraction of the full field of view (FFOV) collected during a single mission. After the aircraft lands, the full frame data can be downloaded, processed, and reviewed forensically to identify and track other entities that were not targeted by video windows during the mission. An example of a full WAMI frame, which covers an entire medium-sized city, is shown in Figure 11.4.

Using full-frame WAMI data downloaded and processed post-mission, analysts forensically backtrack vehicle and entity transactions from an event of interest (time/location) by recreating synthetic video windows from postmission data [14]. The nefarious entities participating in the event were not known a priori (so they could not be targeted with a video window), but because their activities were incidentally collected in the full-frame data, backtracking reveals their origin. This information can be used to task collection and other actions in the present. Forensic analysis of multiple collects may be used to develop pattern of life. Several companies including Reston, Virginia–based PIXIA provide commercial solutions for real-time mission overwatch or forensic pattern-of-life analysis [15].

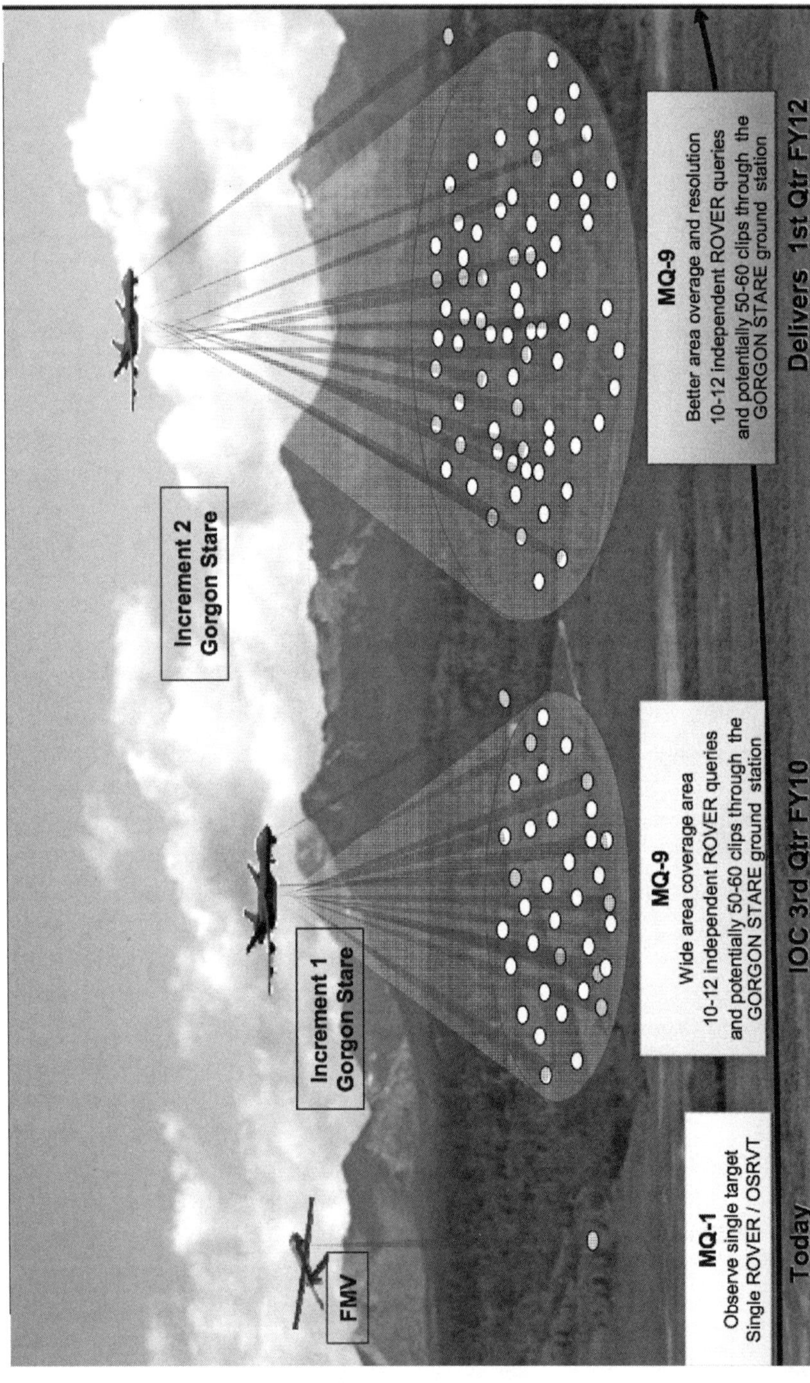

Figure 11.3 Overview of the Gorgon Stare WAAS concept. (Source: United States Air Force [12].)

Figure 11.4 Example of WAMI. [Source: United States Air Force sensor data management system (SDMS) [13].]

Dayton-based Persistent Surveillance Systems operates a commercial 192-megapixel camera system made of 12 commercially available Canon cameras that has been demonstrated for use by police departments, special event security, disaster response, and traffic studies. During a 2012 demonstration flight, Dayton police got reports of an attempted robbery and shooting at a bookstore. By starting with this tip-off event, police used WAMI imagery to backtrack the suspect's car to a residential neighborhood (the suspect's origin), and map his transactions before and after the shooting. Police were able to construct a detailed map used as evidence to arrest him for the crime [16].

First-generation WAMI systems relied on integration of a small number of physical cameras with a maximum pixel density of about 200 megapixels. These cameras are oriented in a housing mounted on the side or top of an aircraft arranged as shown in Figure 11.5.

The central boresight bs_0 is pointed at the center of a circular flight path. As the aircraft holds a steady turn around the aim point, the camera projects a trapezoidal footprint on the collection area. Because targets closer to the aircraft are closer to the camera, the ground resolution is not uniform across the entire field of regard, leading to edge distortion.

In 2007, DARPA initiated the Autonomous Real-Time Ground Ubiquitous Surveillance Imaging System (ARGUS-IS). Comprised of 368 color 5-megapixel cell phone cameras split across four composite focal plane arrays, the 1.8-gigapixel (1,800-megapixel) ARGUS-IS can automatically track vehicles and dismounts over a 36-square-mile area with a resolution of about 15 cm ground sample distance (GSD) [14, 18]. ARGUS was originally designed

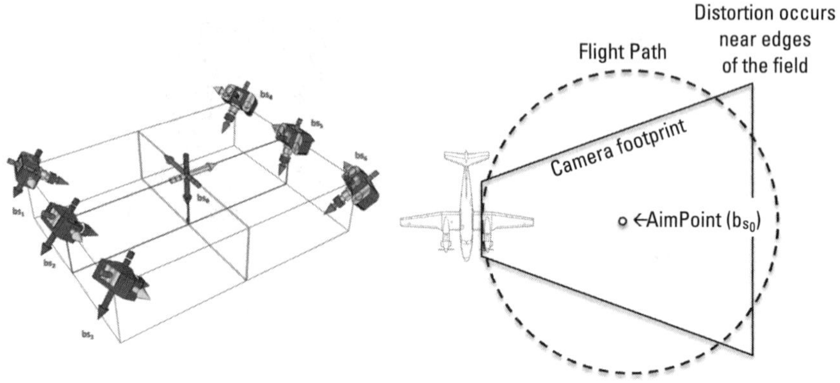

Figure 11.5 Typical layout of first-generation WAMI multisensor camera arrays relative to the sensor ensemble point, bs0. (Source: Motion Imagery Standards Board MISB EG 0810.2 [17].)

with the real-time video window mode in mind—the 65-window baseline for ARGUS-IS is the objective state of the Gorgon Stare program in Figure 11.3; advances in ground processing enabled a hybrid mode where ARGUS high-resolution imagery could be postprocessed to enable forensic backtracking over the FFOV. Constrained only by the limitations of the data downlink, ARGUS video windows are nominally at 10–15 Hz, creating a near-FMV quality picture in both resolution and frame rate. Onboard processing and storage constraints limit post-mission forensic data to approximately 3.3 Hz [14].

ARGUS successfully completed flight testing on a U.S. Army Black Hawk helicopter in 2010 [19]. The system was later planned for deployment on the army's A-160 autonomous helicopter and later on the Air Force's *Blue Devil 2* airship [20]. Unlike side-mounted WAMI systems, because ARGUS-IS is designed to collect from a near-nadir orientation, it experiences significantly less image distortion at the edges of the sensor field. The airborne processing system and associated ground station were equipped to support live exploitation of video windows and forensic recall of incidental collection across the entire field of regard [14].

11.3 MOVINT from Radar

One type of MOVINT sensor widely deployed is radar. Radar is an active sensing modality that uses radio waves to detect and track objects. Although the physics behind radar were observed as early as 1886, the technology did not come into widespread use until the British used it to track aircraft during World War II [21, p. 76]. Radar is also used in ballistic missile defense, space surveillance, weather monitoring, and other applications. While these applications focus on looking "up" for air- and space-based objects, radar can also be mounted on aircraft and spacecraft to provide active sensing of objects on the Earth.

11.3.1 Basic Principles of GMTI

GMTI is the use of radar to detect and track objects on the basis of their reflectivity and velocity. In contrast to synthetic aperture radar (SAR), which produces a static "picture" of the target by processing repeated returns to a moving array, GMTI allows collection of dynamic activity. GMTI works by measuring the Doppler shift of the target to detect the radial component of the target's velocity vector. To detect a moving object, the target must be distinguished from the background clutter, and the radar processor must integrate observations from successive pulses. The minimum detectable velocity (MDV) is a function of transmission wavelength (λ), clutter baseline (B), the speed of the platform, v_p, and the collection geometry [azimuth (α) and elevation (θ)].

$$MDV = \frac{\lambda}{2}\left(\frac{4v_p}{B}\sqrt{(\sin\alpha\sin\theta)^2 + (\cos\alpha\cos\theta)^2}\right)$$

Higher power radars (lower wavelength), slow-moving platforms, and high-clutter baselines improve performance by allowing detection of slower objects. Radars operating in the X-band are desirable because of low atmospheric attenuation and high angular resolution with a small antenna that can fit on an aircraft or spacecraft [22, p. 38].

Long-term tracking of objects with GMTI is a difficult problem. One of the major limitations of radar-based GMTI is that the radar can either be used in GMTI mode or SAR mode, and GMTI mode can only "see" an object when it is moving. Therefore, vehicles that frequently alternate between stopping and moving are difficult to track (although this behavior itself may be detected as anomalous). Sensor obscurations due to buildings, trees, and underpasses cause track breaks. The change in the orientation of the sensed object may change the radar return, leading to an ambiguity about the consistency of the track. Closely spaced objects may not be resolved as unique objects, and objects that closely cross paths lead to track ambiguity. Advanced processing techniques may compensate for each of these shortcomings, but applying multiple sensing modalities is also a solution to obtain high track fidelity.

GMTI data is defined by NATO Standard (STANAG) 4607. This standard includes information about the platform type, velocity, heading, sensor orientation, and minimum detectable velocity. Target information includes the target latitude, longitude, height, radial velocity, signal-to-noise ratio (SNR), classification, classification probability, slant range, and radar cross-section [23].

NATO STANAG 4607 refers to GMTI "dots" while the NATO STANAG 4676 (still in draft) provides for the properties of GMTI-derived tracks.

11.3.2 Evolution of GMTI Collection Systems

Originally implemented as a development system in 1985, the Northrop Grumman E-8C joint surveillance target attack radar system (JSTARS) consists of a 24-foot AN/APY-7 passive electronically scanned array antenna mounted on the underside of a Boeing 707 aircraft. The AN/APY-7 operates in multiple modes including wide area surveillance, GMTI, target classification, and SAR modes. The antenna has a 120° field of view, covers nearly 50,000 km^2, and can track up to 600 targets at a range of 250 km. An example of JSTARS GMTI "dots" as an overlay to reference imagery is shown in Figure 11.6.

The JSTARS was sent to Operation Desert Storm on an accelerated deployment in 1991 where it flew a total of 49 missions. According to Gen. John Jumper, the JSTARS was employed in Operation Iraqi Freedom and Operation

Figure 11.6 Example of imagery with GMTI tracks from JSTARS overlay. (*Source:* NGA (11-040). JSTARS imagery subject to public release per DoD guidance [24].)

Enduring Freedom for convoy overwatch, combat search and rescue, and "building 'pattern analysis' of insurgent movements across hundreds of miles" [25, p. 58].

The evolution of radar-based GMTI advanced significantly since the advent of JSTARS. The latest incarnation is the DARPA-sponsored Vehicle and Dismount Exploitation Radar (VADER). According to *Jane's Defence Weekly*, "VADER scans for vehicles and dismounts at high area rates, provides precision target location, and includes a set of data exploitation tools that allow for long-duration ground vehicle tracking, SAR-coherent change detection, motion pattern analysis, and dismount motion characterization" [26]. In 2009, VADER was tested along a 31-mile stretch of the Arizona border [27]. The VADER Exploitation Ground System (VEGS) includes algorithms from the DARPA NetTrack program for "persistent reconnaissance, surveillance, tracking, and targeting of evasive vehicles and people moving on foot in cluttered environments" [28].

GMTI is an excellent collection capability to identify large-scale patterns of movement and identify regions for further targeting. Because radar returns only provide information about the general size and radial velocity of the target, GMTI may be useful for classifying objects as trucks, motorcycles, or dismounts but is generally not helpful in identifying ground targets. Even when objects can be classified as "slow-moving" and "human-sized," it is difficult to discern whether you are observing a band of insurgents or a herd of livestock.

11.4 Additional Sources of Activities and Transactions

This chapter introduces some sources of collection for georeferenceable data to support ABI methods, but the principle of data neutrality encourages the use of other data sets naturally well conditioned for analysis of activities, transactions, entities, and networks. FININT, widely used in law enforcement, especially to counter money laundering, is naturally transaction-based. Financial analysts trace accounts (a proxy for an entity), money flows (transactions), and other activities. Cyberintelligence is almost always transaction-based as analysts examine the routine of information packets through a computer network. These transactions can be "georeferenced" to physical space but also to alternative physical addresses in cyberspace.

If you lived in the United Kingdom in 2013, there was one surveillance camera for you and each of your 10 friends. The British Security Industry Authority found that there were nearly 6 million closed-circuit television (CCTV) cameras in the island nation [29], and it is believed that the United Kingdom has nearly 20% of the CCTV cameras in the world [30]. While some see this as a disturbing trend, the same *Telegraph* article [29] noted that, "95% of

Scotland Yard murder cases used CCTV footage as evidence." Commercial and civil ground-based sensors have become increasingly important in law enforcement, accident investigation, crime prevention, behavioral analytics, homeland security, customs and border protection, and private home security.

Ground-based traffic sensors like pressure plates and traffic cameras collect MOVINT transactions in cities worldwide. "Every neighborhood in the city walks. We really need to have an idea of what that activity looks like so we can serve our citizens better," said Nicholas O'Brien from New York City's Office of Data Analytics [31]. A start-up called Placemeter integrates video feeds from around the city to detect and count over 10 million people per day, providing pedestrian counts to civil organizations and private businesses. Placemeter augments civil traffic cameras by paying private citizens (so-called meters) to stream auto-processed pedestrian counts from an old smartphone mounted on a window in their own home—a revolutionary albeit invasive method for incidental collection of real-time crowd-sourced activity data over a wide area [32].

Incidental collection of entity activities and transactions is also proliferating for commercial applications. Apple developed the iBeacon system, "a new class of low-powered, low-cost transmitters that can notify nearby iOS 7 devices of their presence" [33]. Using Bluetooth low-energy proximity sensing protocol in the iPhone and many personal electronics to identify a "universally unique identifier," iBeacons collect incidental information about proxies in indoor settings like office buildings or stores [34].

With the proliferation of Internet-connected, location-aware smartphones comes a new class of open-source intelligence including Facebook location postings, Foursquare "checkins," and geolocated Twitter "tweets." These are sources of preconditioned geolocated activities that are mostly accurate in space but extremely precise in time. Foursquare, which bestows the honor of "mayor" to the user with the most checkins at a given location, has operational security–challenged mayors at all of the major U.S. intelligence agency headquarters [35]. Most digital cameras and mobile phones insert GPS tags with the time and precise location of photos uploaded to sharing sites like Instagram, Flickr, and Panoramio.

11.5 Collection to Enable ABI

Traditional collection is targeted, whether the target is a human, a signal, or a geographic location. Since ABI is about gathering all the data and analyzing it with a deductive approach, an incidental collection approach as described in Chapter 9 is more appropriate. Consider the case of a WAMI collect. If the boresight is located on a house of interest (targeted), millions or billions of other pixels of the surrounding area may be collected. These pixels contain incidentally collected activity that may or may not be related to the original collection request.

Radar collection directed at a large segment of a border crossing will incidentally collect the motion of many entities in the area. Analyzing large volumes of georeferenced HUMINT or significant activity (SIGACT) reports through content filtering and entity correlation enables discovery of the unknown. Collecting and indexing electronic signals for subsequent correlation and fusion enables sequence neutral analysis. These techniques are sometimes referred to as cast-the-net collection and are analogous to fishing with a net. Cast the net, reel it in, and see what is inside. Sometimes it's a fish, sometimes it's a tin can...but the fact that it's a can of soup with French writing on the side might be interesting in the context of other information in your database.

ABI collection is most effective when densely applied against an area of interest (AOI) rather than a specific target. Flynn, Juergens, and Cantrell state that "intelligence, surveillance, and reconnaissance are most effective against low-contrast enemies when massed" [36]. This was a major breakthrough in collection discipline and is initially counterintuitive—it requires a shift from collecting the maximum area or number of targets. Late in the Afghanistan conflict, the military shifted from a model of piecemeal allocation of ISR to a mix of colocated assets to "conduct ISR 'soaks,' [and] generate cross-cueing opportunities" [37, 38].

ABI is about going after human-centric, network-based threats whose signature is below an obvious threshold for detection in any single domain of intelligence. Furthermore, elements of the network may be dispersed in space and time or disguised/denied from our collection capabilities. Persistence—of collection, analysis, and knowledge—is an enabler for ABI.

11.6 Persistence: The All-Seeing Eye (?)

For over 2,000 years, military tactics have encouraged the use of the "high ground" for surveillance and reconnaissance of the enemy. From the use of hills and treetops to the advent of military ballooning in the U.S. Civil War to aerial and space-based reconnaissance, nations jockey for the ultimate surveillance high ground. The Department of Defense defines "persistent surveillance" as "a collection strategy that emphasizes the ability of some collection systems to linger on demand in an area to detect, locate, characterize, identify, track, target, and possibly provide battle damage assessment and retargeting in near or real time" [2]. John Stenbit, the former assistant secretary of defense for command, control, communications, and intelligence, said, "Persistence in our context is to match the frequency of revisit with the time stability of the object that you are looking at—the speed with which things change" [39].

Popular culture often depicts persistent collection like the all-seeing "Eye of Sauron" in Peter Jackson's *Lord of the Rings* trilogy, the omnipresent computer in "Eagle Eye," or the camera-filled casinos of *Ocean's Eleven,* but persistence for intelligence is less about stare and more about sufficiency to answer questions.

In this textbook, *persistence* is the ability to maintain sufficient frequency, duration, temporal resolution, and spectral resolution to detect change, characterize activity, and observe behaviors. This chapter summarizes several types of persistent collection and introduces the concept of virtual persistence—the ability to maintain persistence of knowledge on a target or set of targets through integration of multiple sensing and analysis modalities.

11.7 The Persistence "Master Equation"

Persistence, P, can be defined in terms of eight fundamental factors:

$$P \propto [(x, y), z, T, f, \lambda, \sigma, \theta, \Pi]$$

where

(x, y) is the area coverage usually expressed in square kilometers.

z is the altitude, positive or negative, from the surface of the Earth.

T is the total time, duration, or dwell.

f (or Δt) is the frequency, exposure time, or revisit rate.

λ is the wavelength (of the electromagnetic spectrum) or the collection phenomenology. $\Delta\lambda$ may also be used to represent the discretization of frequency for multisensor collects, spectral sensors, or other means.

σ is the accuracy or precision of the collection or analysis.

θ is the resolution or distinguishability. θ may also express the quality of the information.

Π is the cumulative probability, belief, or confidence in the information.

Combinations of these factors contribute to enhanced persistence. Area coverage is enhanced by increasing altitude, usually at the expense of resolution. Dwell time (T) is governed by the platform's endurance, a function of size and fuel. Spectral diversity, confidence, and accuracy are enhanced by implementing advanced sensor suites and multiple, mutually reinforcing sensors. These require increased payload capacity. The trade space requires a layered approach to persistence that includes a portfolio of supporting capabilities.

Counterinsurgency operations in Iraq and Afghanistan during the third age of intelligence accelerated the development and deployment of advanced, multisensor, persistent surveillance aircraft. Some of these concepts are summarized in Figure 11.7.

Medium- and high-altitude aircraft like the MQ-1 Predator, the follow-on MQ-9 Reaper, and the high altitude RQ-4 Global Hawk provide surveillance capabilities for the U.S. Air Force. The army operates smaller, tactical RQ-5 Hunter and RQ-7 Shadow UAVs that provide soda-straw video for situational awareness. Limited by on-board fuel reserves, these platforms fly for up to several hours, with the 32,000-lb, $222-million dollar Global Hawk topping out at just over one day of endurance at 60,000 feet [40].

Advances in structures, aerodynamics, and power technologies introduced the concept of persistent or ubiquitous aircraft for persistent surveillance. A persistent aircraft uses an efficient fuel source like hydrogen to enable extremely long duration powered flight. Boeing's Phantom Works is preparing its Phantom Eye experimental aircraft for high-altitude testing. The objective system carries 2,500 lbs of payload and stays aloft for 7–10 days at over 60,000 feet [41].

A ubiquitous aircraft is one that uses energy regeneration to maintain indefinite loiter over a target area. Under NASA's Environmental Research Aircraft and Sensor Technology (ERAST) program, two aircraft, *Pathfinder* and *Helios* were produced by AeroVironment, Inc. These aircraft use a combination

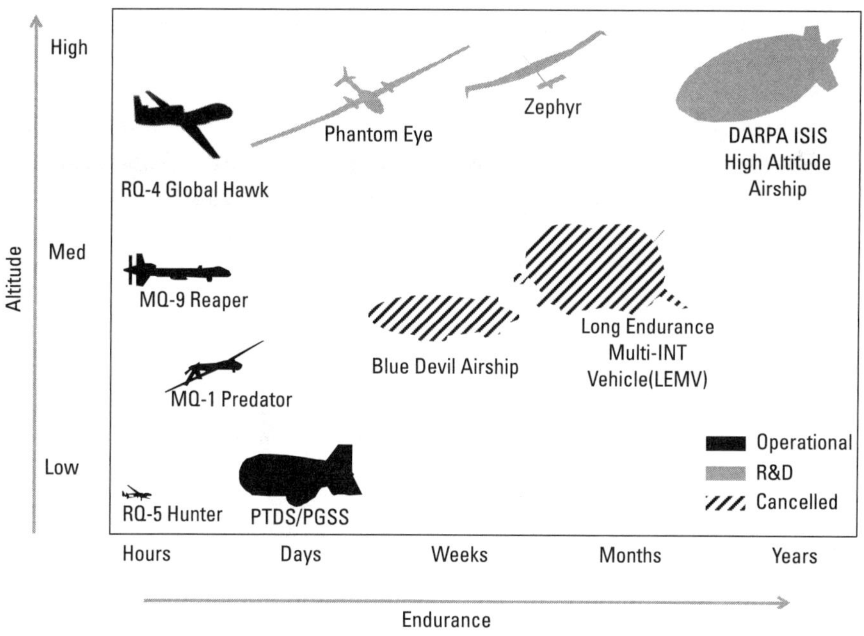

Figure 11.7 Comparison of persistent surveillance air platform concepts.

of solar cells, storage batteries, and hydrogen-air fuel cells. Solar cells power the vehicle during the day and regenerate the batteries and fuel cells that are used at night. These aircraft are typified by extremely long wings that are efficient at high altitudes. The wings and solar panels may share structures to reduce weight. Several firms—including Internet companies like Facebook and Google—are exploring these concepts for persistent data collection and airborne Internet service.

At the other end of the spectrum are aerostats: lighter-than-air craft, they are lofted by means of aerostatic lift—a buoyant force generated by filling the vehicle with a lightweight gas like hydrogen or helium. Aerostats may be tethered or free-flying, although because they are difficult to control, most operationally useful aerostats are tethered.

The Navy deployed 59 tethered aerostats to forward operating bases in Afghanistan from 2009 to 2013. The aerostats contain "electro-optical/infrared sensors, unattended transient acoustic measurement and signatures intelligence sensor, wide-area sensor system, and communications relay system" [42]. A similar system deployed by the army includes a GMTI system capable of detecting dismounts [42]. Other aerostat programs are used for cruise missile defense [43, 44] and border security [45]. Between 2007 and 2012, the military launched 15 aerostat and airship programs at a cost of more than $7 billion [42, 46].

Aerostats are highly susceptible to high winds and weather, which can lead to expensive losses [46]. However, since they do not require fuel to operate they have lower sustainment costs than most air platforms. Because they are tethered, their operations are limited to a single location. The inability to continuously reposition the aerostat also makes it a target for attacks. Aerostats are ideal platforms for "big data" collection of imagery, radar, and other sensors, because the tether can be used to contain a hard-wire data link, reducing the need for bandwidth-limited radio frequency communications.

Rigid and semirigid airships are self-propelled aircraft buoyed by a lightweight gas like helium or hydrogen. A hybrid airship combines this technology with aerodynamic principles for increased lift. Because they travel at slow speeds or maintain position by hovering over a target, airships provide an ideal platform for persistent surveillance. With the right aerodynamic and propulsive controls, airships also provide a stable platform for radar and imagery systems. Their large size offers significant potential to host large antenna arrays for communications or signals intelligence. Airships typically operate at much higher altitudes than tethered aerostats.

Example: Blue Devil Block II

Based on a 2010 urgent operational need request for enhanced ISR capabilities in Afghanistan, the Air Force undertook the Blue Devil II program, a semirigid airship based on a TCOM Polar 1000 concept (a scale-up of the existing Polar

400 model). "At 335 feet in length and a volume of greater than one million cubic feet, Blue Devil Block II was the largest airship in the world" [47, p. 1]. Figure 11.8 shows the Blue Devil II concept.

Designed to support missions of up to nine days, Blue Devil II's 6,000-lb capacity payload module could support communications, a GMTI radar pallet, WAMI, a SAR sensor, multiple high-definition video cameras, and a processing pallet capable of reducing data through on-board processing [48]. Because each sensor was self-contained, a new sensor could be swapped out in hours with no payload integration costs. From an altitude of 20,000 feet, the EO/IR wide aperture camera covers 38,000 mi^2 at 0.5m GSD, the SAR/MTI covers 76,000 mi^2, and the SIGINT system covers 125,000 mi^2 [48, p. 5]. Figure 11.8 also shows "event and geolocation data" from an "UGS [unattended ground sensor] field." UGS are remotely emplaced, ground-based event-detection sensors. Blue Devil II also included the simultaneous integration of wide area EO and IR imaging (see Section 11.3) and a SIGINT sensor.

The DARPA Strategic Technologies Office (STO) and the Air Force Research Laboratory (AFRL) proposed the Integrated Sensor Is the Structure (ISIS) in late 2003. The concept is for an extremely high-altitude, long-endurance, fully autonomous stratospheric airship, summarized in Figure 11.9. ISIS recognizes that significant weight savings can be achieved if the structure of the aircraft is the sensor array rather than designing an aircraft to carry a sensor array. Designed for operation at altitudes above 70,000 feet, the ISIS concept

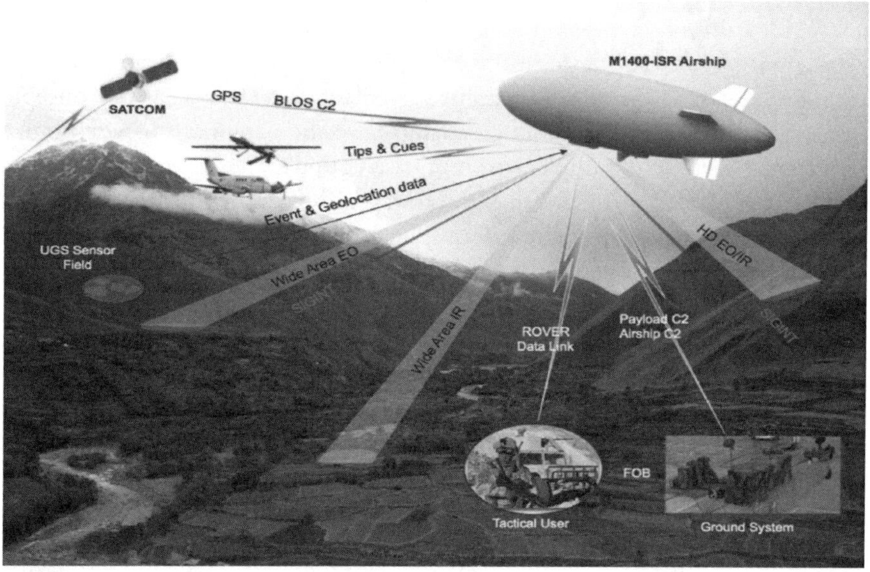

Figure 11.8 Overview of the Blue Devil multi-INT airship concept. (Image courtesy of Mav6, LLC. Reprinted with permission.)

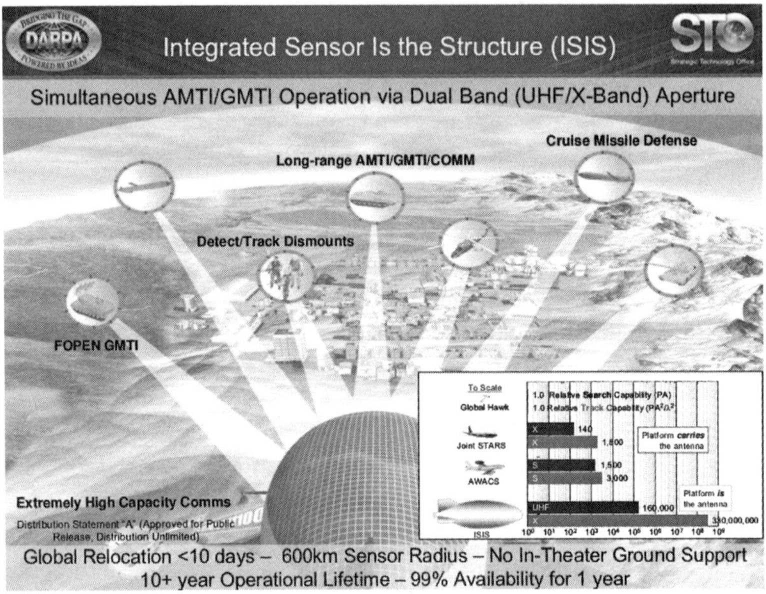

Figure 11.9 Overview of the DARPA ISIS program. (Source: DARPA [49].)

is purported to operate for up to 10 years [49]. "The ISIS concept includes 99% on-station 24/7/365 availability for simultaneous airborne moving target indicator (AMTI) (600 kilometers) and GMTI (300 kilometers) operation" [50, p. 5].

High-altitude, long-endurance aircraft provide the flexibility to operate from multiple worldwide locations, integrate new payload advances as they are developed, and maintain persistent surveillance over a wide area. Concepts like the DARPA ISIS high-altitude airship have the potential to revolutionize persistent collection by implementing quasi-stationary sensing platforms that maintain persistence for years on end. However, the coverage area for persistent surveillance increases with altitude, and the ultimate high ground is space.

11.8 Space-Based Persistent Surveillance

Since the first successful photoreconnaissance satellite, Corona, was launched by the CIA in August 1960, nations have sought intelligence advantage from space. Different orbital regimes, summarized in Figure 11.10, are suited for different types of missions. These regimes are divided by altitude. Increasing the altitude of the orbit increases the orbital period and lowers the velocity with respect to the Earth. Keplerian mechanics define the kinematic properties of a classical two-body system (e.g., a satellite orbiting a planet).

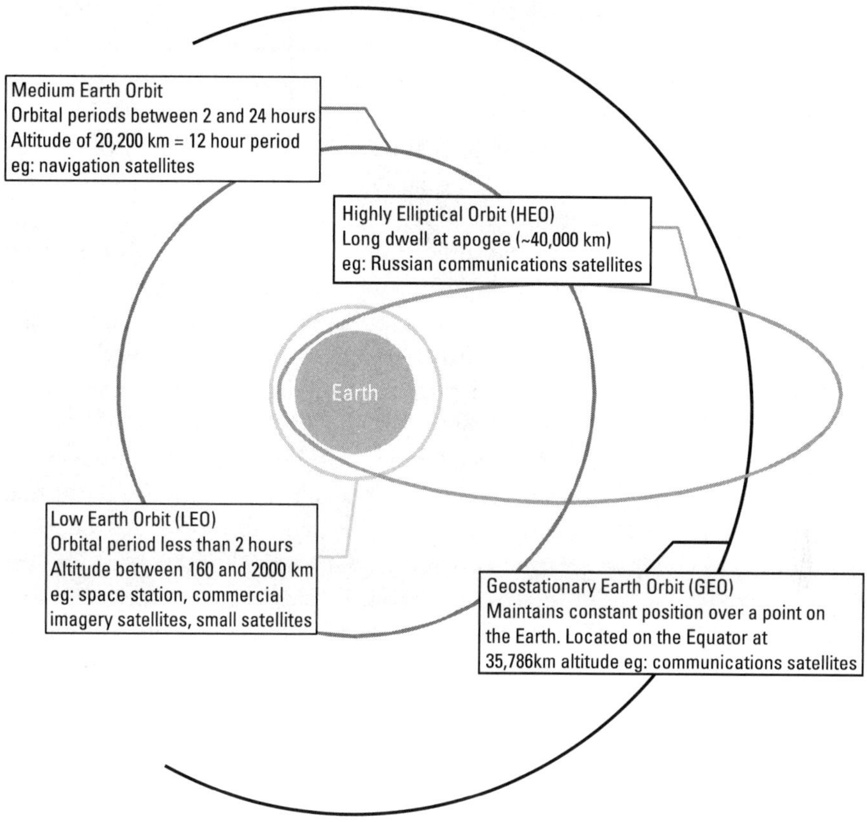

Figure 11.10 Summary of different orbital regimes for persistent surveillance platforms.

Satellites in low Earth orbit (LEO) circle the Earth between 90 minutes and two hours. Depending on the orbital inclination relative to the equator, they may pass over the same point on the ground more than once per day. This is called the revisit rate of the platform. Satellites in LEO are in constant motion relative to a point on the ground and therefore cannot transmit or observe this point for more than about 90 seconds in a single pass.

Medium Earth orbits (MEOs) are those orbits with altitudes between about 2,000 and 35,000 km. A circular orbit with an altitude of 20,200 km has an orbital period of 12 hours. Depending on their observation capabilities and geometry, two satellites (phased on opposite sides of the Earth) can maintain coverage over a given region when placed in a MEO orbit with these characteristics. Navigation systems like the U.S. GPS and the Russian GLONASS use MEO orbits.

Geostationary Earth orbits (GEOs) are equatorial orbits with a period of 24 hours. This means they take one day to circle the Earth, but in this time, the Earth rotates the exact same distance—meaning the satellite maintains a constant longitudinal position. Communications satellites are typically placed in these orbits so ground-based antenna maintain a constant pointing angle to the transmitter.

At high latitudes, the curvature of the Earth blocks a majority of the signal from geostationary satellites. A special type of orbit, a highly elliptical orbit (HEO) is used to maintain persistence over northern latitudes. The Soviet Union first used this orbit for communication satellites called Molniya and later for spy satellites that persisted over North America. Because the orbital velocity slows near apogee and increases near perigee, the platform spends a majority of its orbital period far from the Earth. Although this provides long dwell, the distance from the Earth is variable throughout the elliptical orbit, complicating the design of some surveillance systems.

Higher orbits increase the orbital period and hence the dwell time over a region of the Earth. They also increase the total surface area of the Earth that can be instantaneously accessed. However, the extreme distance from the Earth significantly complicates surveillance tasks. GEO satellites require exceedingly large sensor arrays or imaging apertures. Electromagnetic waves traveling through free space obey the inverse-square law. They lose power proportional to the square of the distance traveled, complicating the ability to collect low-power signals. Also, because satellites in GEO orbit 35,000 km from the Earth, the total round-trip time for a radio signal to reach the satellite and return is 240 ms, complicating efforts for bidirectional real-time data transmission and control.

11.8.1 Space-Based GMTI

In 2004, the Air Force, NRO, and other partners in the Department of Defense embarked on an ambitious program to design a constellation of space-based radar (SBR) satellites. The study proposed up to 21 satellites in LEO at a nominal altitude of 1,000 km to provide persistent global access [51, p. 2]. A study by the Defense Science Board defined metrics for radar-based persistence:

> The concept of persistence does not lend itself to a single metric. It will depend on the nature of the objects of interest, the dynamics of the situation, and the supported task—detect, track, identify, and/or engage. E.g., persistent monitoring of the status of construction of a missile site might demand a weekly revisit using SAR imaging capabilities. Persistent monitoring of a missile launch site in a ready to launch status, using [MTI] track capability might require a revisit every few seconds. Tracking a large moving unit might require an [MTI] revisit every few minutes while tracking a single vehicle might require a revisit each tens of seconds [51, p. 7].

The study also noted that the system must quickly switch between SAR and GMTI modes, because GMTI loses tracking when a vehicle stops and only SAR can verify that a new movement is the same vehicle. Radar systems cannot operate in both modes simultaneously.

One of the major considerations for a space radar constellation is orbital altitude. Higher orbits provide greater coverage of the Earth's surface. Because higher altitudes have lower orbital velocities with respect to the Earth's surface, GMTI satellites in higher orbits can detect objects with a lower velocity. However, the "power-aperture product" (transmit power times aperture size) is inversely proportional to the square of the distance to the target, a radar in MEO could require 100–400x the power of a similarly sized aperture in LEO. A technique known as space-time adaptive processing (STAP), where the statistical properties of clutter-induced radar returns are estimated and removed from returns through postprocessing, improves the ability to distinguish the Doppler shift of moving targets in high-clutter environments.

The Congressional Budget Office (CBO) studied four alternative constellations in an unclassified 2007 report. The study examined five-, nine-, and 21-satellite constellations with apertures varying from 16 × 2.5m (40m^2) to 24 × 4m (100m^2). CBO found that the 40-square-meter arrays "would need to perform near their theoretical optimum level for space radar to be able to detect targets moving more slowly than about 20 miles per hour" [52]. They estimated the program could cost as much as $90 billion in 2007 dollars. The CBO study noted that persistent tracking of ground targets in nearly all environments was not possible. It concluded that to have a 95 percent probability of locating a mobile missile launcher in North Korea before it moved, the Space Radar constellation would require at least 35 and as many as 50 satellites. The mean track life for the largest, most advanced constellation considered by CBO (100-square meters, nine satellites, aggressive signal processing) could only track a 10-m/s moving target in North Korea for six minutes [52]. The Department of Defense cancelled the Space Radar program in 2008 [53].

11.8.2 Commercial Space Radar Applications

Despite the failure of the United States to mount a joint SAR/GMTI radar program through SBR government and commercial entities from Canada, Germany, and Italy operate advanced commercial SAR spacecraft. Some of these systems are multisatellite constellations with C-band or X-band imaging and resolutions under 1m. The basic properties of four systems are shown in Table 11.1.

A notable example is the Canadian RADARSAT-2, designed by MacDonald, Dettwiler and Associates (MDA) for the Canadian Space Agency

Table 11.1
Design Characteristics of Non-U.S. SAR Spacecraft (Source: CBO [52, p. 8])

	RADARSAT-2	TerraSAR-X	SAR-Lupe	COSMO-Skymed
Country	Canada	Germany	Germany (Mil)	Italy
Number of Sats	1	2	5	4
Antenna Size	15 × 1.37m	4.8 × 0.7m	3.3 × 2.7m	5.7 × 1.4m
Mass	2,300 kg	~1,000 kg	770 kg	1,700 kg
Altitude	798 km	514 km	500 km	620 km
Orbital Inclination	98.6	97.4	~90	97.9
Center Frequency	5.4 GHz (C)	9.65 GHz (X)	X-band	9.6 GHz (X)
Solar Cell Power	3,156 W	1,800 W	Unknown	3,600 W
Image Resolution	~1m	~1m	0.5m	<1m
Design Life	7 years	5 years	10 years	5 years

(CSA),which has been upgraded with several advanced capabilities relevant to ABI.

Recognizing the need for enhanced maritime domain awareness, object detection, and pattern-of-life analysis of the arctic mission, the Defense Research and Development Canada (DRDC) began a program called Polar Epsilon. The $64.5-million program involves "arctic surveillance (land), environmental sensing, near-real time ship detection, and satellite-based maritime surveillance with radar" [54]. The program implemented two new ScanSAR beam modes [55]:

- Detection of Vessels, Wide Swath, Far Incidence (DVWF): 450 km swath, single polarization, 1-look azimuth, 5-looks range, large incidence, optimized for ship detection.
- Open Surveillance, Very Wide Swath, Near Incidence (OSVN): 530 km swath, dual polarization, 1-look azimuth, 50looks range, small incidence, improved ship detection.

The performance of these modes for ship detection is summarized in Figure 11.11. Performance varies depending on incidence and range to target, but the two new modes improve significantly on the ability to detect and geolocate smaller vessels at low incidence angles and over wide swaths.

RADARSAT-2 also included an experimental mode called the moving object detection experiment (MODEX), which allows GMTI measurements from postprocessing sequential SAR images with special noise reduction algorithms [56, 57]. MODEX is capable of "picking up a mid-sized car travelling with a 40 km/h radial velocity in a rural area with 90% probability" [58]. Developers

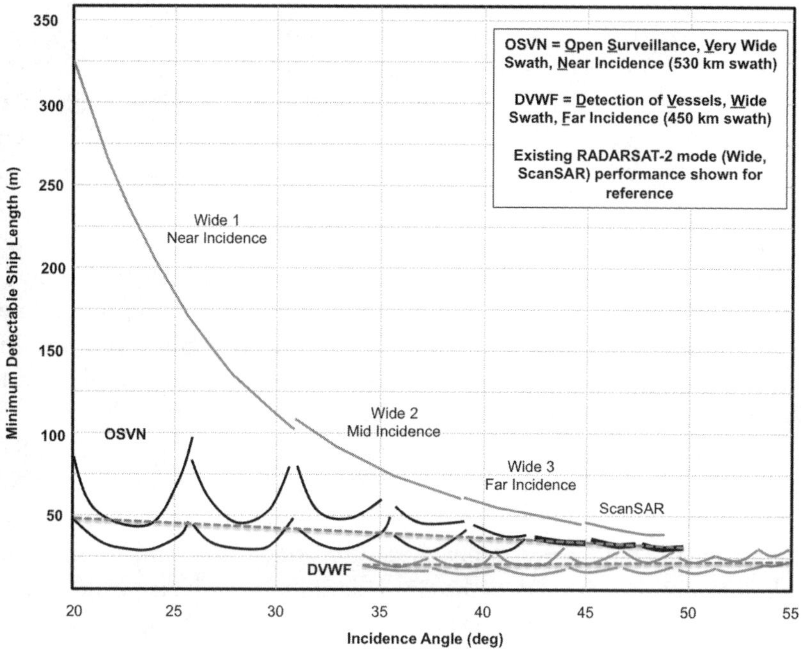

Figure 11.11 Performance of ship detection modes for Canada's RADARSAT-2. (Interpreted and adapted from [55, pp. 6–8]).

believed that the smaller SAR antennas of the RADARSAT follow-on constellation would preclude a useful capability, but researchers at DRDC Ottawa recently proposed MTI processing techniques and antenna reconfiguration schemes that could provide a viable MTI solution for the RADARSAT follow-on [59].

Canada's Polar Epsilon 2 is a $184.6-million program to add global ship identification and tracking, daily Arctic coverage, and four-day revisit with coherent-change detection (CCD) with the launch of the three-satellite Radarsat Continuation Mission in 2018 [54]. This program adds the capability to associate radar imagery and automatic identification system (AIS) detections. AIS is an electronic beacon that broadcasts the vessel name, identification number, position, speed, heading, and navigational status. "The International Maritime Organization's International Convention for the Safety of Life at Sea requires AIS to be fitted aboard international voyaging ships with gross tonnage (GT) of 300 or more, and all passenger ships regardless of size" [60]. Ground-based receiving stations near coastlines monitor AIS broadcasts of vessel position, speed, heading, and navigational status.

While the signals were never designed to be detectable from space, experiments revealed that space-based collectors could gather the signals. Space

Shuttle Mission STS-129 attached an AIS receiver antenna to the International Space Station [61]. Orbcomm, in partnership with the U.S. coast guard, launched a constellation of eight LEO satellites specifically designed to detect AIS signals. On July 14, 2014, Orbcomm launched six second-generation satellites as part of its $230-million network expansion program [62]. The service provides global coverage of AIS signals for maritime domain awareness, collision avoidance, vessel status, and security.

AIS metadata on vessel speed, heading, position, status, and type is visualized and analyzed using a spatiotemporal analysis environment as shown in Figure 11.12. Directional shapes show heading while diamonds indicate nonmoving vessels or navigational aids. A time-slider allows animation over time. Processing integrates multiple vessel positions into tracks for origin-destination and pattern-of-life studies.

Canada's Polar Epsilon program demonstrated the ability to fuse AIS detections and radar returns using RADARSAT-2. The fusion of AIS and SAR provides two data sources on each potential target, improving the probability of detection and allowing a possibility of identification. For example, if the radar return indicates that the ship is the size of a tanker or military vessel, but the user-programmed AIS signal indicates that the vessel is a 12-m pleasure craft, this anomaly is flagged for further investigation. Vachon and Quinn demonstrated fused results from this process in a Google Earth KML visualization in 2012 [54, pp. 30–31].

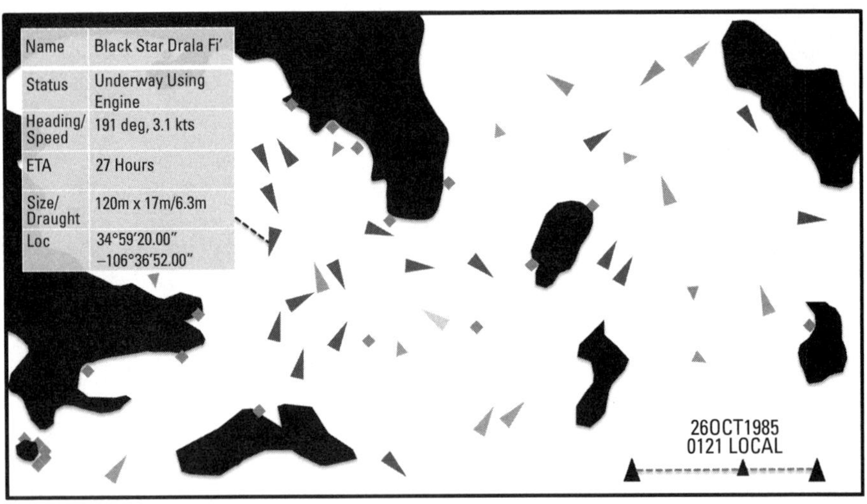

Figure 11.12 Notional visualization and situational awareness of maritime data using AIS.

11.8.3 Space-Based Persistent EO Imagery

Altitude and the physical constraints of optics present significant challenges for space-based persistent EO imaging. A general form of the spatial resolution equation for the sizing of an imaging aperture, *D*, to resolve an object of size *R* from a distance, *h* is:

$$D = 1.22\frac{h\lambda}{R}$$

Where λ is the wavelength of light. For visible imaging (a wavelength of approximately 500 nm), the aperture diameter required varies with height (altitude) as shown in Figure 11.13.

The four curves represent ground reference distances of 5m, 2.5m, 1m, and 0.1m respectively, which are approximately sufficient to detect, classify, characterize, and identify vehicle-sized objects, respectively. The vertical reference lines identify the three orbital regimes (LEO, MEO, and HEO/GEO). Because packaging the spacecraft into a launch vehicle fairing is a major design constraint, the three horizontal reference lines identify the primary mirror size of NASA's Hubble Space Telescope (2.4m), the internal diameter of the largest payload fairing on the United Launch Alliance Atlas V (4.5m), and the

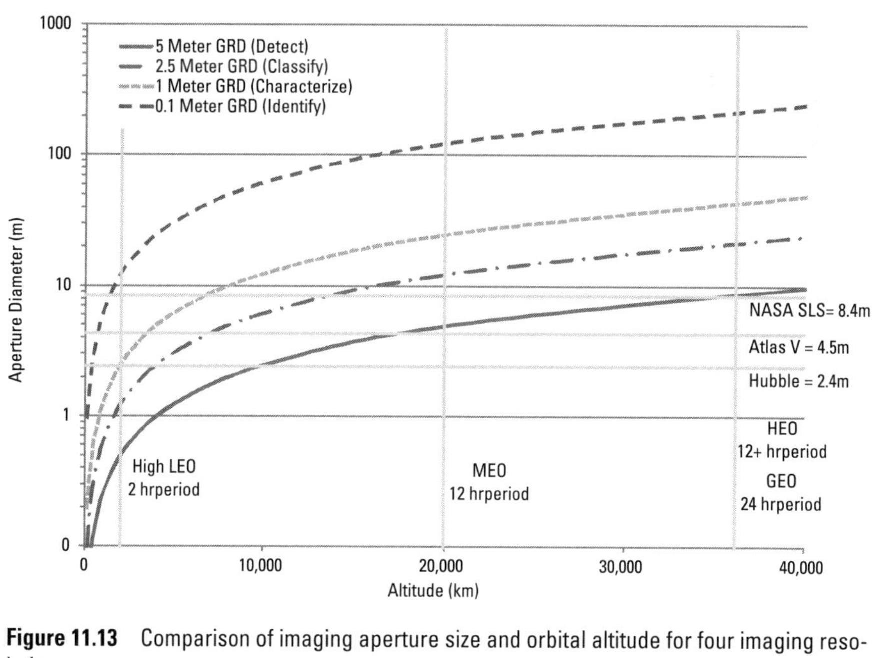

Figure 11.13 Comparison of imaging aperture size and orbital altitude for four imaging resolutions.

proposed diameter of the largest variant of NASA's future Space Launch System (SLS), 8.4m.

Launching a space telescope with 1-m GRD to a MEO orbit with a 12-hr orbital period requires an aperture of over 20m. Packaging such a large and heavy optical instrument into a launch vehicle presents significant challenges. Some alternatives include the foldable mirror system proposed for the James Webb Space Telescope (JWST) [63, 64], sparse imaging optics using the theory of interferometry as proposed by Golay [65], fractionated satellites like DARPA's System F6 [50, p. 14], and inflatable optical membranes [66]. Each of these presents unique structural, operational, and technological challenges.

An alternative architecture proposed by Silicon Valley start-up Skybox Imaging—now part of Google—uses a dispersed constellation of multiple microsatellites in LEO. The microsatellites deployed by Skybox are 60 cm x 60 cm x 95 cm with a mass less than 200 kg, but their patent filing includes concepts twice as large and weighing up to 500 kg [67, p. 2]. *Skysat-1,* launched November 21, 2013, and *Skysat-2,* launched July 8, 2014, are in 607-km and 627-km circular orbits, respectively, with a period of 98 minutes. An example of the Skybox concept is shown in Figure 11.14.

The satellites produce up to 90 seconds of 1.1-m GSD (nadir) panchromatic h.264 MPEG-4 encoded 8-bit video at 30 frames per second [69]. Skysat also captures 4-band multispectral imagery (blue, green, red, and near-IR) and video with a resolution of about 2.0m at nadir [69]. In 2014, Skybox contracted with Space Systems Loral to build 13 satellites out of a plan to eventually loft a constellation of 24 [70]. This increment of the constellation allows revisit of any point on the Earth up to three times per day, albeit for a maximum

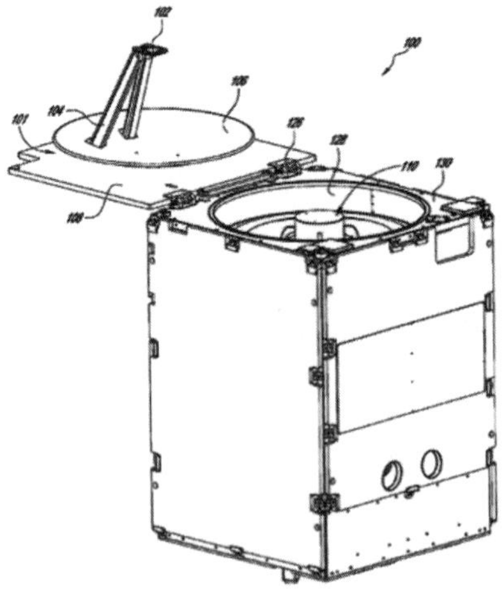

Figure 11.14 Schematic of an imaging microsatellite. (Source: U.S. Patent and Trademark Office [67].)

imaging time of 90 seconds. Despite popular depictions from television shows like "Alias" and movies like "Enemy of the State," it is not feasible to produce persistent high-resolution motion imagery from space.

Another approach to imaging microsatellites implemented by San Francisco-based Planet Labs uses a large constellation of even smaller satellites called "nanosatellites." In 2014, they deployed a "flock" of 28 10×10×30cm satellites from the International Space Station. Their "doves" provide highly frequent 3–5-m resolution imagery of an extremely wide area with a focus on large-scale change detection and environmental monitoring. The planned constellation of 131 satellites is optimized to provide subdaily revisit on midlatitude locations. In contrast, NASA's low-resolution (15-m GSD) Earth-observing Landsat 8 satellite only revisits a location on the Earth every 16 days. Instead of focusing on high resolution, long orbital lifetimes, highly precise collection, and exquisite capabilities, PlanetLabs provides an extremely robust low-cost architecture for low-resolution space-based persistent EO imagery. Other companies pursuing low-cost space-based imaging and video as of 2014 include Urthecast, BlackSky, and Nanosatisfi. While these systems provide high revisit rates for environmental monitoring, since a single pixel represents 3–5m, the intelligence value of these capabilities has yet to be quantified.

A comparison of the Skybox/Google SkySat, the PlanetLabs "Dove," and a standard 1U cubesat is shown in Figure 11.15, alongside a comparison to other imaging satellites.

Without unique analytic methods to narrow the search space based on patterns of life and probabilistic location estimates, catching mobile targets

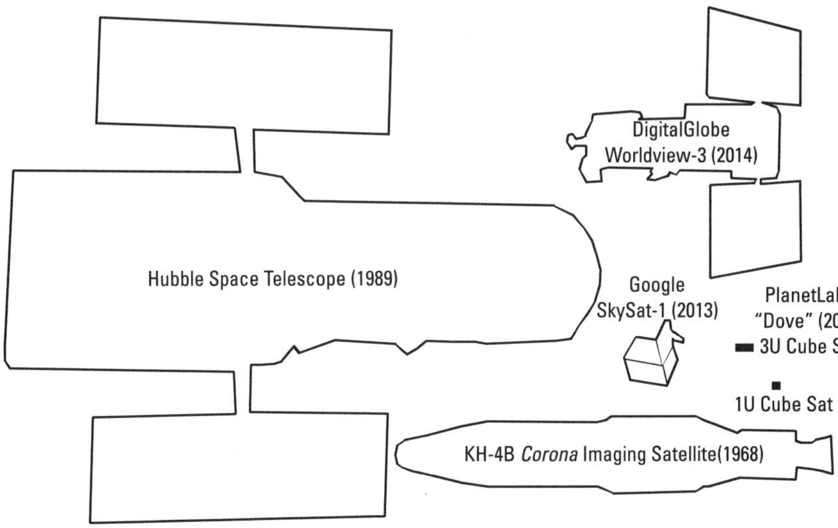

Figure 11.15 Comparison of satellite sizes.

after the act is extremely difficult. As Flynn, Juergens, and Cantrell, in their review of best practices for ISR employment for special forces, noted, "The enemy is so well hidden that it takes multiple sources of intelligence to corroborate one another…without a robust, collaborative intelligence network to guide it, sensors are often used in reactive modes that negate their true power and tend to minimize their full potential" [36].

11.9 Summary

Each concept for persistent surveillance has unique advantages and drawbacks, but large, multisensor persistent constellations are seldom viable. Some persistent concepts introduced and discussed in this chapter are compared and contrasted in Table 11.2.

NGA's Director Cardillo describes persistence not as a system of collection platforms but as "a mindset that encourages analysts to use any combination of methods to answer intelligence questions" [71]. The key to enabling persistent ABI analysis is not a new magical collection system with an advanced focal plane array or a unique waveform. Strategic advantage comes from innovative integration of the data that's all around us. Our ground-, air-, and space-based remote sensing assets are not vacuum cleaners to suck up large volumes of information, but rather scalpels precisely applied to obtain the last missing piece of information once deductive analysis has narrowed the search space.

Table 11.2
Key Attributes of Selected Persistent Surveillance Concepts

Concept	Dwell/ Revisit	Area Coverage	Resolution	Cost	Overall Utility
Ground-Based Cameras	Very High	Very Low	Very High	Very Low	Low/Med
MQ-9 Reaper	Very Low	Low	High	Low	Low
RQ-4 Global Hawk	Low	Low	High	Low/Med	Low/Med
Tethered Aerostat	Medium	Low	High	Low	Low/Med
Blue Devil II/LEMV	Med/High	Medium	Medium	Med	Med/High
Ubiquitous Aircraft	Med/High	High	Medium	Low/Med	Med/High
Near-Space Airship	High	High	Medium	Medium	Med/High
Space-Based SAR	High	Medium	High	Very High	Medium
Space-Based GMTI	Medium	High	Med/High	Very High	Med/High
Satellite-based AIS	High	High	High	Med	Low/Med
Commercial EO Sat.	Low	High	High	Med/High	Medium
Persistent EO Imaging	Very High	Very High	Low	Very High	Med/High
Microsatellites (EO)	Med/High	Low	Medium	Low/Med	Medium
Nanosatellites (EO)	High	Medium	Very Low	Low	Low

Increasingly, persistent information is available from networks of ground-based cameras and the Internet-connected mobile devices required for modern life.

The Chapters 12–14 will describe key advances in analytic methods and approaches to studying data to guide the application of these advanced surveillance tools, including how to apply a persistent mindset to the analysis of increasingly complex multi-INT data sets.

References

[1] NATO, "AAP-6 NATO Glossary of Terms and Definitions," 2004.

[2] "Joint Publication 1-02: Department of Defense Dictionary of Military and Associated Terms." U.S. Department of Defense, 16 July 2014.

[3] 10 US Code, §467.

[4] "Frequently Asked Questions Terms and Acronyms," National Security Agency/Central Security Service, web.

[5] Rosenau, W., "Special Operations Forces and Elusive Enemy Ground Targets: Lessons from Vietnam and the Persian Gulf War," RAND, Santa Monica, CA, 2001.

[6] "Motion Imagery Standards Profile Version 6.6." Department of Defense/Intelligence Community/National System for Geospatial Intelligence (DoD/IC/NSG) Motion Imagery Standards Board, 27 Feb 2014.

[7] "Multi-Spectral Targeting System (MTS)," Raytheon, web, http://www.raytheon.com/capabilities/products/mts/.

[8] Menthe, L., et al., "The Future of Air Force Motion Imagery Exploitation: Lessons from the Commercial World," RAND, Santa Monica, CA, 2012.

[9] "Multi-INT Analysis and Archive System (MAAS®)— Operationally Deployed Full Motion Video (FMV) and Imagery Processing, Exploitation and Dissemination (PED)," General Dynamics Information Systems, web, http://www.gd-ais.com/Products/ISR-Imagery-Analysis/MAAS.

[10] Schanz, M. V., "The Reaper Harvest," *Air Force Magazine,* April 2011.

[11] Nakashima, E., and C. Whitlock, "With Air Force's Gorgon Drone 'We Can See Everything,'" *The Washington Post,* January 1, 2011.

[12] Deptula, D., "Air Force ISR in a Changing World," 30 Mar 2010, web. Available: http://airpower.airforce.gov.au/UploadedFiles/General/Day2_Deptula.pdf.

[13] United States Air Force Sensor Data Management System, " Wright Patterson Air Force Base (WPAFB) 2009 Dataset," web. Available: https://www.sdms.afrl.af.mil/index.php?collection=wpafb2009.

[14] "Rise of the Drones," *NOVA,* Public Broadcasting System, 2013. Television.

[15] "Home," PIXIA Corporation. Web. www.pixia.com.

[16] Timberg, C., "New Surveillance Technology Can Track Everyone in an Area for Several Hours at a Time." *The Washington Post,* 5 Feb 2014.

[17] "MISB Engineering Guideline 0810.2, Profile 2: KLV for LVSD Applications." 11 Jun 2010.

[18] "From Video to Knowledge," Lawrence Livermore National Laboratory, May 2011.

[19] "BAE has Success with ARGUS-IS," UPI.com, February 9, 2010.

[20] Beckhusen, R., "Army Dumps All-Seeing Chopper Drone," *Wired: Danger Room,* June 25, 2012.

[21] Buderi, R., *The Invention That Changed the World: How a Small Group of Radar Pioneers Won the Second World War and Launched a Technological Revolution,* Simon and Schuster, 1996.

[22] Hacker, T. L., "Performance Analysis of a Space-Based GMTI Radar System Using Separated Spacecraft Interferometry," Master of Science, Massachusetts Institute of Technology, Boston, MA, 2000.

[23] "STANAG 4607 JAS (Edition 3)—NATO Ground Moving Target Indicator (GMTI) Format," NATO Standardization Agency, 14 Sep 2010.

[24] "Activity-Based Intelligence," Powerpoint Presentation, National Geospatial-Intelligence Agency. Approved for Public Release. NGA Case #11-040. 2011.

[25] Grant, R., "JSTARS Wars," *Air Force Magazine,* November 2009.

[26] Jennings, G., "US Army to Field King Air-based VADER Special Mission Aircraft," *IHS Jane's 360,* web.

[27] Iaconangelo, D., "Border Patrol VADER: 4 Things To Know About The New Drone Surveillance Radar System," *Latin Times,* June 20, 2013.

[28] "Army Extends Support for UAV Man-Hunting Radar from Northrop Grumman through 2013," *UAS News,* web, Feb 2013.

[29] Barrett, D., "One Surveillance Camera for Every 11 People in Britain, says CCTV Survey," *Telegraph,* web, July 10, 2013.

[30] "UK has 1% of World's Population but 20% of its CCTV Cameras," *Mail Online,*" web, March 27, 2007.

[31] "The City of Tomorrow May Already Be Here," CNN, May 2014, web, http://www.cnn.com/interactive/2014/05/specials/city-of-tomorrow/.

[32] "Placemeter," web, http://placemeter.com.

[33] "Submit your iOS 7 apps today," Apple Corporation, web, https://developer.apple.com/ios7/.

[34] Addey, D., "iBeacons," web, http://daveaddey.com/?p=1252.

[35] "Meet the Foursquare Champs of Top-Secret Washington," *Vocativ,* web. [Online] Available: http://www.vocativ.com/usa/nat-sec/mayor-of-the-nsa-meet-the-foursquare-champs-of-top-secret-washington/. 4 Apr 2013.

[36] Flynn, M. T., R. Juergens, and T. L. Cantrell, "Employing ISR SOF Best Practices," *Joint Forces Quarterly,* No. 50, 3rd Quarter 2008.

[37] Odierno, R., "ISR Evolution in the Iraqi Theater," *Joint Force Quarterly,* No. 50, 3rd Quarter 2008.

[38] Brown, J., "480th ISRW Airmen Decide on, Direct ISR Operations Across the Globe," US Air Force ISR Agency, 09 Jan 2014, web. http://www.afisr.af.mil/news/story.asp?id=123373695.

[39] Ackerman, R. K., "Persistent Surveillance Comes Into View," *AFCEA Signal Magazine,* May 2002.

[40] "RQ-4 Global Hawk," U.S. Air Force. Fact Sheet. Web.

[41] LaBelle, K., and G. Kasper, "Boeing: Status Update: Boeing's Phantom Eye Takes a Huge Step Forward," Boeing Corporation, 12 Feb 2014, web.

[42] "Defense Acquisitions, Future Aerostat and Airship Investment Decisions Drive Oversight and Coordination Needs," Government Accountability Office, GAO-13-81, Oct. 2012.

[43] "JLENS—Joint Land Attack Cruise Missile Defense Elevated Netted Sensor System." United States Army. Approved for Public Release by AMSAM-PA. AMCOM Public Release Case # 246.2012.

[44] Morton, J. F., "Our Warfighters Need JLENS," *The Hill,* web. Available: http://thehill.com/blogs/congress-blog/homeland-security/211459-our-warfighters-need-jlens.

[45] "Tethered Aerostat Radar System," Wikipedia. Accessed: 03 Aug 2014.

[46] Matthews, W., "Aerostats Lost: Weather, Mishaps Take Heavy Toll on Dirigibles," *Defense News,* May 7, 2013.

[47] "Air Force and Army Corps of Engineers Improperly Managed the Award of Contracts for the Blue Devil Block 2 Persistent Surveillance System," Inspector General of the US Department of Defense, DODIG 2013-128, Sep. 2013.

[48] Mav6, "M1400 Lighter-Than-Air Aircraft." Web. http://mav6.com/Mav6-M1400-Overview.pdf.

[49] Defense Advanced Research Projects Agency (DARPA), "Integrated Sensor Is the Structure (ISIS) Overview."

[50] Defense Advanced Research Projects Agency (DARPA), "RDT&E Budget Item Justification Sheet (R-2 Exhibit), Space Programs and Technology PE 0603287E, Project SPC-01," Feb. 2008.

[51] "Defense Science Board Task Force on Contributions of Space Based Radar to Missile Defense," Office of the Under Secretary of Defense For Acquisition, Technology, and Logistics, Washington, DC 20301-3140, Jun. 2004.

[52] "Alternatives for Military Space Radar," Congressional Budget Office, Jan. 2007.

[53] "Space Radar Program Cancelled," *Satellitetoday.com,* 07 March 2008.

[54] Vachon, P. W., and R. Quinn, "Operational Ship Detection in Canada using RADARSAT: Present and Future," *Defense Research and Development Canada,* June 20, 2012.

[55] Vachon, P. W., "New RADARSAT Capabilities Improve Maritime Surveillance," *Defense Research and Development Canada,* October 18, 2010.

[56] Nohara, T. J., et al., "SAR-GMTI Processing with Canada's Radarsat 2 Satellite," in *Adaptive Systems for Signal Processing, Communications, and Control Symposium,* 2000, pp. 379–384.

[57] Chiu, S., et al., "Computer Simulations of Canada's RADARSAT2 GMTI," *RTO SET Symposium on Space-Based Observation Technology,* Samos, Greece, 2000.

[58] Boucher, M., "The Defence and Security Applications of the RADARSAT Constellation Mission," *SpaceRef Canada,* March 19, 2013.

[59] Gierull, C. H., and S. Ishuwa, "Potential Marine Moving Target Indication (MMTI) Performance of the RADARSAT Constellation Mission (RCM)," in Synthetic Aperture Radar, *2012. EUSAR, 9th European Conference on,* Nuremberg, Germany, 2012, pp. 404–407.

[60] "Automatic Identification System," Wikipedia. Accessed: 09 Aug 2014.

[61] "Atlantis leaves Columbus with a Radio Eye on Earth's Sea Traffic," European Space Agency, December 4, 2009.

[62] "OG2: Mission 1 Launched on July 14, 2014." Orbcomm, web, 2014.

[63] Lightsey, P. A., et al., "James Webb Space Telescope: Large Deployable Cryogenic Telescope in Space," *Optical Engineering,* Vol. 51, No. 1, pp. 011003–1–011003–19, 2012.

[64] Nella, J., et. al., "James Webb Space Telescope (JWST) Observatory Architecture and Performance," *Proc. SPIE 5487, Optical, Infrared, and Millimeter Space Telescopes,* October 12, 2004. pp. 576–587.

[65] Chung, S.-J., D. W. Miller, and O. L. deWeck, "Design and Implementation of Sparse Aperture Imaging Systems," *Astronomical Telescopes and Instrumentation,* 2002, pp. 181–192.

[66] Soh, M., J. H. Lee, and S.-K. Youn, "An Inflatable Circular Membrane Mirror for Space Telescopes," *Proceedings of SPIE—The International Society for Optical Engineering,* No. 5638, 10 February, 2005, pp. 262–271.

[67] Miranda Do Carmo, H.,"Integrated Antenna System for Imaging Microsatellites." Google Patents, 2014.

[68] "SkySat-1." Web. http://www.geoimage.com.au/satellite/skysat-1.

[69] "Imagery and Video Data Sheet." Skybox Imaging, 08 Oct 2013.

[70] "Skybox Imaging Selects SSL To Build 13 Low Earth Orbit Imaging Satellites," web, February 10, 2014.

[71] Cardillo, R., "Remarks as Prepared for Robert Cardillo, Director, National Geospatial-Intelligence Agency for AFCEA/NGA Industry Day 2015," Approved for Public Release. NGA Case #15-281." Web. https://www1.nga.mil/MediaRoom/SpeechesRemarks/Documents/2015/031615_AFCEA_As_Prepared_FINAL.pdf.

12

Automated Activity Extraction

The New York Times reported that data scientists "spend from 50 to 80 percent of their time mired in this more mundane labor of collecting and preparing unruly digital data, before it can be explored for useful nuggets" [1]. Pejoratively referred to in the article as "janitor work," these tasks, also referred to as data wrangling, data munging, data farming, and data conditioning, inhibit progress toward analysis [2]. Conventional wisdom and repeated interviews with data analytics professionals support the "80%" notion [3–5]. Many of these tasks are routine and repetitive: reformatting data into different coordinate systems or data formats; manually tagging objects in imagery and video; backtracking vehicles from destination to origin; and extracting entities and objects in text. Ac to NGA's 2020 Analysis Technology Plan, "Automation may be in the form of tools and algorithms that reduce data dimensionality, provide investigative cues based on data correlation and object/change detection, and help the analyst transition from a forensic-based to a model-based approach to GEOINT analysis" [6]. ABI-enabling technologies for automated object, event, activity, and transaction extraction replace rote tasks with algorithms to free up more time to analyze the subsequent data. This chapter introduces a number of automated techniques for human-machine teaming and support to the analytic process.

12.1 The Need for Automation

Chapter 10 introduced the three V's of "big data"—the exploding volume, velocity, and variety—and Chapter 11 described new persistent, wide-area, multi-INT sensors. Although users are "drowning in data," they are thirsting for knowledge and insights. Intelligence consumers require highly confident judgments on a wider range of threats on increasingly tight timelines.

A 2003 study by DARPA in collaboration with several U.S. intelligence agencies found that analysts spend nearly 60% of their time performing research and preparing data for analysis [7]. The so-called bathtub curve, shown in Figure 12.1, shows how a significant percentage of an analyst's time is spent looking for data (research), formatting it for analysis, writing reports, and working on other administrative tasks. The DARPA study examined whether advances in formation technology such as collaboration and analysis tools could invert the "bathtub curve" so that analysts would spend less time wrestling with data and more time collaborating and performing analytic tasks, finding a significant benefit from new IT-enhanced methods.

As the volume, velocity, and variety of data sources available to intelligence analysts explodes, the problem of the "bathtub curve" gets worse. This chapter introduces several concepts for automated processing and transformation of data to support large-scale analysis for ABI that seeks to invert this curve, allowing analysts to spend more time doing analysis.

12.2 Data Conditioning

Data conditioning is an overarching term describing the preparation of data for analysis and is often associated with "automation" because it involves automated processes to prepare data for analysis. BAE Systems notes that ABI-enabling capabilities "employ advanced software analysis tools integrated with commercial, off-the-shelf computing infrastructure to automate the ingestion, storage and processing of petabytes of data as it is received" [8].

Historically, the phrase "extract, transform, load" (ETL) referred to a series of basic steps to prepare data for consumption by databases and data services.

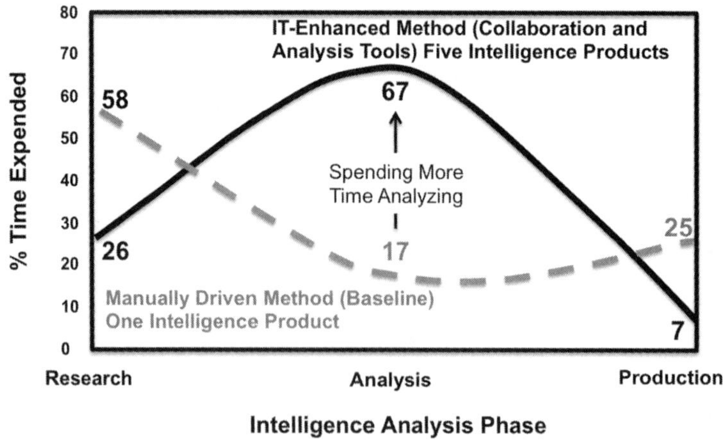

Figure 12.1 IT-enhanced versus manually driven intelligence analysis. (Adapted from [7].)

Often, nuanced ETL techniques were tied to a specific database architecture. Data conditioning includes the following:

- Extracting or obtaining the source data from various heterogeneous data sources or identifying a streaming data source (e.g., RSS feed) that provides continuous data input;
- Reformatting the data so that it is machine-readable and compliant with the target data model;
- Cleaning the data to remove erroneous records and adjusting date/time formats or geospatial coordinate systems to ensure consistency;
- Translating the language of data records as necessary;
- Correcting the data for various biases (e.g., geolocation errors);
- Enriching the data by adding derived metadata fields from the original source data (e.g., enriching spatial data to include a calculated country code);
- Tagging or labeling data with security, fitness-for-use, or other structural tags;
- Georeferencing the data to a consistent coordinate system or known physical locations;
- Loading the data into the target data store consistent with the data model and physical structure of the store;
- Validating that the conditioning steps have been done correctly and that queries produce results that meet mission criteria.

ABI accentuates the importance of georeferencing the data and highlights the need to condition the data against a spatial reference even if that reference is not inherent in the source data. One example is the extraction of place names from text files and the subsequent georegistration of the data to the coordinates of those place names—essentially a nested data conditioning problem. Data conditioning of source data into a spatiotemporal reference frame enables georeference to discover. Table 12.1 summarizes some common ABI examples for data conditioning.

While the principle of data neutrality promotes data conditioning from multiple sources, this chapter focuses on a subset of automated activity extraction techniques including automated extraction and geolocation of entities/events from text, extraction of objects/activities from still imagery, and automated extraction of objects, features, and tracks from motion imagery.

Table 12.1
Common Data Conditioning Techniques Applied to Support ABI

Source	Typical Extraction Activities
Text reports	Entities, events, coordinates, locations
Still imagery	Buildings, roads, geographic features, vehicles (e.g., tanks), changes between frames
Motion imagery	Vehicle motion (tracks), human activities
Hyperspectral imagery	Materials
Infrared imagery	Warm objects (e.g., concealed soldiers), operating equipment
Financial transactions	Account numbers, identities, amounts

12.3 Georeferenced Entity and Activity Extraction

While many applications perform automated text-parsing and entity extraction from unstructured text files, the simultaneous automated extraction of geospatial coordinates is central to enabling the ABI tradecraft of georeference to discover.

Marc Ubaldino, systems engineer and software developer at the MITRE Corporation, described a project called Event Horizon (EH) that "was borne out of an interest in trying to geospatially describe a volume of data—a lot of data, a lot of documents, a lot of things—and put them on a map for analysts to browse, and search, and understand, the details and also the trends" [9]. EH is a custom-developed tool to enable georeference to discover by creating a shapefile database of text documents that are georeferenced to a common coordinate system. Working files are stored as Esri geodatabases, the common storage and management framework for ArcGIS software. MITRE also developed a series of data conditioning tools for the government with names like Oxygen, Cesium, and Boron. These simple tools are chained together to orchestrate data conditioning and automated processing steps. According to MITRE, these tools have "decreased the human effort involved in correlating multi-source, multi-format intelligence" [10, p. 47]. This multimillion records corpus of data was first called the "giant load of intelligence" (GLINT). Later, this term evolved to geolocated intelligence.

One implementation of this method is the LocateXT software by ClearTerra, a "commercial technology for analyzing unstructured documents and extracting coordinate data, custom place names, and other critical information into GIS and other spatial viewing platforms" [11]. The tool scans unstructured text documents and features a flexible import utility for structured data (spreadsheets, delimited text). The tool supports all Microsoft Office documents (Word, PowerPoint, Excel), Adobe PDF, XML, HTML, Text, and more. Some of the tasks performed by LocateXT are described as follows [12]:

- Extracting geocoordinates, user-defined place names, dates, and other critical information from unstructured data;
- Identifying and extracting thousands of variations of geocoordinate formats;
- Creating geospatial layers from extracted locations;
- Configuring place name extraction using a geospatial layer or gazetteer file;
- Creating custom attributes by configuring keyword search and extraction controls.

LocateXT's extraction process is shown in Figure 12.2. The technology is capable of running as a desktop tool or a web geoprocessing service, or through an application programming interface (API) as an "engine" for geoprocessing and georegistering many data files as part of a larger workflow.

The resultant ArcMap view of a LocateXT document extraction in Esri geodatabase format is shown in Figure 12.3. The feature attributes contain the coordinates extracted as well as a "pretext" and "posttext" field. These represent 255 characters (the maximum number of characters allowed in the most basic Esri format, the shapefile) before and after the extracted match, which usually

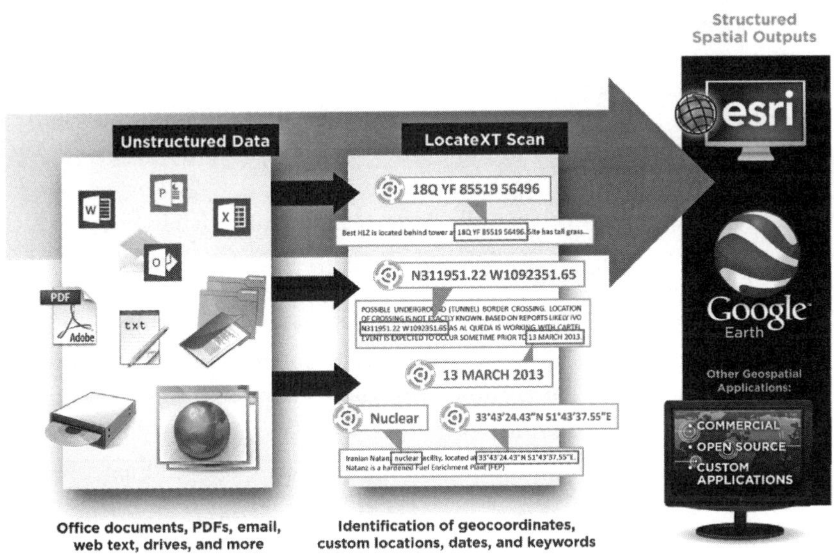

Figure 12.2 LocateXT scan and extract process. (©2014 ClearTerra. Reprinted with permission.)

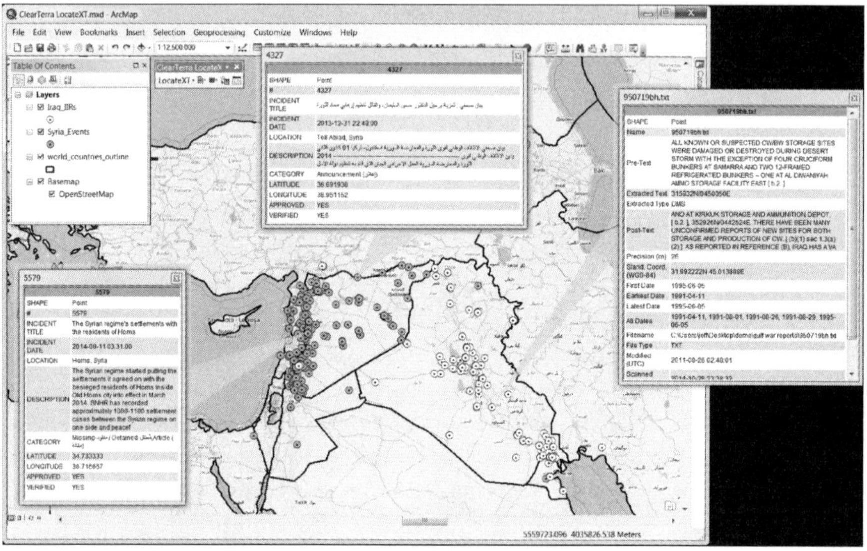

Figure 12.3 Example of LocateXT Running in ArcMap. (©2014 ClearTerra. Reprinted with Permission.)

contains contextual information about the source report. For example, an event in Figure 12.3 contains pretext:

> All known or suspected CW/BW [chemical weapon/biological weapon] storage sites were damaged or destroyed during DESERT STORM with the exception of four cruciform bunkers at Samarra and two 12-framed refrigerated bunkers—one at Al Diwaniyah Ammo Storage Facility East [13].

Imagery analysts accessing this event marker immediately know that the event is related to weapons of mass destruction. They are also on the lookout for cruciform bunkers and related equipment. They may cross-reference this event to the specific ammo storage facility mentioned. Links to the source data file are also included to maintain provenance throughout the analysis process.

LocateXT software integration with the widely used ESRI suite of GIS tools, makes it attractive to military and intelligence customers. LocateXT also integrates with ArcGIS Online, the server-based capability by ESRI. "This capability not only addresses the time-consuming manual process of finding and plotting geocoordinates but brings it into the social networking era by immediately sharing with others," said Jeff Wilson of ClearTerra [14]. In a demonstration to the United States Geospatial Intelligence Foundation ABI working group, Wilson showed how thousands of documents in a local directory could be georeferenced with a drag and drop interface.

12.4 Object and Activity Extraction from Still Imagery

Extraction of objects, features, and activities from imagery is a core element of GEOINT tradecraft and central to training as an imagery analyst. A number of tools and algorithms have been developed to aid in the manual, computer-assisted, and fully automated extraction from imagery. Feature extraction techniques for geoprocessing buildings, roads, trees, tunnels, and other features are widely applied to commercial imagery and used by civil engineers and city planners.

A number of methods are available for automated extraction and geoprocessing. Low-level feature extraction refers to those that are automatically extracted from an image without any shape information [15, p. 115]. A common technique used is called edge detection, which produces a vectorized line drawing with the salient features in the image. Portraits by caricature artists are an example of salient feature detection, as the individual is immediately recognizable in a highly downsampled sketch.

Edge detection relies on discontinuities or large changes in the brightness of the image over a spatial region. Because many man-made objects have defined edges and because sharp linear features are usually less prevalent in nature, edge detection is widely used in image processing and computer vision to extract man-made features and objects such as roads and buildings from their surroundings. These discontinuities may also be caused by changes in object orientation, color, or variations in material properties [16].

The applications of edge detection are diverse, and thousands of technical papers have been written on the topic. Raytheon BBN developed a system called VISER that "incorporates a large set of low-level features that capture appearance, color, motion, audio, and audio-visual co-occurrence patterns in videos" [17]. Edge detection techniques are embedded in many automated algorithms including those that perform tracking and fusion.

Higher level feature extraction techniques are object-based and rely on classifiers and machine learning. "An object is a region of interest with spatial, spectral (brightness and color), and/or texture characteristics that describe the region" [18]. The classification-based approach calibrates an automated algorithm by providing a range of training samples of the desired object. Multilevel neural networks are a common technique used to classify the training set in this approach [19].

A widely used approach is called scale-invariant feature transform (SIFT), published by David Lowe in 1999 [20, 21]. SIFT produces a "feature description" of an object in imagery, identifying significant points primarily defined by high-contrast regions of the image (a variation on edge detection). The scale-invariant aspect of the algorithm refers to the fact that characteristic features of the image should be consistent between one vantage point and another (e.g., across multiple images). The significant points are referred to as "key points." Lowe summarizes some of the key techniques applied to recognition for this class of problems in Table 12.2 [22].

High-level feature extraction is also applied to facial recognition [25]. Social-networking giant Facebook pioneered advancements in facial recognition, with some sites claiming that performance approaches or exceeds that of a human [26]. Most facial recognition approaches follow a four-stage model: Detect → Align → Represent → Classify. Much research is aimed at the classify step of the workflow. Facebook's approach improves performance by applying three-dimensional modeling to the alignment step and deriving the facial representation using a deep neural network [27]. While Facebook's research applies to universal face detection, classification in the context of the problem set is significantly easier. When the Facebook algorithm attempts to recognize individuals in submitted pictures, it has information about the "friends" to which the user is currently linked (in ABI parlance, a combination of contextual and relational information). It is much more likely that an individual in a user-provided photograph is related to the user through his or her social network. This property, called local partitioning, is useful for ABI. If an analyst can identify a subset of the data that is related to the target through one or more links (for example, a history of spatial locations previously visited), the dimensionality of the wide area search and targeting problem can be exponentially reduced.

Example: SPAWAR RAPIER

The Navy's Space and Naval Warfare Systems Command (SPAWAR) developed the Rapid Image Exploitation Resource (RAPIER®) ship detection system, a

Table 12.2
Significant Aspects of SIFT

Problem	Technique	Advantage
Key localization/Scale/ Rotation	Difference-of Gaussians/ Scale-space pyramid/ Orientation assignment	Accuracy, stability, scale & rotational invariance
Geometric distortion	Blurring/Resampling of local image orientation planes	Affine invariance
Indexing and matching	Nearest neighbor/Best Bin First search [24]	Efficiency/speed
Cluster identification	Hough Transform voting	Reliable pose models/
Model verification/Outlier detection	Linear least squares	Better error tolerance with fewer matches
Hypothesis acceptance	Bayesian Probability analysis	Improved reliability

Source: [23]

patented technology developed over five years [28]. RAPIER "uses a suite of advanced image processing algorithms to automatically detect ships from high-resolution commercial satellite imagery" and other sources [29]. Algorithms reduce false alarms by masking land, detecting/removing clouds, and removing noise due to glints, waves, and other environmental phenomena. Algorithms extract, classify, geolocate, and measure ships. They also determine their approximate heading by estimating the location of the bow (usually curved) and stern (usually flat) and a wake if present. A typical output is an HTML "tip sheet" that contains a chip of the detected ship, and the metadata from the detection algorithm is shown in Figure 12.4.

Figure 12.4 is classified as a liquefied gas carrier with 99% confidence. These ships are characterized by large cylindrical gas tanks along the length of the vessel. Characteristics of ships are defined through models, which are trained through the system by repeated exposure to example cases. Figure 12.5 shows edge detection and classification algorithms applied to a container ship.

To calibrate RAPIER, designers imported the geometric properties of thousands of vessels from publically available databases of vessel registries. An initial classification step groups ships by their length and length/width ratio, coarse estimates that can also be used to estimate the gross tonnage of the vessel.

Different algorithms are used depending on the nature of the imagery and background noise caused by the water's surface. Calm water (lower sea state) appears black in panchromatic imagery whereas choppy water (high sea state) appears to have a gray texture on average when the intensity of pixels is examined across the image. RAPIER uses Fourier transforms and binary logic to apply different detection thresholds depending on the detected sea state.

Detection confidence is determined using a weighted sum:

$$C = W_{\mu}\mu \times W_{\sigma}\sigma \times W_{A}A \times W_{I}I_{max}$$

Where μ and σ are the mean and standard deviation of the pixel intensity, A is the area of the detected object, and I_{max} is the maximum pixel intensity in the detection region. W_i are empirical weights assigned to each factor, which are nominally 0.25 per factor [31].

Figure 12.4 Extracted ship and associated metadata from RAPIER®. (Source: SPAWAR Systems Center [29]. Approved for public Release. Distribution unlimited.)

Figure 12.5 Application of edge detection and feature extraction algorithms to a candidate ship. (Source: SPAWAR Systems Center [30], U.S. patent.)

In addition to HTML tip sheets, RAPIER also outputs a Google Earth .KMZ file containing geolocated ship detections and associated metadata. Visualization tools highlight detected objects by putting a box around them as shown in Figure 12.6.

Traditionally, broad area search is a monotonous, time-consuming task that requires analysts to scan raster imagery line by line for candidate objects. RAPIER's approach to ship detection focuses human analysts on potential objects of interest. Analysts can filter by metadata like confidence, size, and heading to further filter objects. By aggregating RAPIER results over many subsequent collections, analysts develop an understanding of maritime patterns. In a related application, SPAWAR developed a method for fusing overhead imagery with automatic vessel reporting systems—specifically the AIS—to fuse imagery-derived position/classification data with a unique identifier associated with each vessel [32]. This process integrates "georeference to discover" with entity resolution through proxies to identify the pattern of life of individual vessels.

SPAWAR continues development of RAPIER and recently extended the algorithms to maritime target detection, tracking, and identification from full-motion video and synthetic aperture radar [33, 34]. Developers also applied the RAPIER framework to fire and flood detection [35, 36].

Ongoing research in this field seeks to extend methods for object detection to activity detection. The relative position of objects, contextual information

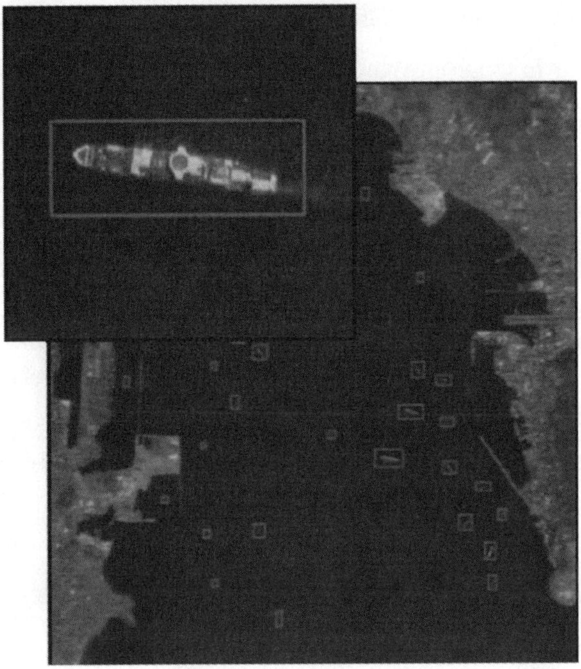

Figure 12.6 Example of ship detections from RAPIER®. (Source: SPAWAR Systems Center [29]. Approved for public release. Distribution unlimited.)

about the scene, or object positions/behaviors in the context of past observations may allow characterization of activity.

Another approach relevant to ABI is the concept of change detection from a sequence of images of the same area of interest. A raster image is a series of intensity values across a Cartesian plane. If two images are taken in the same plane at two different times, they can be "subtracted" to detect the change from one image to another. In practice, the rate of sampling of the pair of images must be greater than the rate of change of the target activity. For this reason, automated change detection and activity extraction from motion imagery demonstrates significant utility in ABI analysis.

12.5 Object and Activity Extraction from Motion Imagery

With the proliferation of motion imagery in the early 2000s, many of the fundamental algorithms used for computer vision and object extraction were extended to motion imagery. Object and activity extraction from ground-based, airborne, and space-borne motion imagery is a continuing area of intense research and development focus across the intelligence community and DoD.

12.5.1 Activity Extraction from Video

The Automated Low-Level Analysis and Description of Diverse Intelligence Video (ALADDIN) program, from the Intelligence Advanced Research Projects Agency's (IARPA) Office of Incisive Analytics, "seeks to combine the state-of-the-art in video extraction, audio extraction, knowledge representation, and search technologies in a revolutionary way to create a fast, accurate, robust, and extensible technology that supports the multimedia analytic needs of the future" [37]. ALADDIN develops technology to rapidly search a very large collection of video clips for specific events of interest. In ALADDIN, an "event" has the following attributes:

- It is a complex activity occurring at a specific place and time;
- It involves people interacting with other people and/or objects;
- It consists of a number of human actions, processes, and activities that are loosely or tightly organized and that have significant temporal and semantic relationships to the overarching activity;
- It is directly observable.

The technology development program defined content descriptors (metadata about events in the video) and semantic relationships to define events. As a performer on ALADDIN, IBM developed the IBM Multimedia Analysis and Retrieval System (IMARS), a "novel visual feature-based machine-learning framework for large-scale semantic modeling and classification of image and video content" [38]. IMARS categorizes and classifies video based on descriptors and their research represents attributes in a compact matrix format that is learned through offline processing but is capable of executing at scale on streaming video [39].

Activity discovery and exploitation in motion imagery from airborne surveillance platforms like the U.S. Air Force Predator UAV is a time-consuming, manual process. The DARPA Video and Image Retrieval and Analysis Tool (VIRAT), a 2008 program led by Dr. Mita Desai, develops "the ability to quickly search large volumes of existing video data and monitor real-time video data for specific activities or events" [40]. VIRAT-developed tools alert operators to the occurrence of specific events and activities and enable content-based contextual searches of large motion imagery stores to aid in multisource correlation. According to a Broad Agency Announcement (BAA) on the topic, "DARPA is seeking innovative algorithms for activity representation, matching and recognition which can support both indexing and retrieval" for dynamic activity-based information to support activity analysis [41]. "Bad guys do bad things, such as all the actions involved in burying an IED (improvised explosive device)—so it is activity that matters," VIRAT and PerSEAS program manager Mita Desai said in a DARPA statement [41].

By 2014, the military's fleet of UAVs logged more than 4,000,000 flight hours, producing millions of hours of high-resolution motion imagery [42]. Analysts exploit only a small fraction of this imagery. It resides in petabyte-scale data stores, tagged with limited metadata and is seldom reviewed after the fact. VIRAT-developed algorithms offer an opportunity for forensic reprocessing of these holdings to spatially and temporally tag specific activities to enable large-scale activity-based discovery and correlation. Analysts will be able to query existing motion imagery stores for all past occurrences of events. As new events of interest or extraction algorithms are defined, existing holdings can be reprocessed to geotemporally reference new events, enriching ABI metadata holdings and conditioning additional data for discovery and correlation. Figure 12.7 describes the objective VIRAT system, which processes motion imagery and derives a series of "content descriptors" based on "primitives" that represent behaviors, activities, events, and transactions [43].

VIRAT also developed a taxonomy of the types of activities that should be automatically extracted from video clips. The taxonomy, shown in Table 12.3, was focused on human activities, which are generally much harder to recognize than particular objects. VIRAT performers demonstrated numerous techniques against "surveillance videos of realistic scenes with different scales and resolution, each lasting 2 to 15 minutes and containing up to 30 events" [44].

Figure 12.8 illustrates one possible operational concept for VIRAT. An analyst wants to query (or receive an alert) for all the instances where cars make a U-turn in a 4 km^2 area over a 12-hour period. The "U-turn" behavior corresponds to primitives in track data including radial velocity, heading, and acceleration. Track data is processed both from streaming real-time motion imagery and an indexed data store to find all vehicles that change heading by 180° coincident with 0 radial velocity. The VIRAT system queues up all video clips that match the desired behavior and geolocates the events on a map with spatial and temporal coordinates.

According to Rimey, Hoff, and Lee, "Recognizing activities requires observations over time, and recognition performance is a function of the discrimination power of a set of observational evidence relative to the structure of a specific activity set" [45]. They highlight the importance of increasingly proliferating persistent surveillance sensors and focus on activities identified by a critical geospatial, temporal, or interactive pattern in highly cluttered environments. One of the keys to their approach is integration of the contextual environment—other multi-INT information that improves the probability of correct association and improves reasoning based on likely activities in that context.

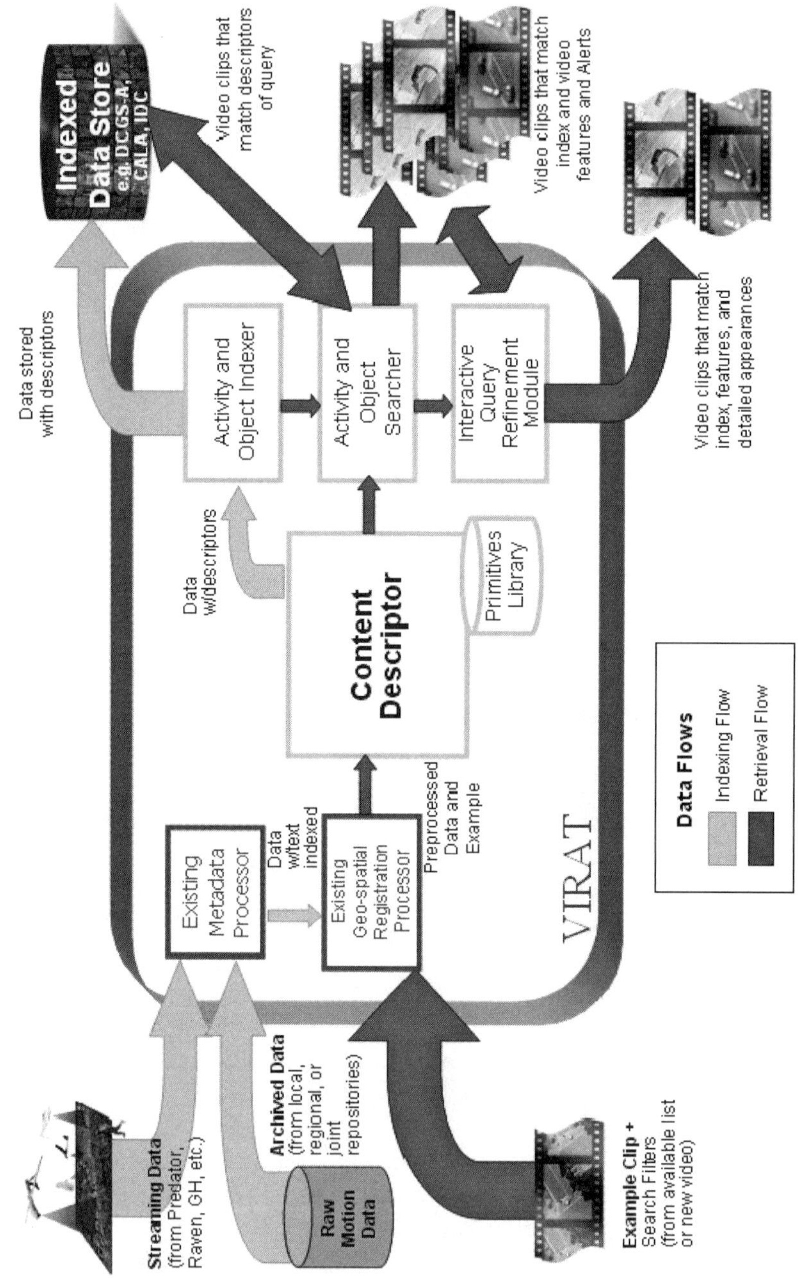

Figure 12.7 Description of the DARPA VIRAT system. (Source: [40], DARPA.

Table 12.3

Example of the Types of Activities Extracted by the VIRAT System (Source: [40], DARPA)

Activity Category	Candidate Activities
Single Person	Digging, loitering, picking up, throwing, exploding/burning, carrying, shooting, launching, walking, limping, running, kicking, smoking, gesturing
Person-person	Following, meeting, gathering, moving as a group, dispersing, shaking hands, kissing, exchanging objects, kicking, carrying together
Person-vehicle	Driving, getting-in (out), loading (unloading), opening (closing) trunk, crawling under car, breaking window, shooting/launching, exploding/burning, dropping off, picking up
Person-facility	Entering (exiting), standing, waiting at checkpoint, evading checkpoint, climbing atop, passing thru gate, dropping off
Vehicle	Accelerating (decelerating), turning, stopping, overtaking/passing, exploding/burning, discharging, shooting, moving together, forming into convoys, maintaining distance
Other	VIP activities (convoy, parade, receiving line, troop formation, speaking to crowds), riding/leading animal, bicycling, etc.

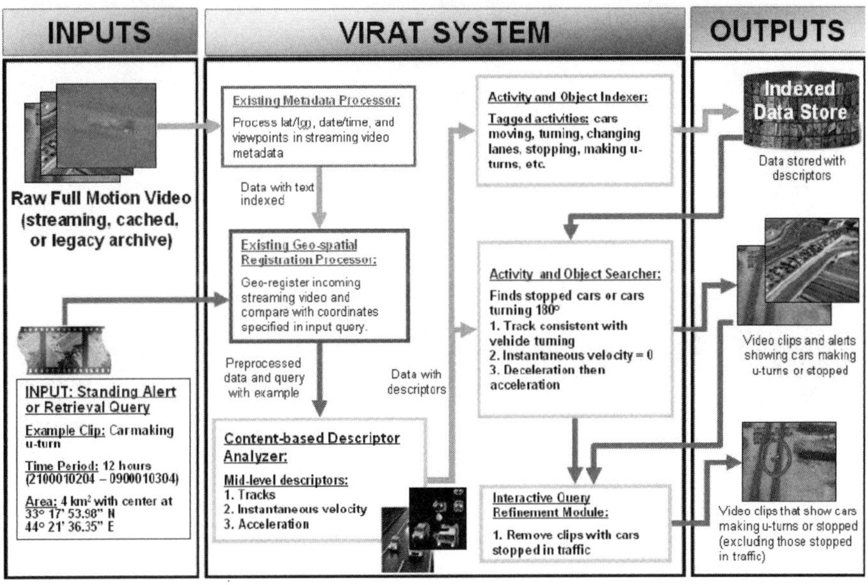

Figure 12.8 Operational concept for the VIRAT program. (Source: [40], DARPA.

Applications to Military Remote Sensing

In 2013, RAND produced a detailed report on motion imagery processing and exploitation (MIPE) that examined techniques for exploiting the exploding volumes of FMV and WAMI produced by military collection systems. RAND noted that automated algorithms "cannot completely replace all human analysts" searching for objects and activities but added that automation played a significant role in focusing analyst attention to "a particular video frame or subframe" [46]. The concept of narrowing the analyst's search area—not replacing the human-driven workflow and analytic processes—is a key enabler for automation in support of ABI, especially for wide area multi-INT collections and other large data sets.

Analysts manually exploited the Army's Constant Hawk WAMI system using the Massachusetts Institute of Technology Lincoln Labs APIX viewer beginning around 2006 [47, p. 11]. While one set of data went to ground-based analysts for immediate exploitation, a second set of disks was processed to stitch images together and sharpen, color-correct, and compress them for forensic analysis. Forensic analysis required analysts to begin from an event of interest—usually the explosion of an IED—and rewind the video to backtrack participating entities to their point of origin. Working with the Army's Task Force ODIN to produce intelligence products, analysts painstakingly documented the destination/origin tracks by advancing the video one frame at a time and clicking on the target object [48]. Exploitation of linked tracks often took longer than mission time (i.e., it takes longer than one hour to exploit one hour of motion imagery with forensic backtracking). The meticulous, laborious process was further confounded in urban environments dominated by occlusions and the standard for a successful track was high: Military forces were preparing to raid the targeted origin destinations, and false alarms would cause significant danger and negative public reaction.

Over time, change detection algorithms were implemented to identify differences in video over the same locations "such as disturbed earth... that would indicate a possible land mine" [49]. Change detection algorithms enable the "focus of attention" tradecraft identified by the RAND study, but more advanced algorithms are needed to fully analyze activities and transactions across multiple sensors and increasingly large areas. The RAND MIPE report also noted that a high-payoff, low-hanging fruit technology that benefits WAMI analysis is background subtraction, which "automatically indicates whether the foreground of a video feed has changed significantly between frames, thus directing an analyst to return his or her attention to the changing scene," noting that such a technology would be especially helpful in reviewing large unpopulated areas for low-signature change [46].

12.5.2 Activity and Event Extraction from WAMI

WAMI collectors like Constant Hawk, Gorgon Stare, and Blue Devil introduced a new capability for detecting events over a city-sized area due to their relatively long dwell time with reasonable frame rate and resolution.

Start-Stop Detection

Tracking objects across the full field of view (FFOV) of a WAMI platform is an extremely computationally intensive task. When WAMI platforms were first deployed, forward-deployed hardware configurations lacked sufficient power, and algorithms lacked sufficient confidence to enable full-frame automated tracking of objects. Analysts and technologies integrated the concept of discreteness with automated motion-detection algorithms to perform human-machine teaming for exploitation of WAMI. Motion-detection calculations—essentially change detection between two or more frames of motion imagery—are exponentially less computationally intensive than multitarget tracking. It is relatively easy to identify that something moved between frames, but associating the moving object with the correct object in the previous frame requires another computational step.

Analysts realized that identifying only the starts and stops of transactions focused their attention on locations at which to begin manual tracking and backtracking in WAMI data. Furthermore, by correlating unusually timed starts and stops with discrete locations for a suspected entity, analysts could quickly test hypotheses related to that entity's pattern of life and quickly map the geospatial network of that entity using the methods discussed in Chapters 5 and 22.

Although many R&D efforts focused resources on developing tracking algorithms, many experienced analysts were comfortable with start/stop detection and temporal analysis of those starts/stops as a clue to being analysis at that location. Further development efforts refined the types of metadata documented for each event and improved the confidence of detections by integrating other sources of data.

Event Detection

While VIRAT focused on identification and extraction of activities in focused, high-resolution motion imagery typified by the Predator UAV, lower-resolution WAMI also provides potential for automated activity extraction. DARPA's Desai also managed the Persistent Stare Exploitation and Analysis System (PerSEAS) program, which focused on WAMI and activity recognition from large-format, low-resolution systems by characterizing kinematic behaviors in context. Figure 12.9 shows the concept for PerSEAS. PerSEAS developed "algorithmic solutions to [associate] track fragments in order to identify localized events" [50]. PerSEAS automatically observes a large area over time to identify patterns and alert analysts for further analysis. According to Desai, "The objectives of VIRAT and PerSEAS are not to replace human analysts, but to make them more effective and efficient by reducing their cognitive load and enabling them to search for activities and threats quickly and easily" [41].

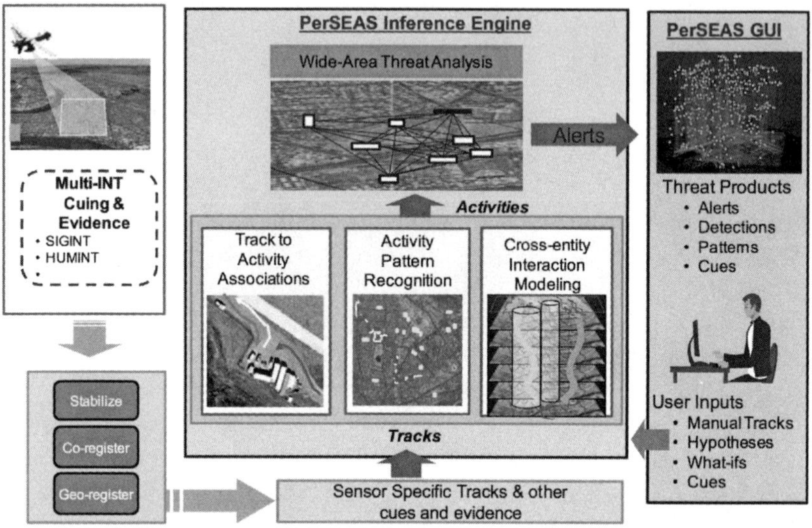

Figure 12.9 Concept for DARPA's persistent stare exploitation and analysis system (PerSEAS). (Source: [50], DARPA.)

Kitware, one of the performers on PerSEAS, developed software for activity and threat recognition to "efficiently find long-duration, spatially distributed, multi-agent activities and suspicious behaviors despite the shortcomings of computer tracks, which rarely maintain entity identity continuously from its origin to its destination" [51]. Kitware's approach to event recognition and correlation across fragmented tracks is based on dynamic Bayesian networks [52].

12.6 Tracking and Track Extraction

In addition to detecting events and activities over a wide area, airborne motion imagery and radar sensors provided the capability to document complete transactions over a city-sized area for the first time. While early exploitation efforts focused on manual documentation of object tracks from origin to destination, increasingly capable automated algorithms for tracking from video allowed full-field-of-view processing of the entire take to form complete transactions. These are generally referred to as "tracks."

Tracking involves the mathematical solution of two related problems: motion detection and object matching. Motion detection, described earlier in this chapter in terms of change detection, is the annotation of change from one observation to the next. In the case of high-frame-rate imagery, two or more successive frames are taken from almost exactly the same vantage point, incidence angle, lighting condition, and altitude, which improves the consistency of change detection. Changes are calculated by comparing the returns—energy in the form of radar and intensity in the form of imagery—between two (or more) successive frames.

Object matching requires estimating the likely position of the object in the next observation and correctly associating the two observations with each other. The association step increases in computational complexity as the density of the target environment increases—that is, when the algorithm has to compare the new position with multiple possible objects, there are multiple opportunities to introduce errors.

12.6.1 The Role of Sampling Rate and Resolution

In tracking, the sampling rate (in motion imagery, also called the frame rate) and resolution play a significant role in tracker performance. The frame rate must be high enough so that the tracked object is close enough to the position in the subsequent frame that it can be reliably associated with its previous position. However, at very high frame rates, sometimes motion cannot be detected across two adjacent frames because the object has not traveled far enough to cross the detection threshold at a given resolution. Paradoxically, humans move slowly enough that extremely high-frame-rate imagery may actually degrade motion detection and tracking performance.

Resolution also plays a significant role in resolving the details of objects for appearance-based tracking and for extracting moving objects out of a noisy background. At 0.5-m ground sample distance (GSD), each pixel in an image represents .5meter. A typical sedan is about 2m across and 5m long. At 0.5-m GSD, a sedan is comprised of approximately 20 pixels. Features like a windshield or pickup truck bed may be barely resolvable. From a perfectly vertical orientation, humans are not wide and deep enough to comprise a single pixel; although depending on the time of day, their shadows may be trackable with this approach.

Because of the trade-offs involved, successful tracking approaches typically use a mix of algorithms and aggregate the results to increase the probability of detection and correct association while accounting for noise and false alarms.

12.6.2 Terminology: Tracks and Tracklets

A track is a continuous representation of the movement of an object over time, but in practice, occlusions, sensor errors, artifacts, noise, intersections with other objects, and stops/starts of objects confound the ability to continuously track an object across an entire transaction. Figure 12.10 illustrates the contiguous nature of the real-world track (top). The middle of the figure shows a ground-truthed track (squares) as recorded by a reliable sensor at seven discrete points. These are the real-world positions of the object along the track. In the world of

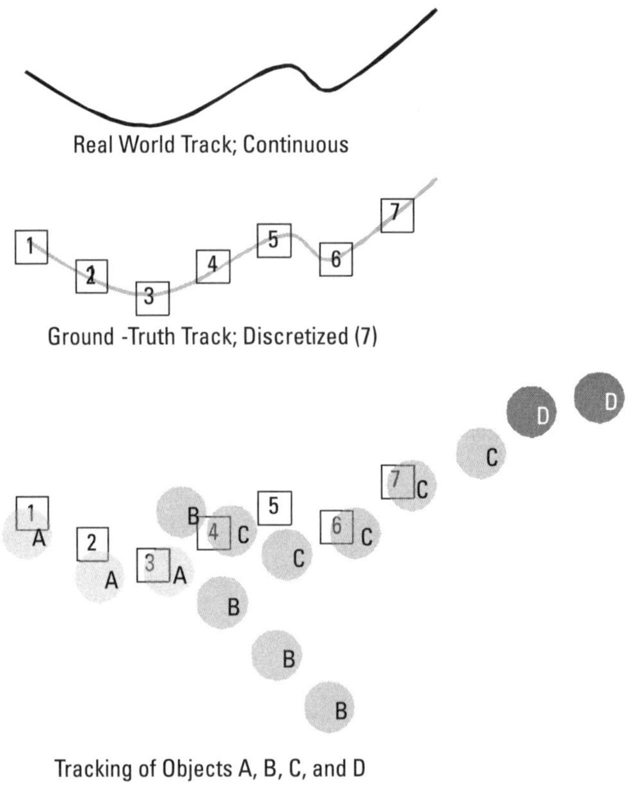

Figure 12.10 Basic track terminology. (Adapted from [53].)

remote sensing, tracks are never continuous because they are sampled discretely in time.

The bottom of Figure 12.10 shows how sensed objects A, B, C, and D line up with the ground-truth discretized observations. When the circle corresponding with each object lines up with the square, it means that the object detection lines up with a real-world position of the object. Note at position (4), both objects B and C are detected. Three of the observations of object B do not correspond to a real-world track. None of the observations line up with position (5). Track C continues after the seventh observation, and D appears to be a continuation of C but was resolved as a separate object by the sensor.

Tracklets are tracks of more than one observation that are associated with a single objects. Tracklets occur frequently during occlusions or other geometric/environmental obscurations (e.g., caused by the motion of clouds). Figure 12.11 shows how individual sensor observations from Figure 12.10 are associated into tracklets.

Tracklets can be stitched to form tracks. The association between tracklets is determined from observation metadata. Kinematic metadata—for example,

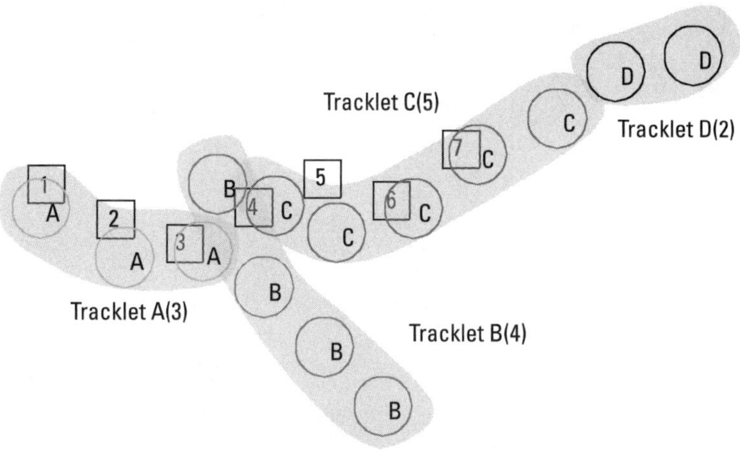

Figure 12.11 Tracklets formed from sensor data. (Adapted from [53].)

the speed and heading of the object—are widely used to anticipate the object's position at the time of the next sample. Appearance metadata, which includes the object's radar cross-section, color, shape, hue, reflection, and other identifying characteristics, is used where possible. Many tracking algorithms combine kinematics and appearance to improve tracker performance.

Figure 12.12 shows an example of tracklet stitching to form complete tracks. In the first instance, tracklet AB is formed by stitching tracklet A and tracklet B, ignoring one of the conflicting observations of object B. Tracklet CD is resolved as a separate object moving in a different direction and beginning around point (4). It is evident from the ground-truthed data representing the real path of the object that this conjunction is incorrect.

The correct stitching is shown in the lower half of Figure 12.12, where tracklet AC is formed from eight observations of the object. Note that point (5) was not accurately sampled and that tracklet D may be a continuation of AC if the kinematic and appearance metadata are similar.

12.6.3 The Kalman Filter

A Kalman filter is a "computational algorithm that processes measurements to deduce an optimum estimate of the past, present, or future state of a linear system by using a time sequence of measurements of the system behavior, plus a statistical model that characterizes the system and measurement errors, plus initial condition information" [54]. The basic concept of the Kalman filter is shown in Figure 12.13. The term "filter" is confusing to novice analysts and engineers. The process is used to find the best estimate of the future state of the object from measurements containing a bias—or noise. The Kalman Filter can

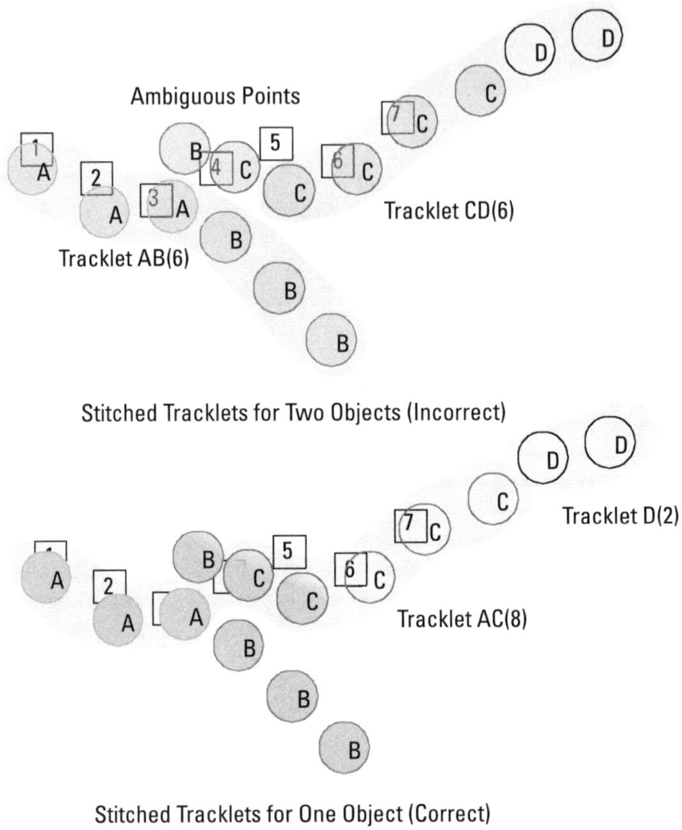

Figure 12.12 Track stitching and the resolution of object motion. (Adapted from [53].)

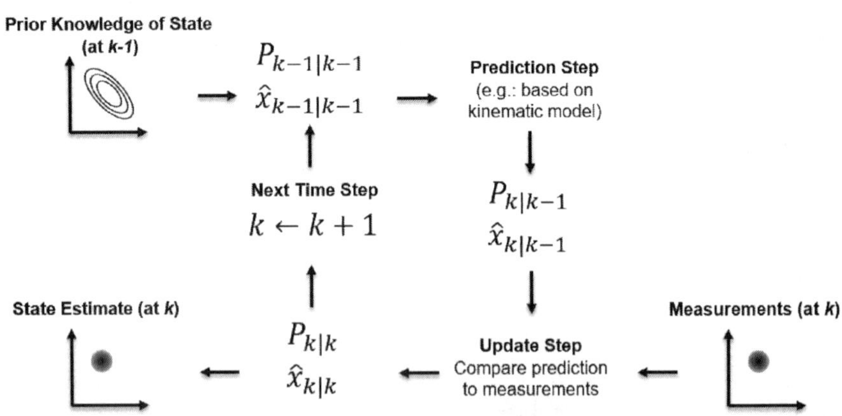

Figure 12.13 Basic concept of Kalman filtering [55]. (Public domain image.)

be thought of as "filtering out the noise" to produce a reliable estimate of the future state of an object.

Kalman filtering is a two-step process. As shown in Figure 12.13, the Kalman filter begins with a prior state of knowledge and predicts the next likely state (e.g., future position of the object) based on a physical model. In the case of tracking algorithms, the prediction step is usually based on the kinematic properties of the object and information about the current state. In the case of a sedan observed on a road driving at 60 miles per hour (88 ft/sec), the prediction step would assume that the sedan is 88 feet directly ahead one second later. Measurements of the object are taken at the next timestep, k. The update step compares the detected state of the object with the predicted state of the object and updates the state estimate with newly observed information. The equations governing the Kalman filter for estimating the state of an object, $x \in \Re_n$ are [56, p. 20].

$$x_k = Ax_{k-1} + Bu_k + w_{k-1} \tag{12.1}$$

with a measurement, $y \in \Re_m$

$$y_k = Hx_k + v_k \tag{12.2}$$

where:

A is an $n \times n$ matrix relating the state at the previous time, k–1 to the state at time k (irrespective of any control input or noise in the system);

x_{k-1} represents the state of the system at the previous time, k–1;

B is an $n \times 1$ matrix relating the control input, $u \in$ Rl, to the state x;

u is the input to the system causing change from time k–1 to k (in addition to any forecasted change irrespective of u);

w_{k-1} is the Gaussian process noise associated with the state estimate, x;

y_k is the measurement of the state, with associated Gaussian process noise v_k. v_k and w_{k-1} are assumed to be independent;

H is an $m \times n$ matrix that relates the state, x_k, to the measurement, y_k.

In simple terms, the Kalman filter takes known information (x), projects it into the future based on known relationships (A, B), corrects with measurements (y, H), and accounts for noise (w, v). This approach and its derivatives (including nonlinear, continuous time, frequency-weighted, and hybrid filters) form the basis of most tracking, trajectory estimation, guidance and navigation, and signal processing applications. Many of the automated algorithms used to estimate states and positions of objects and to project activities and transactions into the future feature a variant of the Kalman filter inside their code.

12.6.4 Probabilistic Tracking Frameworks

Signal Innovations Group (SIG), headquartered in Research Triangle Park, N.C., developed the Tracking Analytics Software Suite (TASS), an enterprise software for motion imagery tracking that creates "geolocated, time-stamped tracks and activities for all moving targets in WAMI and FMV data" [57]. TASS was identified by the Undersecretary of Defense for Intelligence in the 2010 RFI as a viable tracking framework and was "used as a basis for NGA ABI requirements" as part of the TASER ABI program [58, 59].

SIG uses a tracking framework based on sequential Bayesian inference where each statistical model is represented by an abstract class, and the track likelihood is given by the equation:

$$p\left(\mu_k|Z^k\right) \propto p_f(z_k \mid \mu_k, F_k) \cdot p_K(z_k \mid \mu_k) \cdot \\ p_D(\mu_k, F_k) \cdot \int p\left(\mu_k|\mu_{k-1}\right) p\left(\mu_{k-1}|Z^{k-1}\right) d\mu_{k-1} \tag{12.3}$$

where

$p(\mu_k|Z^k)$ is the probabilistic position of the object at future time k;

$p_f(z_k|\mu_k, F_k)$ is the likelihood based on features, F at position ;

$p_K(z_k|\mu_k)$ is the likelihood based on kinematics at position ;

$p_D(\mu_k, F_k)$ is the likelihood based on the detectability of the object;

$p(\mu_k| \mu_{k-1})$ is the motion model for prediction of the position from k–1 to k;

$p(\mu_{k-1}|Z^{k-1})d\mu_{k-1}$ is the previous posterior, with μ representing the state space as [x y speed heading]$_T$.

This approach allows for plug-and-play tracking using different models for each of the terms in (12.3).

Kinematic tracking can be improved by "reducing prediction uncertainty via traffic flow–modeling and self learning of target behaviors and scene dynamics" [60, p. 6]. Figure 12.14 shows an example of flow-based predictions—an adaptation to the traditional Kalman filtering approach—where objects are treated as particles projected along potential paths. In Figure 12.14, road constraints, vehicle interactions, and the state of traffic signals contribute to the state estimation problem.

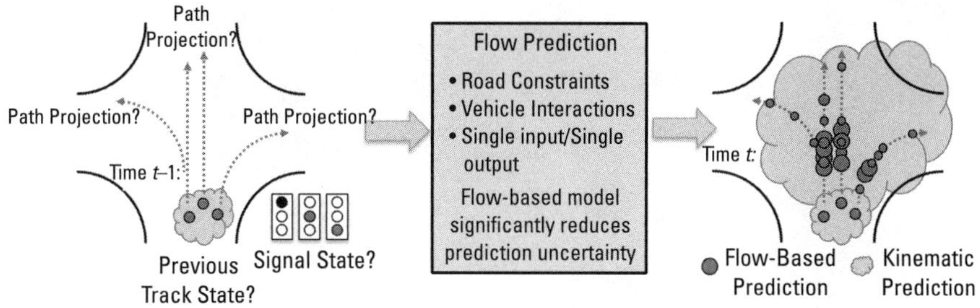

Figure 12.14 Quantification of uncertainty in urban tracking. (©2014 Signal Innovations Group (SIG). Reprinted with permission [60].)

SIG's approach also includes a variable probability of detection (VPD) model where the detectability is a function of the features of the object and the properties of the environment. Radial velocity sampling is assigned a separate probability of detection (PD) based on the predicted speed/heading relative to the sensor, relying on observed velocity-dependent PD characteristics rather than a constant minimum detectable velocity alone. Variable PD relative to obscuration is also included. For example, obscurations reduce the detectability of the object. The model allows integration of GIS data and other geospatial information to account for occlusions. Obscuration information can be incorporated through 3D scene/GIS models where appropriate, or learned based on repeated observations of tracks around a region. This approach improves track quality, track contiguity, and reduces the number of broken tracks in the same regions, leading to global improvement in tracking performance.

12.6.5 Clustering, Track Association, and Multihypothesis Tracking (MHT)

In single-target tracking, the object matching problem is defined by the ability to correctly associate the object through subsequent observations. In a dense tracking environment, the probability of correct association drops significantly with this approach. MHT is an approach that maintains a probabilistic estimate of possible states for all possible objects. As future observations are made that disprove a hypothesis—for example, a subsequent observation shows the appearance or size of the suspected object has changed significantly—possibilities are "pruned" from the forecasted model. Unfortunately, tracking multiple hypotheses in dense environments is extremely computationally intensive, because each detected object must be tested for a match to every other detected object in every subsequent frame.

One approach to addressing this problem is track association through clustering. Clustering partitions the tracking problem into regions to reduce the number of potential matches that must be considered. Figure 12.15 shows an example of clustering where observations, z, are divided into two clusters.

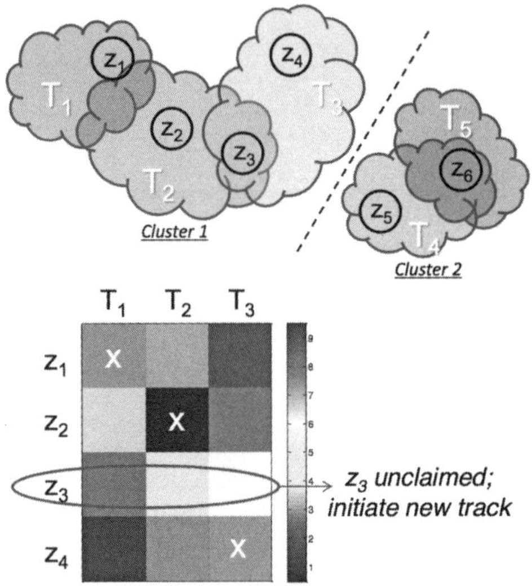

Figure 12.15 Clustering and track association. (©2014 Signal Innovations Group (SIG). Reprinted with permission [60].)

Because z_5 and z_6 are significantly far from the other observations, they are unlikely to be associated with tracks T_1, T_2, or T_3. A "cost" is associated with each detection-to-track assignment using a Munkres assignment algorithm, which minimizes the total cost for each assignment within a cluster [61]. Figure 12.15 shows how observation z_1 is paired with track T_1, z_2 is paired with T_2, and z_4 is paired with T_3 within cluster 1. According to the color scheme, z_3 has a cost of [8, 5, 6] when associated with [T_1, T_2, T_3] whereas [z_1, z_2, z_4] have costs of [3.5, 1, and 3.5]. Since all the costs for z_3 are higher than the already assigned costs, a new track is initiated for z_3.

Figure 12.16 illustrates the concept of MHT. At time *t*, two detections, are received in the nearby vicinity of flow-based predictions (dots in the left of

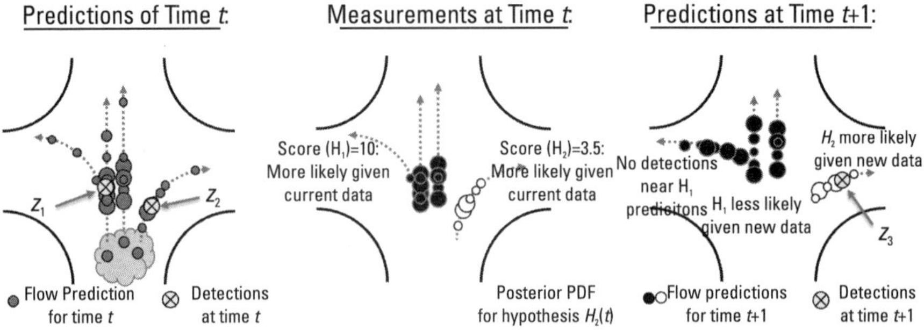

Figure 12.16 Overview of MHT. (©2014 SIG. Reprinted with permission [60].

Figure 12.16). Because detection z_1 is closer to the majority of the prediction samples than detection z_2, it has a higher "track score." As shown in the middle of the figure, the score of hypothesis H_1 (10) is greater than hypothesis H_2 (3.5). In single target tracking, the track would be associated with H_1 and the observation z_2 would be dropped as a track. At time t+1 (right), a third detection z_3 is observed closer to H_2 than H_1. Using MHT with delayed decisions, H_2 is associated with the original prediction, and H_1 is maintained as less likely or dropped as a track if the probability of correct association falls below a defined threshold.

12.6.6 Detecting Anomalous Tracks

Another automation technique that can be applied to wide area data is the detection of anomalous behaviors—that is, "individual tracks where the track trajectory is anomalous compared to a model of typical behavior" [62]. Again, the "focus of attention" CONOP is facilitated by this automated process, which focuses an operator to an area of interest in a scene that is too large and complex to monitor without automated assistance. Kennedy, Wang, and Brandes implemented a technique where they subdivided a scene into 9x9 pixel blocks and identified the most likely track behaviors throughout a scene (Figure 12.17).

Figure 12.17 shows the results of the traffic pattern characterization, highlighting the 50 most common track behaviors in the data set. The upper left corner represents the most common action and the lower right corner represents the rarest action. The most likely behaviors correspond to "normal" traffic patterns over the collected area and are not necessarily extensible to other areas (for example, median U-turn crossovers are a standard feature of many thoroughfares in Michigan but are rare in other parts of the United States) [63]. Anomalous tracks are isolated and highlighted from the larger data set for further identification as shown in Figure 12.18.

In this instance, the vehicle drives along a main street and turns into a parking lot—which itself is not an anomalous activity. Then, at a track time of 40 seconds, the vehicle makes a series of loops around the parking lot. This behavior may be classified as normal if the vehicle is simply searching for a parking spot during a busy time of the day, but the learned normalcy model flags the behavior as unusual relative to the other tracks in the data set.

Using automated algorithms to triage large WAMI and GMTI data sets frees up analysts' time to investigate flagged tracks, understand "normal" behavior according to a statistically learned model, and correlate anomalous tracks with suspected discrete locations. Reliable automated tracking algorithms also speed the process of developing patterns of life and save time in laborious manual track extraction. They also help process very large multicollect data sets to identify new discrete locations and resolve unusual entity behaviors.

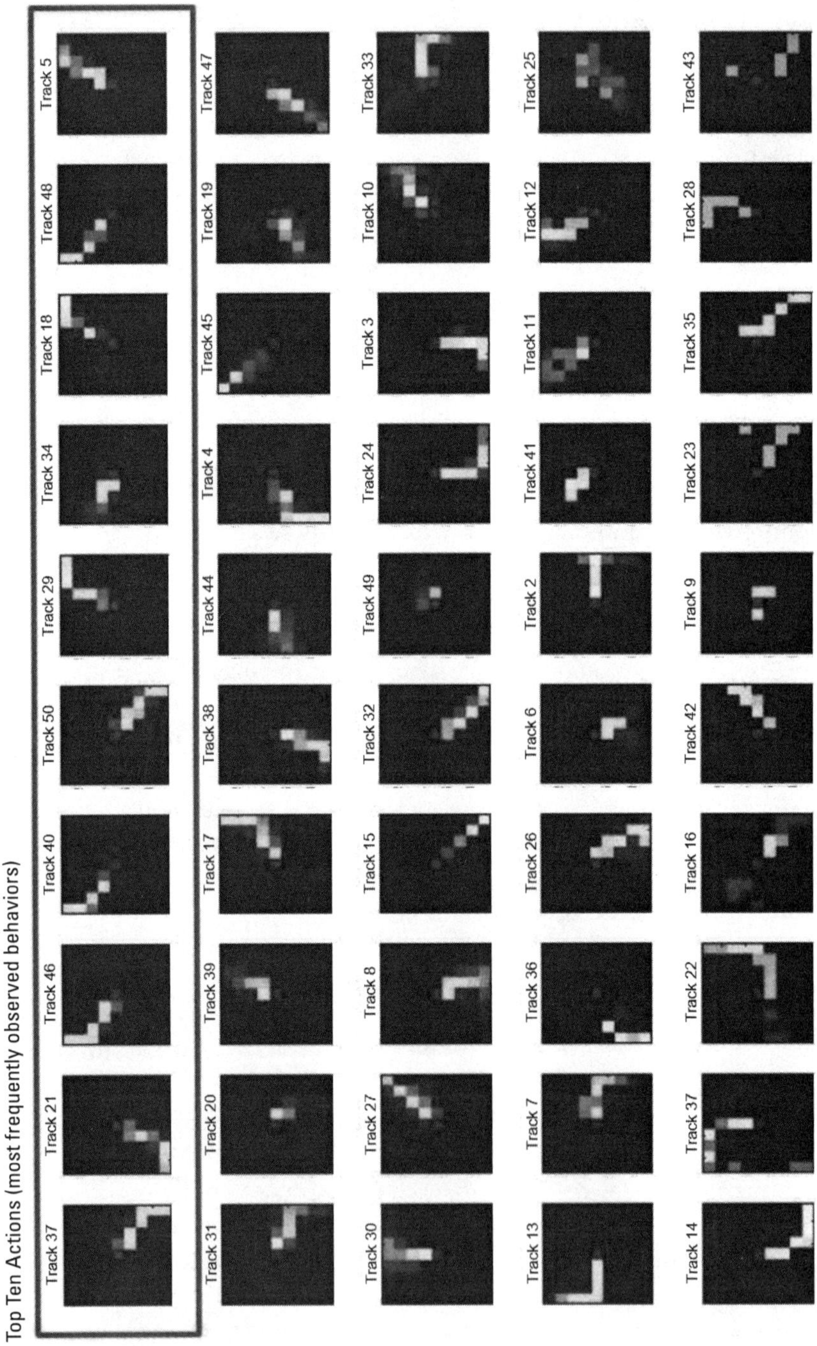

Figure 12.17 Matrix of common traffic patterns. (Source: [62]. Reprinted with permission.)

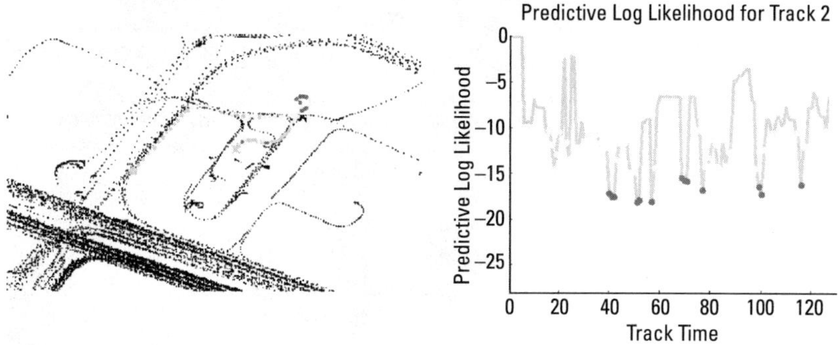

Figure 12.18 Example of an anomalous track. (Source: [62]. Reprinted with permission.)

Increasingly, these algorithms are being deployed as part of analytic support systems connected to large graphics processing units (GPU)-based computing systems so analysts can leverage increasingly sophisticated hardware and software to spend more time performing analysis.

12.7 Metrics for Automated Algorithms

One of the major challenges in establishing revolutionary algorithms for automated activity extraction, identification, and correlation is the lack of standards with which to evaluate performance. DARPA's PerSEAS program introduced several candidate metrics that are broadly applicable across this class of algorithms, as summarized in Table 12.4.

12.8 The Need for Multiple, Complimentary Sources

In signal processing and sensor theory, the most prevalent descriptive plot is the receiver operating characteristic (ROC) curve, a plot of true positive rate or probability of detection versus FAR. An example of a ROC curve for a generic process is shown in Figure 12.19. The ROC curve describes the properties of a sensor (or process) as the detection sensitivity is varied. The central property of the curve is that the rate of false detections always goes up as the sensitivity is increased, and the shape of the curve is defined by the properties of the sensor or process.

Although analysts abhor false alarms (they slow the analysis process because they require extra time to evaluate and discriminate) but as Figure 12.19 shows, they are an inevitable result of any detection process. Most analysts want to eliminate false alarms, but for a given sensor or process, false alarms cannot

Table 12.4
Example Metrics Used to Evaluate Automated Activity Extraction Algorithms (Adapted from [50])

Metric	Description
P_d	For a given threat activity, the number of activities correctly returned divided by the total instances of that activity in the candidate collection. Higher P_d is better. This metric can be extended to other probabilities including the probability of correct association and probability of correct entity origin.
False alarm rate (FAR) per km^2-hr	For a given threat activity, the number of nonmatching threat activities that are reported per square kilometer per hour. Lower FAR is better.
Normalized time to alert (TA)	ratio of the time remaining from the time of alert to the conclusion of the threat activity divided by the total time of the threat activity. Higher TA is better.
Time to exploit (TE)	Reduction in time to exploit, given an n hour mission and one named area of interest. How much quicker can an analyst perform a standard exploitation task with an algorithm-assisted system versus doing it manually? Higher percentage reduction is better.
Number of activity types	This is the number of activity/event types that can be discerned with PD and FAR for a given data source.
Resources required per unit area per unit time	The time needed to perform activity extraction over a given space/time range for a given amount of CPU capability available. Alternatively, this metric can be represented as the number of CPUs (cores) required to process an area in a certain time or the ratio of processing time to mission time for a given number of processors.
Scaling (data volume)	Change in processing speed for increases in space-time volumes. This involves measuring the processing time for known volumes and comparing speed versus size. The speed should grow less than linearly as compared to the increase in volume.
Scaling (density)	change in processing speed for increases in entity and/or event density. This involves measuring processing time for known densities and comparing speed versus density. The speed should grow less than linearly as compared to the increase in density.

be reduced without sacrificing sensitivity. A reduction in false alarms for a given process means that true positives will be missed as well.

Kaiser Fung describes this paradox in *Numbers Rule Your World* using the example of drug testing of professional athletes. Fung quoted Mike Lowell, third baseman of the Boston Red Sox:

> [Human Growth Hormone testing] has to be 100 percent accurate, because if it's 99 percent accurate, there are going to be seven false positives in big league baseball, and what if any of those names is one of the major names? You've scarred that person's career for life. You can't come back and say 'Sorry, we've made a mistake,' because you just destroyed that person's career [64, p. 100].

Fung notes that in the case of drug testing of professional athletes, a tester might dial the ROC curve in favor of a low FAR because the consequences of falsely accusing a player may be career-ending. False alarms are important in intelligence, too. A false detection of a submarine-launched ballistic missile

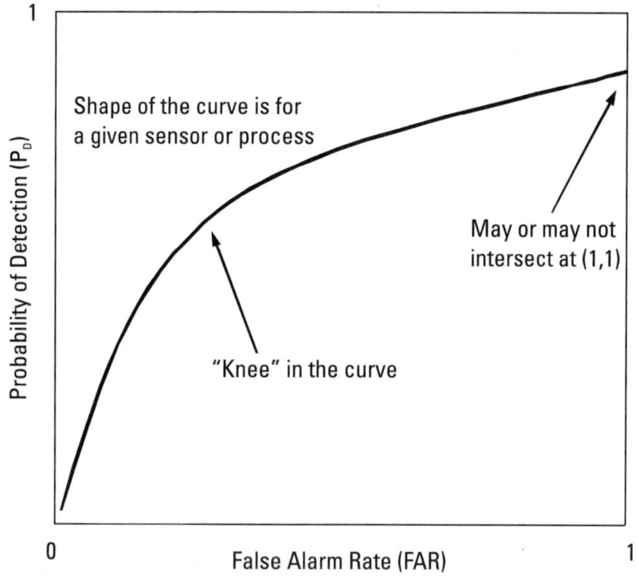

Figure 12.19 The ROC curve.

from the polar ice cap might cause a world-destroying counterstrike. However, in ABI, the stakes are lower because any one single detection seldom decides the outcome of the war. Because ABI analysts aggregate data over many sensors to look for correlations, they prefer false detections to missed detections. The fusion of additional sources of information compensates for the weakness of any one data source in a single domain of collection. That is another statement of the principle of data neutrality in action.

The ROC curve is inherent to a single sensor or process, but combining the results of multiple sensors shifts the ROC curve to the left as shown in Figure 12.20. While single-source processes move along a given ROC curve, multisource fusion moves to a different ROC curve. When the sensors/processes are higher quality or complementary, the FAR decreases for a given probability of detection and vice versa. Techniques used to calculate the resultant probability are described in Chapter 14.

12.9 Summary

Speaking at the Space Foundation's National Space Symposium in May 2014, DNI James Clapper said "We will have systems that are capable of persistence: staring at a place for an extended period of time to detect activity; to understand patterns of life; to warn us when a pattern is broken, when the abnormal happens; and even to use ABI methodologies to predict future actions" [65,

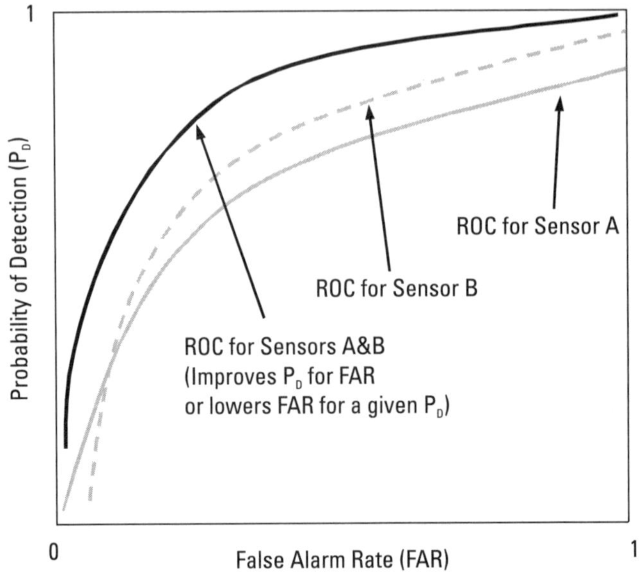

Figure 12.20 ROC curve for a multisource process.

66]. The increasing volume, velocity, and variety of "big data" introduced in Chapter 10 requires implementation of automated algorithms for data conditioning, activity/event extraction from unstructured data, object/activity extraction from imagery, and automated detection/tracking from motion imagery.

On the other hand, "Deus ex machina," Latin for "god from the machine," is a term from literature when a seemingly impossible and complex situation is resolved with irrational or divine means. Increasingly sophisticated "advanced analytic" algorithms provide the potential to disconnect analysts from the data by simply placing trust in the "magical black box." In practice, no analyst will trust any piece of data without documented provenance and without understanding exactly how it was collected or processed.

Automation also removes the analyst from the art of performing analysis. Early in the development of the ABI methodology, analysts were forced to do the dumpster diving and "data janitorial work" to condition their own data for analysis. In the course of doing so, analysts were close to each individual record, becoming intimately familiar with the metadata. Often, analysts stumbled upon anomalies or patterns in the course of doing this work. Automated data conditioning algorithms may reformat and "clean" data to remove outliers—but as any statistician knows—all the interesting behaviors are in the tails of the distribution.

12.10 Acknowledgments

Jonathan Woodworth, Levi Kennedy, and Paul Runkle of SIG contributed significantly to Section 12.6 on tracking and track extraction. Jeff Wilson of ClearTerra contributed to Section 12.3 on georeferenced entity and activity extraction. Heidi Buck of SPAWAR contributed material on RAPIER.

References

[1] Lohr, S., "For Big-Data Scientists, 'Janitor Work' Is Key Hurdle to Insights," *The New York Times,* August 17, 2014.

[2] Holtman, J., "Data Munging with R." Powerpoint Presentation. Web. Available: http://datatable.r-forge.r-project.org/JimHoltman.pdf

[3] Romano, D., "Data Mining Leading Edge: Insurance & Banking," presented at the *Proceedings of Knowledge Discovery and Data Mining,* Brunel University, 1997.

[4] Dasu, T., and T. Johnson, *Exploratory Data Mining and Data Cleaning,* Hoboken, NJ: Wiley-IEEE, 2003.

[5] Steinberg, D., "How Much Time Needs to be Spent Preparing Data for Analysis?," Web. Available: http://1.salford-systems.com/blog/bid/299181/How-Much-Time-Needs-to-be-Spent-Preparing-Data-for-Analysis. 9 Jun 2013.

[6] "2020 Analysis Technology Plan," National Geospatial-Intelligence Agency, Approved for Public Release. NGA Case #14-472. 12 Nov 2014.

[7] Popp, R., et al., "Countering Terrorism Through Information Technology," *Communications of the ACM,* Vol. 47, No. 3, March 2004, p. 36.

[8] "Threat Prediction." BAE Systems. Web. Available: http://www.baesystems.com/our-company-rzz/our-businesses/intelligence-&-security/capabilities-&-services/geospatial-intelligence.

[9] "Engineer's Early Fascination with Geography and Language Evident in MITRE's Georeferencing Toolkit," MITRE Corporation, Mar 2012. Web. Available: https://www.youtube.com/watch?v=zr6_HJEPnJg.

[10] "2013 Annual Report." MITRE Corporation. 2013.

[11] LocateXT Overview. ClearTerra. Web. Available: http://www.clearterra.com/locatext/. Accessed: 26 Oct 2014

[12] "Extract Locations, Unstructured Data, Geoparsing," ClearTerra.- ClearTerra." Web. Available: http://www.clearterra.com/locatext-software. Accessed: 26 Oct 2014.

[13] "Report on Questions Regarding CW Production Capabilities, Pathfinder Record Number 9763 (Declassified Record as part of the study of Gulf War Illness)." Defense Intelligence Agency (DIA). Aug 1991.

[14] "Esri Announces ArcGIS Online Hosting of ClearTerra Enhanced Documents." Web.

[15] Nixon, M. S., and A. S. Aguado, *Feature Extraction & Image Processing, 2nd Edition,* Oxford: Academic Press, 2008

[16] Lindeberg, T., "Edge Detection," in *Encyclopedia of Mathematics,* Springer, 2002, https://www.encylopediaofmath.org.

[17] Natarajan, P., et. al., "BBNM VISER TREVCID 2011 Multimedia Event Detection System," NIST. TRECVID Workshop, Vol. 62, 2011.

[18] "Feature Extraction in ENVI EX Using DigitalGlobe Multispectral Imagery." DigitalGlobe. Information Sheet. 2013.

[19] Mao, J., and A. K. Jain, "Artificial Neural Networks for Feature Extraction and Multivariate Data Projection," *IEEE Transactions on Neural Networks,* Vol. 6, No. 2, March 1995, pp. 296–317.

[20] Lowe, D. G., "Object Recognition from Local Scale-Invariant Features," in *Proceedings of the Seventh IEEE International Conference on Computer Vision,* Vol. 2, 1999, pp. 1150–1157

[21] Lowe, D. G., "Local Feature View Clustering for 3D Object Recognition," in *Proceedings of the IEEE Computer Society Conference on Computer Vision and Pattern Recognition,* 2001, Vol. 1, pp. I–682–I–688.

[22] Lowe, D. G., "Distinctive Image Features from Scale-Invariant Keypoints," *International Journal of Computer Vision,* Vol. 60, No. 2, pp. 91–110, November 2004.

[23] "Scale-Invariant Feature Transform," Wikipedia. Accessed: 24 Oct 2014.

[24] Beis, J. S., and D. G. Lowe, "Shape Indexing Using Approximate Nearest-Neighbour Search in High-Dimensional Spaces," in *Proceedings of the IEEE Computer Society Conference on Computer Vision and Pattern Recognition,* 1997, pp. 1000–1006.

[25] Zhao, W., et al., "Face Recognition: A Literature Survey," *ACM Computer Surv.,* Vol. 35, No. 4, December 2003, pp. 399–458.

[26] "Facebook's Facial Recognition Software is Now as Accurate as the Human Brain, But What Now?," *ExtremeTech,* web. Available: http://www.extremetech.com/extreme/178777-facebooks-facial-recognition-software-is-now-as-accurate-as-the-human-brain-but-what-now. Accessed: 01 Nov 2014.

[27] "DeepFace: Closing the Gap to Human-Level Performance in Face Verification," in *IEEE Conference on Computer Vision and Pattern Recognition (CVPR),* 2014, pp. 1701–1708.

[28] "RAPIER® (RAPid Image Exploitation Resource) Ship Detection System." SPAWAR Systems Center Pacific, web.

[29] "RAPIER (RAPiD Image Exploitation Resource) Ship Detection System, Technology Transfer, SD 959." SPAWAR Systems Center Pacific, April 2011.

[30] Joslin, E., et al., "Method for Classifying Vessels Using Features Extracted from Overhead Imagery." U.S. Patent No. US8170272. 2012.

[31] Buck, H., et al., "Ship Detection System and Method from Overhead Images." U.S. Patent No. US8116522. 2012.

[32] Joslin, E., et al., "Method for Fusing Overhead Imagery with Automatic Vessel Reporting Systems." U.S. Patent No. US8411969, 2013.

[33] "RAPIER Full Motion Video (FMV)." SPAWAR Systems Center Pacific. Web. Available: http://www.public.navy.mil/spawar/Pacific/TechTransfer/ProductsServices/Pages/RAPIERFullMotionVideo(FMV).aspx.

[34] Stastny, J. C., et al., "Adaptive Automated Synthetic Aperture Radar Vessel Detection Method with False Alarm Mitigation." U.S. Patent No. US8422738, 2013.

[35] RAPIER (Rapid Image Exploitation Resource). SPAWAR Systems Center Pacific. Video Recording. https://www.youtube.com/watch?v=sZq2ngxqeAk. 2010.

[36] Bagnall, B., E. Sharghi, and H. Buck, "Algorithms for the Detection and Mapping of Wildfires in SPOT 4 and 5 imagery," in *Proc. SPIE 8515, Imaging Spectrometry XVII,* October 25, 2012, p. 85150A.

[37] Automated Low-Level Analysis and Description of Diverse Intelligence Video (ALADDIN), Intelligence Advanced Projects Research Agency (IARPA), IARPA-BAA-10-01.

[38] "IBM Multimedia Analysis and Retrieval System (IMARS)," IBM," web. Available: http://researcher.watson.ibm.com/researcher/view_group.php?id=877. 22 Mar 2013.

[39] Yu, F. X., et al., "Designing Category-Level Attributes for Discriminative Visual Recognition," *IEEE Conference on Computer Vision and Pattern Recognition (CVPR),* 2013, pp. 771–778.

[40] "Video and Image Retrieval and Analysis Tool (VIRAT)," DARPA, BAA-08-20, October 2008.

[41] Kenyon, H., "DARPA Develops New Tools to Help Process Video Data," *Defense Systems,* June 29, 2011.

[42] Whitlock, C., "When Drones Fall from the Sky," *The Washington Post,* June 20,2014.

[43] Oh, S., et. al. "A Large-Scale Benchmark Dataset for Event Recognition in Surveillance Video," in *IEEE Conference on Computer Vision and Pattern Recognition (CVPR),* 2011.

[44] Zhu, Y., N. M. Nayak, and A. K. Roy-Chowdhury, "Context-Aware Activity Recognition and Anomaly Detection in Video," *IEEE Journal of Selected Topics in Signal Processing,* Vol. 7, No. 1, pp. 91–101, February 2013.

[45] Rimey, R. D., W. Hoff, and J. Y. Lee, "Recognizing Wide-Area and Process-Type Activities," *10th International Conference on Information Fusion,* 2007.

[46] Cordova, A., et al., "Motion Imagery Processing and Exploitation (MIPE)," RAND Corporation, Santa Monica, CA, 2013.

[47] Ratches, J. A., R. Chait, and J. W. Lyons, "Some Recent Sensor-Related Army Critical Technology Events," National Defense University, *Defense & Technology 100 Paper,* Center for Technology and National Security Policy, February 2013.

[48] "Night Eyes for the Constant Hawk," *Defense Update,* 17 Sep 2009.

[49] "Walking Back the Cat: The US Army's Constant Hawk," *Defense Industry Daily,* web, Oct. 2, 2011.

[50] "Persistent Stare Exploitation and Analysis System (PerSEAS)." DARPA, BAA-09-55, Nov 2009.

[51] "Computer Vision." Kitware. Information Sheet. Web. Available: http://www.kitware.eu/products/archive/ComputerVisionFlyer.pdf .

[52] Swears, E., et al., "Complex Activity Recognition using Granger Constrained DBN (GCDBN) in Sports and Surveillance Video," *IEEE Conference on Computer Vision and Pattern Recognition (CVPR),* 2014.

[53] Collins, R., "PPAML WAMI Evaluation Metrics," web, June 25, 2014.

[54] "Kalman Filter," Institute for Telecommunications Sciences, 23 Aug 1996.

[55] Aimonen, P., "File:Basic concept of Kalman filtering.svg" Wikimedia Commons. Web.

[56] Welch, G., and G. Bishop, "An Introduction to the Kalman Filter," *SIGGRAPH 2001,* web. Available: http://www.cs.unc.edu/~tracker/media/pdf/SIGGRAPH2001_CoursePack_08.pdf.

[57] "ISR Products," Signal Innovations Group. Web. Available: https://siginnovations.com/products/.

[58] "Company," Signal Innovations Group. Web. Available: https://siginnovations.com/company/.

[59] "BAE Systems Selected to Provide Activity-Based Intelligence Support for National Geospatial-Intelligence Agency," *Business Wire,* December 20, 2012.

[60] Shargo, P., "Detection-Based Tracker Overview Materials," Powerpoint Presentation.

[61] Kuhn, H. W., "Variants of the Hungarian Method for Assignment Problems," *Naval Research Logistics Quarterly,* Vol. 3, No. 4, December 1956, pp. 253–258.

[62] Kennedy, L., E. Wang, and S. Brandes, "Activity Recognition in Wide Area Motion Imagery," Jul 2010.

[63] "Michigan Highways: In Depth: The Michigan Left." Web. Available: http://www.michiganhighways.org/indepth/michigan_left.html.

[64] Fung, K., *Numbers Rule Your World: The Hidden Influence of Probability and Statistics on Everything You Do,* New York: McGraw-Hill, 2010.

[65] Clapper, J., "Remarks at the National Space Symposium," May 2014.

[66] "DNI Clapper Teases 'Revolutionary' Intel Future; Big Cost Savings From Cutting Contractors," *Breaking Defense,* May 2014.

13

Analysis and Visualization

Analysis of large data sets increasingly requires a strong foundation in statistics and visualization. This chapter introduces the key concepts behind data science and visual analytics. It demonstrates key statistical, visual, and spatial techniques for analysis of large-scale data sets. The chapter provides many examples of visual interfaces used to understand and analyze large data sets.

13.1 Introduction to Analysis and Visualization

Analysis is defined as "a careful study of something to learn about its parts, what they do, and how they are related to each other" [1]. The core competency of the discipline of intelligence is to perform analysis, deconstructing complex mysteries to understand what is happening and why. Figure 13.1 highlights key functional terms for analysis and the relative benefit/effort required for each.

This chapter focuses on the center of Figure 13.1: the use of visualization, query/drill down, and statistical analysis to answer higher value questions. In contrast to basic analysis techniques that focus on alerting and reporting on known events after the fact, our approach to analysis introduces techniques to identify causality, understand patterns of-life, and determine why things happen.

Because most real-world problems are highly dimensional, visualization of multivariate data is increasingly important in data science and intelligence analysis. This chapter describes several basic and advanced techniques for analysis and visualization of large, multidimensional data sets and identifies some of the emergent challenges of deploying and applying these techniques at massive scale.

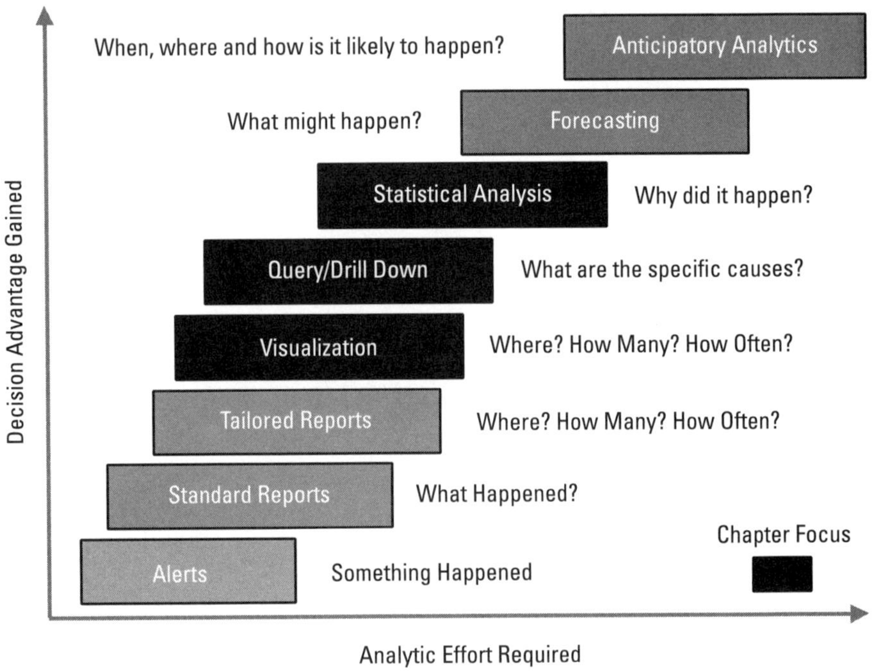

Figure 13.1 Hierarchy of analysis functions and outputs.

13.1.1 The Sexiest Job of the 21st Century...

Big-budget motion pictures seldom glamorize the careers of statisticians, operations researchers, and intelligence analysts. Analysts are not used to being called "sexy," but in a 2012 article in the Harvard Business Review, Thomas Davenport and D. J. Patil called out the data scientist as "the sexiest job of the 21st century" [2]. The term was first coined around 2008 to recognize the emerging job roles associated with large-scale data analytics at companies like Google, Facebook, and LinkedIn. Combining the skills of a statistician, a computer scientist, and a software engineer, the proliferation of data science across commercial and government sectors recognizes that competitive organizations are deriving significant value from data analysis. Today we're seeing an integration of data science and intelligence analysis, as intelligence professionals are being driven to discover answers in those giant haystacks of unstructured data. According to Law, Greenbacker, and Eberhardt of Altamira, the emergent job role of data scientist is defined by two types of skills as shown in Figure 13.2.

Data scientists demonstrate mastery in all of the core skills and usually specialize in one or more of the specialized skills like natural language processing and machine learning. Because data sets are seldom well-conditioned, most data scientists have to write code to manipulate data prior to analysis. Analysts

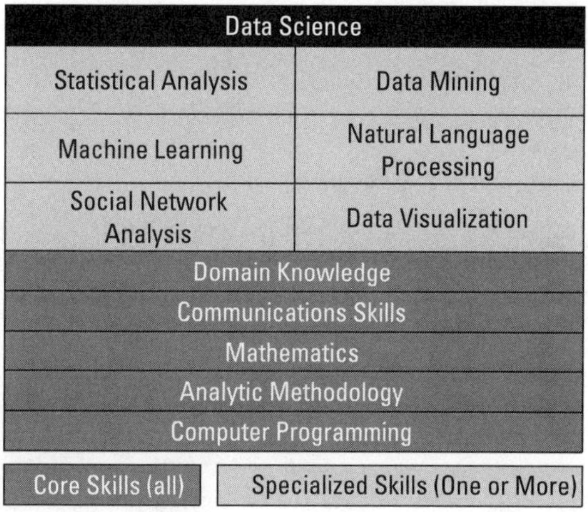

Figure 13.2 Skills mix requirements for a data scientist. (Adapted from [3], Law, Greenbacker, and Eberhardt. Used with permission.)

benefit from knowledge of scripting languages like Python or statistical processing languages like SAS or R.

According to Leek, Peng, and Caffo, the key tasks for data scientists are the following [4, p. 2].

- Defining the question;
- Defining the ideal data set;
- Obtaining and cleaning the data;
- Performing exploratory data analysis;
- Performing statistical prediction/modeling;
- Interpreting results;
- Challenging results;
- Synthesizing and writing up and distributing results.

Each of these tasks presents unique challenges. Often, the most difficult step of the analysis process is defining the question, which, in turn, drives the type of data needed to answer it. In a data-poor environment, the most time-consuming step was usually the collection of data; however, in a modern "big data" environment, a majority of analysts' time is spent cleaning and conditioning the data for analysis. Many of the data sets—even publically available ones—are seldom well-conditioned for instantaneous import and analysis. Often column headings, date formats, and even individual records may need reformatting before the data can even be viewed for the first time. Messy data is almost always an impetus to rapid analysis, and decision makers have little understanding of the chaotic data landscape experienced by the average data scientist.

13.1.2 Asking Questions and Getting Answers

The most important task for an intelligence analyst is determining what questions to ask. The traditional view of intelligence analysis places the onus of defining the question on the intelligence consumer, typically a policy maker. In the military, a "request for information" defines tasking for tactical analysts. These approaches can limit the quality of intelligence by cementing institutional biases, encouraging inattention blindness, and narrowly focusing analysis on known-knowns.

Asking questions from a data-driven and intelligence problem–centric viewpoint is the central theme of this textbook and the core analytic focus for the ABI discipline. Sometimes, collected data limits the questions that may be asked. Unanswerable questions define additional data needs, either through collection of processing. In a nonlinear intelligence cycle, sometimes the output of analysis is yet another question. The analytic techniques in this chapter demonstrate how questions can be quickly asked and answered using cognitive assistance software and advanced visualization.

Analysis takes several forms, described as follows:

- Descriptive: Describe a set of data using statistical measures (e.g., census).
- Inferential: Develop trends and judgments about a larger population using a subset of data (e.g., exit polls).
- Predictive: Use a series of data observations to make predictions about the outcomes or behaviors of another situation (e.g., sporting event outcomes).
- Causal: Determine the impact on one variable when you change one more more variables (e.g., medical experimentation).
- Exploratory: Discover relationships and connections by examining data in bulk, sometimes without an initial question in mind (e.g., intelligence data).

Because the primary focus of ABI is discovery, the main branch of analysis applied in this textbook is exploratory analysis. The primary software interface used is this chapter is JMP® 11 by the SAS Institute. JMP® is a desktop application for visual, interactive statistical discovery. First launched in 1989, JMP® was one of the first statistical tools to make use of the GUI of the Macintosh computer. It is widely applied in engineering, science, genomics, biology, and other fields.

13.2 Statistical Visualization

ABI analysis benefits from the combination of statistical processes and visualization. This section reviews some of the basic statistical functions that provide rapid insight into activities and behaviors.

A histogram represents the distribution of data graphically, tabulating frequencies as adjacent bars [5]. Histograms are useful in representing trends over time or in discrete groupings. Groupings can be equally distributed or quantized into groupings based on sample size or frequency. Groups may also be defined by discrete categories.

Figure 13.3 shows three different histograms created in JMP 11. The left-most histogram is a distribution of speed during a bike ride (4292 data points). The mean is calculated at about 4 m/s. For this type of behavior, the speed varies according to an approximately normal distribution about the mean (sometimes the rider is faster and sometimes slower). The second histogram shows the average duration of telephone calls over a six month period (725 calls). This histogram approximates a power law since the user averages about 8.6 minutes per call. Only a small number of calls are longer than one hour. The right-most

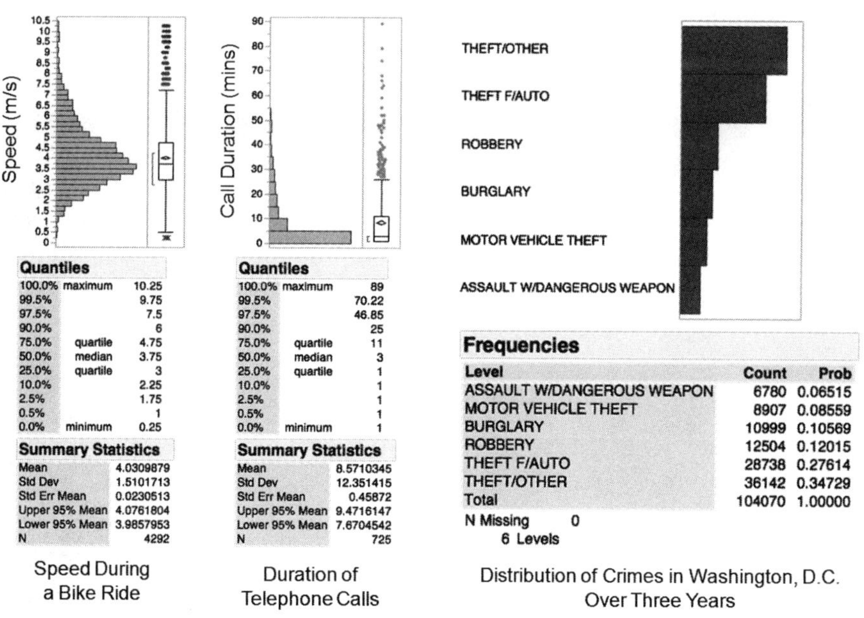

Figure 13.3 Examples of three different histograms.

plot shows the histogram aggregation against Washington, D.C. crime data over a three year period (104,070 points). The variable offense takes one of six values. The histogram bins are defined by the allowable settings of the discrete variable. All three cases provide rapid insight into the data set.

Another property of histograms is that the sum of all bars totals 100% of the data. Cumulative histograms add the value of the current bin to all previous bins, allowing analysts to quickly find the "break point" or the "knee in the curve."

13.2.1 Scatterplots

One of the most basic statistical tools used in data analysis and quality engineering is the scatterplot or scattergram, a two-dimensional Cartesian graph of two variables. Independent variables are usually plotted on the X-axis. Dependent variables are plotted on the Y-axis. In many modern data applications, the analyst does not have control over either parameter and simply "discovers" the data set. In this case, either variable can be assigned to either axis.

Correlation, discussed in detail in Chapter 14, is the statistical dependence between two variables in a data set. Scatterplots are useful to visually inspect multiple relationships for correlation. In intelligence analysis, correlations are useful because they describe a relationship in one dimension that can be exploited with knowledge of the variable in the other dimension. This is useful when many behaviors and signatures cannot be directly observed. In 1997, frozen pizza maker Schwann's wanted to know the production capacity of a new Kraft factory preparing to launch a revolutionary product called DiGiorno. Since the company could not directly ascertain the production volume of the plant, it estimated production by counting the number of cardboard boxes produced by a subcontractor [6, p. 2]. Correlation between two seemingly random variables in a data set is one of the most useful techniques early in the analysis process. An example of a two-dimensional scatterplot and several types of correlations is shown in Figure 13.4.

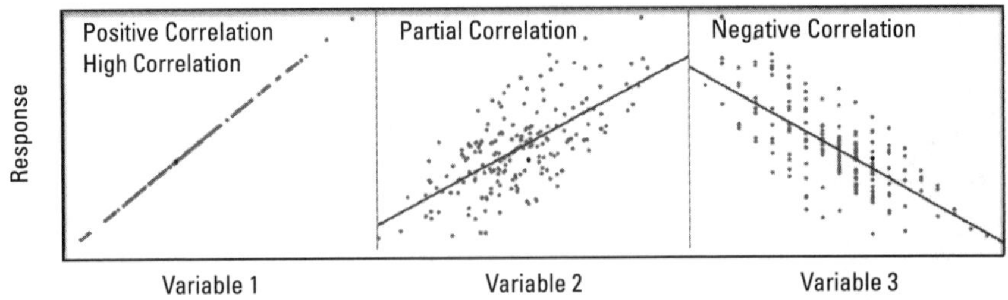

Figure 13.4 Three examples of two-dimensional scatterplots and correlations.

A line of fit drawn through the data points in the scatterplot enables calculation of the correlation, a number between -1 and 1 where zero means no correlation and (-1,1) is perfect correlation in the negative and positive directions respectively. Positive correlation means that an increase in one variable tends to imply an increase in the other variable. Sometimes, patterns appear in the scatterplot, which implies that the correlation between two variables follows a nonlinear mathematical relationship that can be quantified through regression of one or more variables and interactions.

13.2.2 Pareto Charts

Joseph Juran, a pioneer in quality engineering, developed the Pareto principle and named it after Italian economist Vilfredo Pareto. Also known as "the 80/20 rule," the Pareto principle is a common rule of thumb that 80% of observations tend to come from 20% of the causes. In mathematics, this is manifest as a power law, also called the Pareto distribution whose cumulative distribution function is given as:

$$\bar{F}(x) = \Pr(X > x) = \begin{Bmatrix} \left(\frac{x_m}{x}\right)^{\alpha} & x \geq x_m \\ 1 & x < x_m \end{Bmatrix} \tag{13.1}$$

Where α, the Pareto index, is a number greater than 1 that defines the slope of the Pareto distribution. For an "80/20" power law, $\alpha \approx 1.161$. The power law curve appears in many natural processes, especially in information theory. It was popularized in Chris Anderson's 2006 book *The Long Tail: Why the Future of Business is Selling Less of More* [7].

A Pareto chart, shown in Figure 13.5, is a visual manifestation of the Pareto principle and defines the dominant independent variables impacting a process. Many statistical analysis tools including JMP include Pareto charts during model analysis or analysis of variance (ANOVA). Figure 13.5 shows the

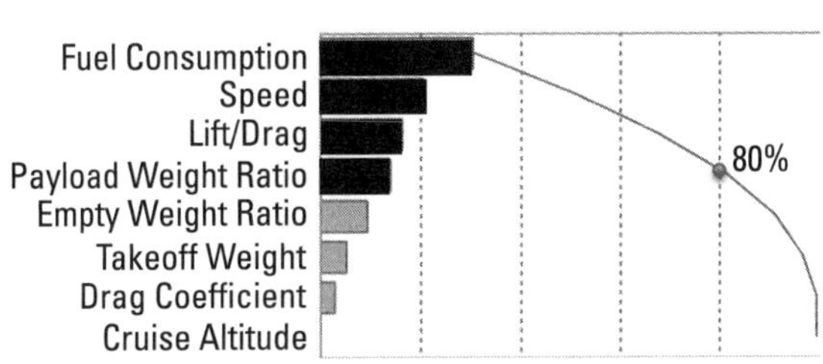

Figure 13.5 Pareto chart showing the dominant design variables in the range calculation for a jet aircraft.

results of an aircraft design tool based on the Breguet range equation, a widely used rule of thumb in aircraft design. The Pareto chart shows the significance of each independent variable against the dependent variable of aircraft range. Black shaded bars highlight the four design variables that dominate 80% of the variability of the range response.

The significance is a function of the mathematical relation between independent and dependent variables and the range of the independent variables in the data set. This is an important distinction when applying this analysis tool to pre-existing data sets for which the underlying physical relations are not known (e.g., analysis of ABI data sets). The most dominant parameters in one data set may not necessarily be extensible to a different data set, region of the Earth, or intelligence issue.

A variation on the Pareto chart, called the "tornado chart," is shown in Figure 13.6. Like the Pareto chart, bars indicate the significance of the contribution on the response but the bars are aligned about a central axis to show the direction of correlation between the independent and dependent variables.

Figure 13.6 includes the *t-ratio,* a statistical measure of the significance of the variable on the response, mathematically related to the length and size of the bars in the tornado chart. The t-ratio provides for statistical hypothesis testing and quantifies the methods in Chapter 4 when applied to analysis of large data sets.

Pareto charts are useful in formulating initial hypotheses about the possible dependence between two data sets or for identifying a collection strategy to reduce the standard error in a model. Statistical correlation using Pareto charts and the Pareto principle is one of the simplest methods for data-driven discovery of important relationships in real-world data sets.

13.2.3 Factor Profiling

Factor profiling examines the relationships between independent and dependent variables. The profiler in Figure 13.7 shows the predicted response (de-

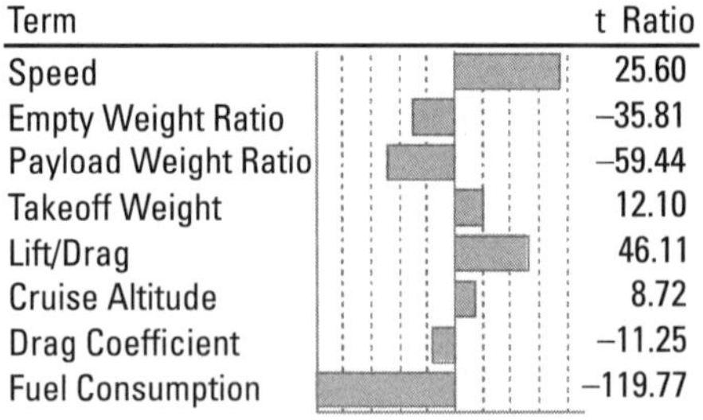

Figure 13.6 "Tornado" chart showing the relative significance and direction of impact for aircraft design parameters.

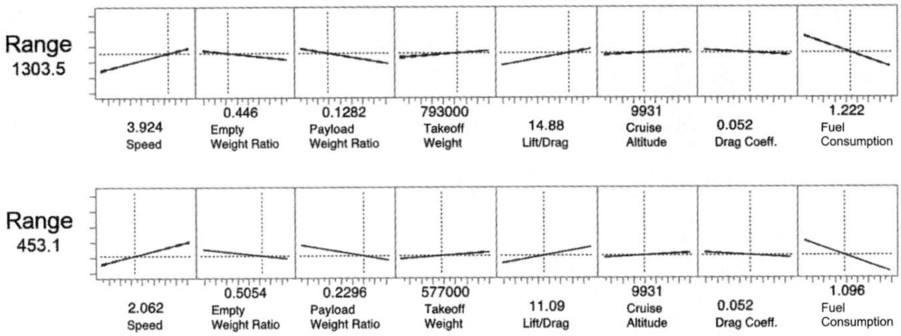

Figure 13.7 Prediction profiler for the range of an aircraft.

pendent variable) as each independent variable is changed while all others are held constant. A series of one-dimensional relationships, plotted as solid lines in Figure 13.7, represents the partial derivatives of the effect of each independent variable at the current factor settings for all other independent variables. The upper and lower depictions in Figure 13.7 show how the range response changes from 1303.5 to 453.1 as the prediction traces are moved in the JMP 11 interactive factor profiler.

Factor profiling is a useful technique for interrogating a simple (or complex) regression model from an existing data set. Because the profiler is recalculated each time the independent variable settings are updated, sometimes the shape of the prediction traces change as different parts of the data set are graphically explored. This type of visual, statistical exploration of data sets allows trends and patterns of be discovered, which then drives an analyst to ask additional questions of the data set.

Real-time factor profiling and visualization of large multidimensional, multidisciplinary problems became feasible with advances in desktop computing power and graphics cards around the year 2001. Since that time, a number of increasingly powerful graphical methods for graphical analysis of large data sets have been popularized. These techniques are now commonly referred to as visual analytics.

13.3 Visual Analytics

Visual analytics was defined by Thomas and Cook of the Pacific Northwest National Laboratory in 2005 as "the science of analytical reasoning facilitated by interactive visual interfaces" [8]. The emergent discipline combines statistical analysis techniques with increasingly colorful, dynamic, and interactive presentations of data. Intelligence analysts increasingly rely on software tools for visual analytics to understand trends, relationships and patterns in increasingly large and complex data sets. These methods are sometimes the only way to rapidly resolve entities and develop justifiable, traceable stories about what happened and what might happen next.

Large data volumes present several unique challenges. First, just transforming and loading the data is a cumbersome prospect. Most desktop tools are limited by the size of the data table that can be in memory, requiring partitioning before any analysis takes place. The a priori partitioning of a data set requires judgments about where the break points should be placed, and these may arbitrarily steer the analysis in the wrong direction. Large data sets also tend to exhibit "wash out" effects. The average data values make it very difficult to discern what is useful and what is not. In location data, many entities conduct perfectly normal transactions. Entities of interest exploit this effect to effectively hide in the noise.

As dimensionality increases, potential sources of causality and multivariable interactions also increase. This tends to wash out the relative contribution of each variable on the response. Again, another paradox arises: Arbitrarily limiting the data set means throwing out potentially interesting correlations before any analysis has taken place. Including all possible variables makes analysis cumbersome and time-consuming. In the worst case, the presence of all variables washes out the impacts so much that no viable trends can be observed and entities cannot be resolved.

Analysts must take care to avoid visualization for the sake of visualization. Sometimes, the graphic doesn't mean anything or reveal an interesting observation. Visualization pioneer Edward Tufte coined the term "chartjunk" to refer to these unnecessary visualizations in his 1983 book *The Visual Display of Quantitative Information,* saying:

> The interior decoration of graphics generates a lot of ink that does not tell the viewer anything new. The purpose of decoration varies—to make the graphic appear more scientific and precise, to enliven the display, to give the designer an opportunity to exercise artistic skills. Regardless of its cause, it is all non-data-ink or redundant data-ink, and it is often chartjunk [9].

Creating compelling and useful visuals that convey information quickly and effectively is a major challenge. Michelle Borkin and Hanspeter Pfister of the Harvard School of Engineering and Applied Scientists studied over 5,000 charts and graphics from scientific papers, design blogs, newspapers, and government reports to identify characteristics of the most memorable ones. "A visualization will be instantly and overwhelmingly more memorable if it incorporates an image of a human-recognizable object—if it includes a photograph, people, cartoons, logos—any component that is not just an abstract data visualization," says Pfister. "We learned that any time you have a graphic with one of those components, that's the most dominant thing that affects the memorability" [10].

Everything happens somewhere, and things happen in places. Maps are therefore a natural backdrop upon which to study activities, events and transactions. Section 13.4 introduces increasingly complex—yet sometimes intuitive—statistical and spatial visualizations based on this premise.

13.4 Spatial Statistics and Visualization

In a world dominated by handheld maps, self-driving cars, and three-dimensional spinning high-resolution globes, the concept of putting data on a map to improve situational awareness and understanding may seem trite, but the first modern geospatial computer system was not proposed until 1968. While working for the Department of Forestry and Rural Development for the Government of Canada, Roger Tomlinson introduced the term "geographic information system" (now GIS) as a "computer-based system for the storage and manipulation of map-based land data" [11]. According to a report by Global Industry Analysts, Inc., the GIS industry is expected to grow to $10.6 billion worldwide by 2015 [12].

While numerous texts review GIS and its diverse uses in detail, the primary interest for ABI is a common backdrop for analyzing, visualization, and associating activities and transactions based on contextual geospatial information. The combination of maps with visual analytic tools and rich spatially enabled data sets enables georeference to discover.

The founding event of the modern study of epidemiology was one of the first uses of maps in the study of public health. London physician John Snow mapped the locations of an 1854 cholera outbreak by talking to local residents (inset of Figure 13.8).

Based on the geographic pattern of disease, Snow hypothesized that the pump on Broad Street (now Broadwick Street) was the source of the outbreak [13]. After disabling the well by removing its handle, the outbreak subsided.

This is a striking example of georeference to discover, aided by spatial data analysis and hypothesis testing. A modern version of the famous map by Snow is shown using a JMP bubble plot in Figure 13.8 where shading and size indicate the number of cases at a given address. The map shows no correlation with pump sites away from Broad Street.

This example also demonstrated a key principle of urban planning adopted by ABI analysts: the "gravity" or "draw" of key spatial locations based on

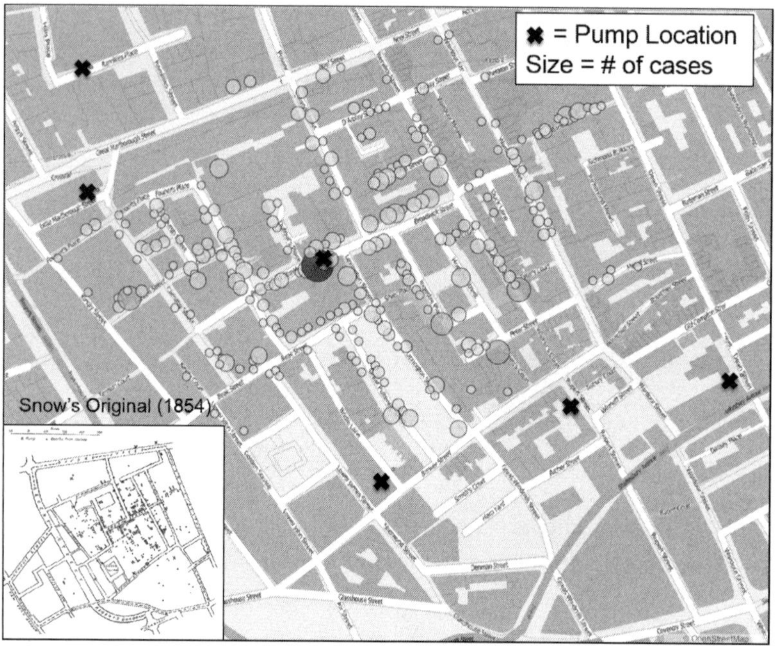

Figure 13.8 Example of spatial mapping of data. Original data by John Snow (1854). Digitized by Robin Wilson (robin@rtwilson.com). (Map data source: OpenStreetMap.)

distance. Local residents don't want to walk very far to gather water so they tend to use the pump closest to their residence. The original plot by Snow (1854) did not use scaled bubbles but rather individual dots. The ability to aggregate and disaggregate data to improve clarity and test hypotheses is a key technique in visual analytics.

13.4.1 Spatial Data Aggregation

A popular form of descriptive analysis using spatial statistical is the use of subdivided maps based on aggregated data. Typical uses include visualization of census data by tract, county, state, or other geographic boundaries. An example of aggregated travel data—the origin-destination (work/home) pairings of workers in Reston, Virginia—is shown in Figure 13.9.

This data aggregates origin-destination pairings (a type of transaction) over a large number of citizens and bins them by geographic region. Urban planners and sociologists study these patterns over long time periods to discern trends and changes. They attempt to associate these trends with broader socioeconomic trends or correlate them to other variables.

Despite the wealth of data available from the U.S. census, it demonstrates a key challenge in ABI: sampling rate. Census data is only sampled once every

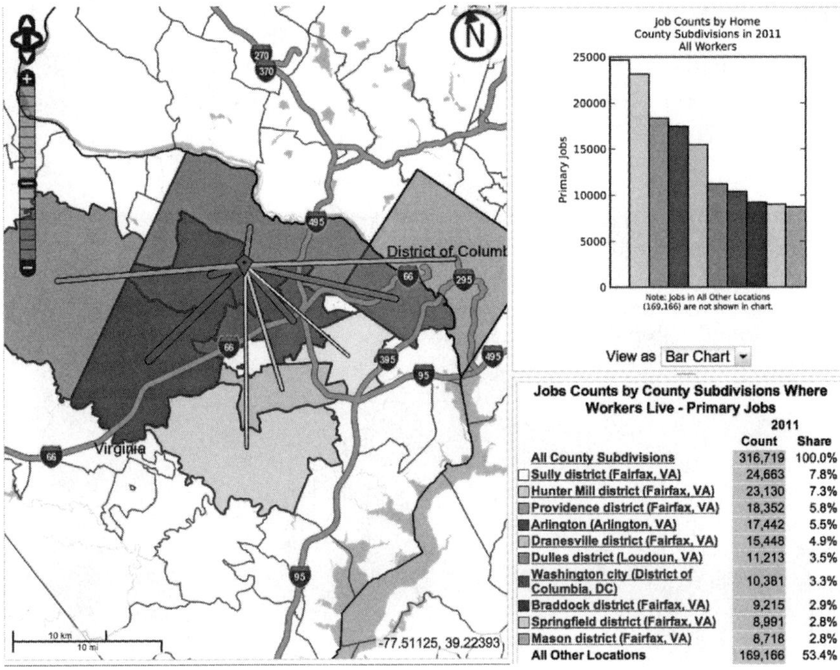

	2011	
	Count	Share
All County Subdivisions	316,719	100.0%
Sully district (Fairfax, VA)	24,663	7.8%
Hunter Mill district (Fairfax, VA)	23,130	7.3%
Providence district (Fairfax, VA)	18,352	5.8%
Arlington (Arlington, VA)	17,442	5.5%
Dranesville district (Fairfax, VA)	15,448	4.9%
Dulles district (Loudoun, VA)	11,213	3.5%
Washington city (District of Columbia, DC)	10,381	3.3%
Braddock district (Fairfax, VA)	9,215	2.9%
Springfield district (Fairfax, VA)	8,991	2.8%
Mason district (Fairfax, VA)	8,718	2.8%
All Other Locations	169,166	53.4%

Figure 13.9 Interactive visualization of 2012 American community survey data—residence locations for citizens who work in Reston, Virginia. (*Data Source:* U.S. Census Bureau.)

10 years. The data in Figure 13.9 comes from the American Community Survey (ACS), an annual survey conducted over a smaller population set. Planners use the ACS data to adjust the delivery of social services based on population trends. Using a subset of data to made judgments about a larger population is called inferential analysis.

Aggregating statistical map data into predefined geographic regions may be useful for understanding broad socioeconomic trends or mapping the "human geography" of a large region, but unless the data is finely discretized in space down to the neighborhood, block, or residence, it is impossible to use this data to resolve entities, understand networks, or determine patterns of life.

This technique is applied to activity databases for violent and nonviolent crimes, termed crime mapping (See Chapter 17).

Spatially aggregated data can be used to focus analytic effort around clusters of "interesting" patterns. Because humans are biased to discover visual patterns (e.g., recognizing faces and predators), the concept of "interesting" is difficult to quantify, but the repeated visualization of aggregated and disaggregated information allows an analyst to "drill down" to regions of interest for investigation.

13.4.2 Tree Maps

Figure 13.10 shows a tree map of spatial data related to telephone call logs for a business traveler.[1] A tree map is a technique for visualizing categorical, hierarchical data with nested rectangles [14]. The technique was developed in

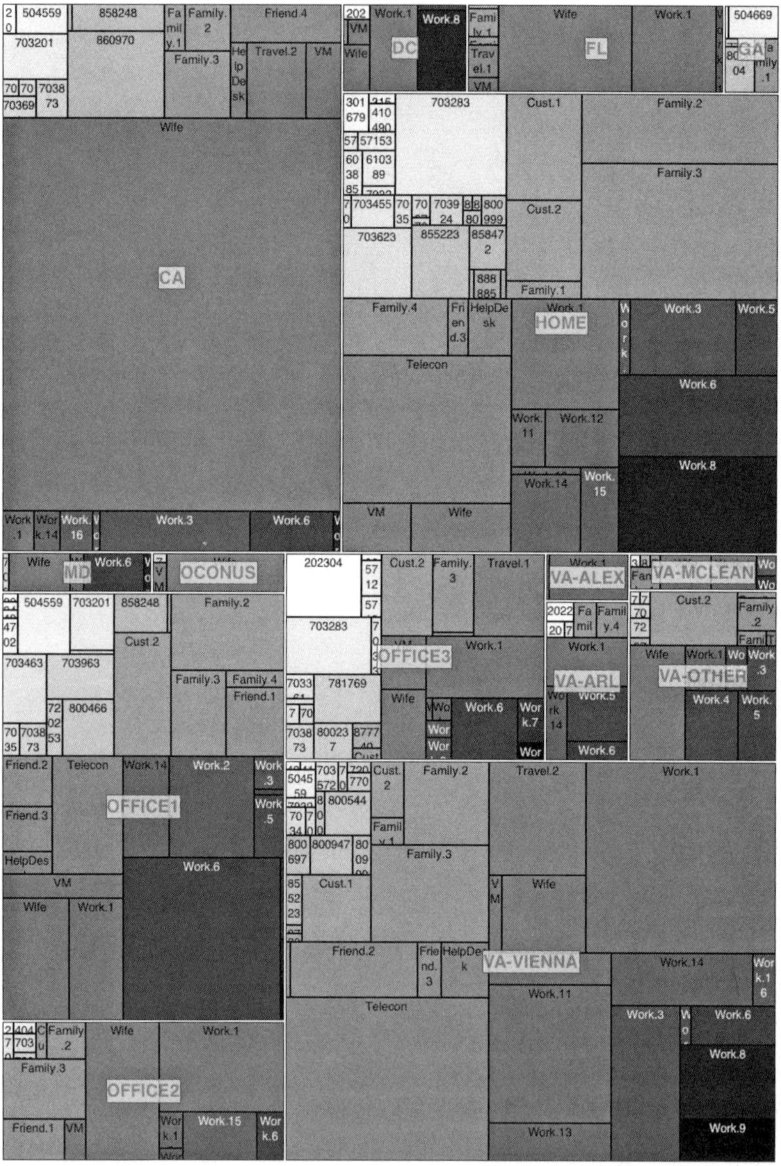

Figure 13.10 Tree map of phone calls grouped by location of origin.

1. Telephone call logs used with permission.

the early 1990s but found widespread popularity after IBM researcher Martin Wattenberg released the Map of the Market, a multidimensional visualization of stock market activity credited as one of the first interactive visualizations on the World Wide Web in 1999 [15].

The tree map divides categorical data into regions, sizes individual boxes by a variable, and colors the boxes based on another variable. In the Map of the Market, the boxes are categorized by industry, sized by market capitalization, and colored by the change in the stock price. A financial analyst can instantly see that consumer staples are down and basic materials are up. The entire map turns red during a broad sell-off. Variations on the Map of the Market segment the visualization by time so analysts can view data in daily, weekly, monthly, or 52-week increments.

The visual in Figure 13.10 groups phone calls by the caller's location at the origination of the call as reported in Verizon Wireless call record logs over a six-month sampling period. The originating location is likely determined based on the location of the cell tower the user is connected to during the origination of the call. The boxes are then grouped by identity of the other party—anonymized by dropping the last four digits of the phone number or using a pseudonym. The size of the box is the sum of the number of minutes the person was called from each location over the six-month period. Boxes are colored by the identity of the second party. Based on the size of the partitioning, we determine that the caller spends about as much time on the phone at home, in California, and across one of three office locations. Other calls originate from areas in Virginia, Washington, D.C., and Maryland. A small number of calls take place from outside the continental United States (OCONUS), indicating the caller took at least one international trip during the sampling period.

The longest duration of calls to wife occur when the caller is in California, leading us to believe the caller frequently travels to a business or customer there. Many long-duration calls to family members take place from VA-Annandale. Long-duration calls to telecon indicate that this individual may work from home (either temporarily or permanently). Because the data were preconditioned to indicate three office locations, analysts would be biased to exclude that hypotheses. An analyst might postulate that this is the user's home location based on that pattern of transactions. Caller Work.1 is frequently called from all locations. Perhaps this person is a close coworker or boss.

The tree map is a useful visualization for patterns—in this case transactional patterns categorized by location and recipient. The eye is naturally drawn to anomalies in color, shape, and grouping. These form the starting point for further analysis of activities and transactions, postulates of relationships between data elements, and hypothesis generation about the nature of the activities and transactions as illustrated above. While tree maps are not inherently spatial, this application shows how spatial analysis can be incorporated and how the spatial component of transactional data generates new questions and leads to further analysis.

This type of analysis reveals a number of other interesting things about the entity (and related entities) patterns of life elements. If all calls contain only two entities, then when entity A calls entity B, we know that both entities are (likely) not talking to someone else during that time. Also, the tree map contains several unique entities that reveal targetable information: Travel.2 is the phone number for United Airlines. Other interesting nodes like the entity's dentist, car mechanic, and neighbors also pop out of the call database as less frequent (and shorter duration) transactions.

13.4.3 Three-Dimensional Scatterplot Matrix

Three-dimensional colored dot plots are widely used in media and scientific visualization because they are complex and compelling. Although it seems reasonable to extend two-dimensional visualizations to three dimensions, these depictions are often visually overwhelming and seldom convey additional information that cannot be viewed using a combination of two-dimensional plots more easily synthesized by humans. Also, three-dimensional plots projected onto a two-dimensional monitor usually have to be rotated for the brain to process the third dimension of information. Three-dimensional models of aircraft are typically projected onto three-view drawings (three two-dimensional views from each orthogonal viewpoint) to aid in clarity.

Figure 13.11 shows a two-dimensional visualization of travel data—a bike ride in northern Virginia. Shading indicates speed; X's indicate stops. The lower half of the route has a number of stops between A and C, including a wrong turn (the cyclists were unfamiliar with this route). Points D indicate stops for street crossing or position checks. The segment near E shows a low velocity with several stops (possibly a grueling hill), while the clockwise circuit back to A is dominated by high speed perhaps indicating a high-quality paved trail.

From Figure 13.11, the viewer cannot tell the direction of the route. Many spatial representations use three dimensions to convey latitude, longitude, and altitude; however, a unique presentation of spatial data is to use the third dimension to instead visualize time. The same information is projected with a time component in three dimensions in Figure 13.12, where the start time is shown in the upper right corner.

Because the beginning and end of the route are now distinct on the chart due to the use of the Z-axis for time, the overlapping part of the route is now visible. Uphill, muddy, and unpaved sections of the road are shown with dim shading, frequent stops, and large changes in Z (long time). In contrast, shallow portions of the graph with few stops show when the cyclists are going downhill or making good progress along level, paved trails.

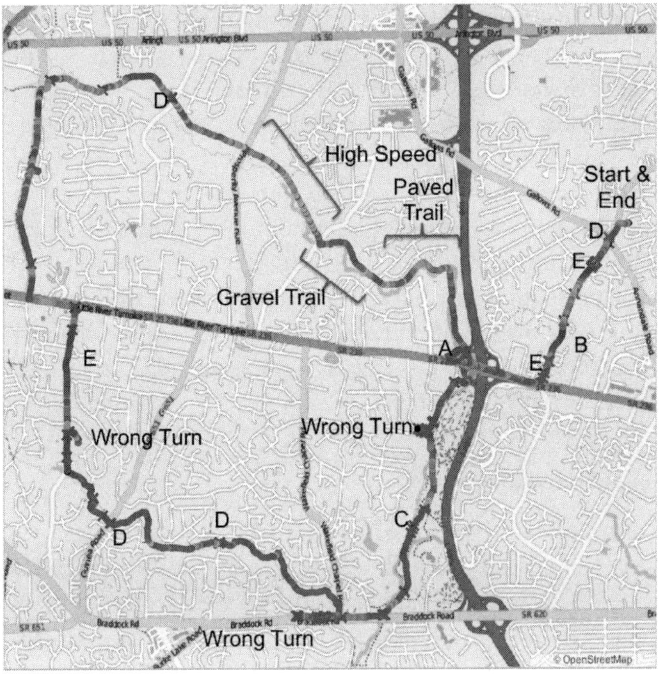

Figure 13.11 Two-dimensional visualization of a bike trip.

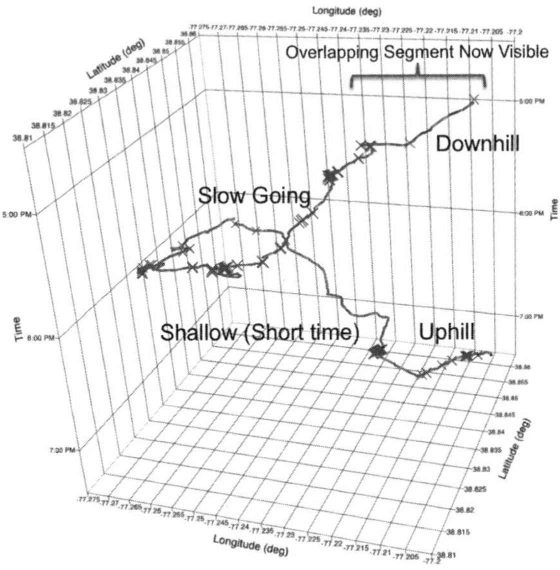

Figure 13.12 Three-dimensional visualization of the same bike trip.

This type of visual is the basis of a commercial analysis tool called GeoTime, shown in Figure 13.13. GeoTime is a spatial/temporal visualization tool that plays back spatially enabled data like a movie. It allows analysts to watch entities move from one location to another and interact through events and transactions. Patterns of life are also easily evident in this type of visualization.

Investigators and lawyers use GeoTime in criminal cases to show the data-driven story about an entity's pattern of movements and activities. In Figure 13.14, the intersection of two lines from two different origins indicate the meeting of suspect 1 and suspect 2. The map shows the meeting took place on the west bank of a river. The timeline lets investigators animate through the sequence of events. The use of interactive graphics and charts to tell stories on a map is called spatial storytelling.

13.4.4 Spatial Storytelling

The latest technique incorporated into multi-INT tradecraft and visual analytics is the aspect of spatial storytelling: using data about time and place to animate a series of events. Several statistical analysis tools implemented storytelling or sequencing capabilities. Figure 13.15 shows a flow map published in 1869

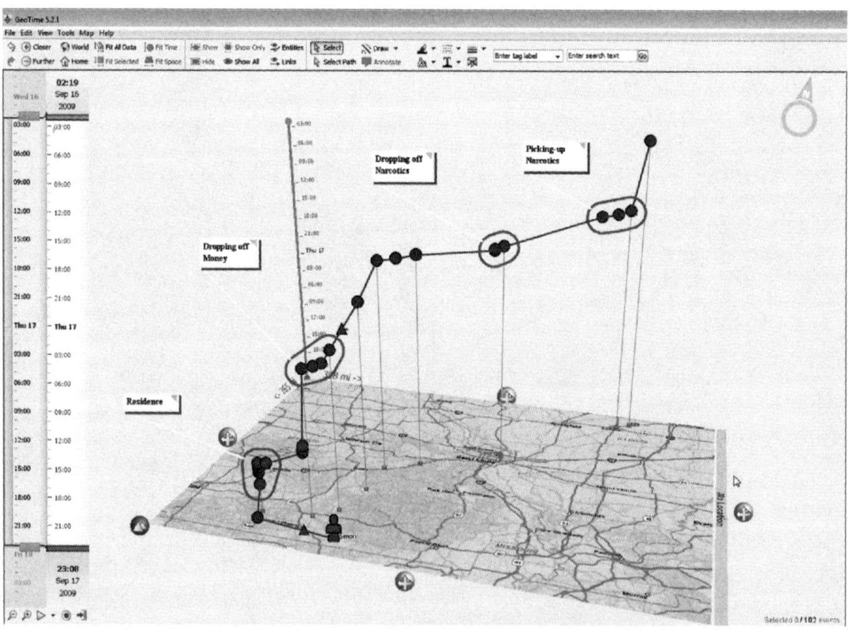

Figure 13.13 Example of oculus GeoTime spatiotemporal analysis events along a transaction. (Screenshots courtesy of Oculus Info, Inc. GeoTime® is a registered trademark of Oculus Info, Inc.)

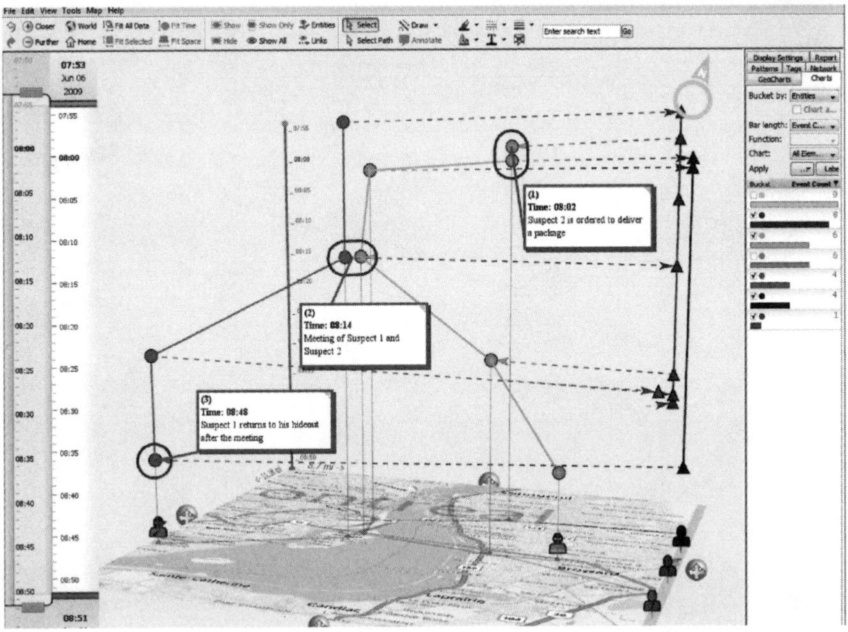

Figure 13.14 Example of Oculus GeoTime spatiotemporal analysis of two suspects [16]. (Screenshots courtesy of Oculus Info, Inc. GeoTime® is a registered trademark of Oculus Info, Inc.)

by French civil engineer Charles Joseph Minard. Visualization pioneer Edward Tufte said this figure "may well be the best statistical graphic ever drawn" [9].

The graphic shows the size of the army, the path taken, and the major events along the path—providing a visually striking example of how the army dwindled in size through many battles. Figure 13.15(a) shows the decreasing temperature along the path of retreat, contributing to even greater losses. The same plot, recreated as flow graph in JMP 11 is shown in Figure 13.15(b).

The Minard depiction of Napoleon's march can also be reimagined using JMP's bubble plot feature. The bubble plot uses size, shape, color, x, y, and time to show the relationship of multiple variables simultaneously. In Figure 13.15(c), the size of the bubble is the number of troops and the shading indicates the direction (advance or retreat).

Bubble plots are useful for animating spatial stories involving statistical change. A 2006 TED talk by statistics guru Hans Rosling used the bubble plot to depict the change in birth rate and income using United Nations data [17]. His talk accelerated the use of bubble charts in visual analytics. Animated bubble charts were a primary feature in the dynamic analysis platform GeoIQ (formerly FortiusOne), which was acquired by Esri in 2012.

Online spatial storytelling communities have developed as collaborative groups of data scientists and geospatial analysts combine their tradecraft with

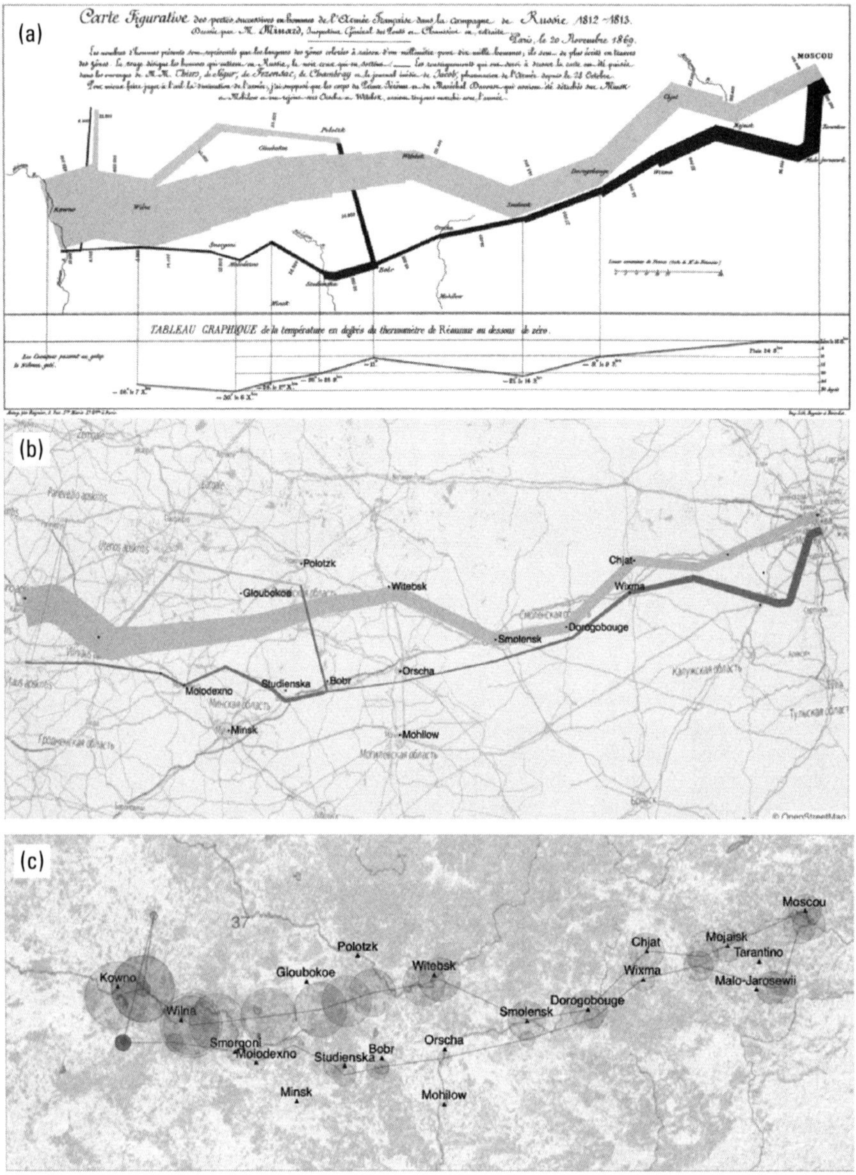

Figure 13.15 (a) Napoleon's Moscow, by Charles Minard; (b) Napoleon's march, as rendered in JMP 11, Data digitized by SAS/JMP (map data source: OpenStreetMap); and (c) Napoleon's March, rendered in JMP 11 as a time-animated bubble-plot (data digitized by SAS/JMP).

increasingly proliferated spatially-enabled data. The MapStory Foundation, a 501(c)3 educational organization founded in 2011 by social entrepreneur Chris Tucker developed an open, online platform for sharing stories about our world and how it develops over time. Stories demonstrate the spread of technology, the evolution of the post office, migration and immigration, election results, the spread of bike trails in Oregon, and trends in Internet access [18]. MapStory is built on the OpenGeo software stack, a widely used open geospatial platform for managing data and creating geospatial applications.

Visual analytics, spatial analysis, and statistical discovery are powerful tools for exploring data sets to understand trends and patterns, but also to tell stories about individual entities and their patterns of life.

13.5 The Way Ahead

Analytics and visualization have advanced dramatically over the past 20 years and co-developed with new commercial hardware and software technologies. At the GEOINT Symposium in April 2014, NGA's former Director Long presented a futuristic view for the integration of data and analysts, termed intelligence immersion. "By immersion, I mean living, interacting and experimenting with the data in a multimedia, multisensory experience with GEOINT at its core," said Long [19]. These immersive experiences sometimes require specialized visualization environments like the Georgia Tech Aerospace Systems Design Laboratory (ASDL) Collaborative Visualization Environment (CoVE) [20] and immersive visualization environments like the Applied Research Laboratory at the Pennsylvania State University.

Visualizing relationships across large, multidimensional data sets quickly requires more real estate than the average desktop computer. NGA's 24-hour operations center, with a "knowledge wall" comprised of 56 eighty-inch monitors, was inspired by the television show "24" [21, p. 19].

There are several key technology areas that provide potential for another paradigm shift in how analysts work with data. Some of the benefits of these technological advances were highlighted by former CIA chief technology officer Gus Hunt at a 2010 forum on big data analytics [22]:

- Elegant, powerful, and easy-to-use tools and visualizations;
- Machines to do more of the heavy lifting;
- Intelligent systems that learn from the user;
- A move to correlation, not search;
- A "curiosity layer"—signifying machines that are curious on your behalf.

Advances like in-memory processing of extremely large data sets, novel three dimensional visualizations, widespread adoption of ultra-high-resolution displays, and the ubiquity of tablet and wearable computing devices will catalyze further revolutions in real-time information processing, personalized analytics, and inspirational visualization. Many of these advances can be adopted and adapted to the intelligence discipline. Chapters 17–24 highlight some of these in the context of enabling technologies and methods for ABI.

References

[1] "Analysis," Merriam-Webster Dictionary.

[2] Davenport, T. H., and D. J. Patil, "Data Scientst: The Sexiest Job of the 21st Century," *Harvard Business Review,* October 2012.

[3] Law, D., and J. Eberhardt, "Do You Know Big Data?," June 9, 2014, web. Available: http://www.ctovision.com/download/know-big-data/.

[4] Leek, J., R. D. Peng, and B. Caffo, "The Data Scientist's Toolbox," Powerpoint presentation, The Johns Hopkins University, 2014.

[5] "Histogram," Wikipedia.

[6] Penenberg, A. L., and M. Barry, "Corporate Spies; The Pizza Plot," *The New York Times,* December 3, 2000.

[7] Anderson, C., *The Long Tail: Why the Future of Business Is Selling Less of More,* New York: Hyperion, 2006.

[8] Thomas, J. J., and K. A. Cook, "Illuminating the Path: The R&D Agenda for Visual Analytics," National Visualization and Analytics Center, Pacific Northwest National Laboratory, 2004.

[9] Tufte, E., *The Visual Display of Quantitative Information,* Cheshire, CT: Graphics Press, 1983.

[10] "What Makes a Data Visualization Memorable?" Harvard School of Engineering and Applied Sciences," web.

[11] Tomlinson, R., "A Geographic Information System for Regional Planning," Department of Forestry and Rural Development for the Government of Canada, August 1968.

[12] "Geographic Information Systems (GIS)," Global Industry Analysts, Inc., GOS-146, January 2012.

[13] Johnson, S., *The Ghost Map: The Story of London's Most Terrifying Epidemic—and How it Changed Science, Cities, and the Modern World,* New York: Riverhead Books, 2006.

[14] "Treemapping," Wikipedia.

[15] Wattenberg, M., "Map of the Market." Web. Available: http://www.bewitched.com/marketmap.html.

[16] "Geotime Overview," Oculus, Inc., Information Sheet. 2013.

[17] Rosling, H., "The Best Stats You've Ever Seen," TED Talks, web, February 2006.

[18] "MapStory: Search Stories." MapStory Foundation. Web. Available: http://mapstory.org/mapstories/search.

[19] Karpovich, J., "NGA's Long Touts GEOINT Immersion," *Earth Imaging Journal*, July 21, 2014.

[20] "Collaborative Visualization Environment." Georgia Institute of Technology. Web. Available: http://www.asdl.gatech.edu/Collaborative_Visualization_Environment.html.

[21] Clark, K., "Seamless Info Stream Key to Success of NGA Operations Centers," *Pathfinder*, Spring 2014.

[22] Hunt, G., "Big Data Operational Excellence: Ahead in the Cloud," Presented at the 2011 Amazon Web Services Summit, 18 Oct 2011.

14

Correlation and Fusion

Correlation of multiple sources of data is central to the integration before exploitation pillar of ABI and was the first major analytic breakthrough in combating adversaries that lack signature and doctrine. Multisensor data fusion techniques enhance the automated activity extraction capabilities discussed in Chapter 12 and provide mathematical structure to the ABI analytic tradecraft. One of the major outcomes of ABI is entity resolution through multi-INT analysis. Fusion, whether accomplished by a computer or a trained analyst, is central to this task. The suggested readings for this chapter alone fill several bookshelves. Correlation is a topic widely covered in mathematics, statistics, and popular psychology. Data fusion has evolved over 40 years into a complete discipline in its own right. This chapter provides a high-level overview of several key concepts in the context of ABI processing and analysis while directing the reader to further detailed references on this ever evolving topic.

14.1 Correlation

Correlation is the tendency of two variables to be related to each other. ABI relies heavily on correlation between multiple sources of information to understand patterns of life and resolve entities. The terms "correlation" and "association" are closely related. In statistics, the term association means that any two variables have a statistical dependence. In knowledge management, association means "relation" or "relationship." Correlation means that there is a more precise, measurable relationship between the two variables defined by a correlation coefficient. Although there is a nuanced mathematical difference between these two terms, we treat them as synonyms. A central principle of ABI is the need to correlate data from multiple sources—data neutrality—without a priori regard for the significance of data. In ABI, correlation leads to discovery of significance.

Scottish philosopher David Hume, in his 1748 book *An Enquiry Concerning Human Understanding,* defined association in terms of resemblance, contiguity [in time and place], and causality Hume says, "The thinking on any object readily transports the mind to what is contiguous"—an eighteenth-century statement of georeference to discover [1].

14.1.1 Correlation Versus Causality

One of the most oft-quoted maxims in data analysis is "correlation does not imply causality." The Internet is rife with amusing correlations, for example, Physicist Bobby Henderson's chart showing a correlation between a rise in global average temperature and the number of pirates. Statistician Nate Silver provides this vivid explanation of the correlation/causality argument:

> Just because two variables have a statistical relationship with each other does not mean that one is responsible for the other. For instance, ice cream sales and forest fires are correlated because both occur more often in the summer heat. But there is no causation; you don't light a patch of the Montana brush on fire when you buy a pint of Häagen-Dazs [2].

Many doubters of data science and mathematics use this sentence to deny any analytic result, dismissing a statistically valid fact as "pish posh." Correlation can be a powerful indicator of possible causality and a clue for analysts and researchers to continue an investigative hypothesis.

In *Thinking, Fast and Slow,* Kahneman notes that we "easily think associatively, we think metaphorically, we think causally, but statistics requires thinking about many things at once," which is difficult for humans to do without great effort [3, p. 13]. We prefer to develop causal links where there are none. We associate concepts and make judgments based on availability and familiarity. The trader-philosopher Nassim Taleb, author of *The Black Swan,* attributes the human propensity to form causal stories to an evolutionary desire to see patterns and discern threats in real time because prehistoric man's failure to do so meant that he was eaten [4].

Unfortunately, in modern times, we extend this evolution to stock picking, scientific experiments, and political forecasts—even though decades of research show this is wrong [5]. The only way to prove causality is through controlled experiments where all external influences are carefully controlled and their responses measured. The best example of controlled evaluation of causality is through pharmaceutical trials, where control groups, blind trials, and placebos are widely used.

In the discipline of intelligence, the ability to prove causality is effectively zero. Subjects of analysis are seldom participatory. Information is undersampled, incomplete, intermittent, erroneous, and cluttered. Knowledge lacks persistence. Sensors are biased. The most important subjects of analysis are intentionally trying to deceive you. Any medical trial conducted under these conditions would be laughably dismissed.

Remember: correlations are clues and indicators to dig deeper. Just as starts and stops are clues to begin transactional analysis at a location, the presence of a statistical correlation or a generic association between two factors is a hint to begin the process of deductive or abductive analysis there. Therefore, statistical analysis of data correlation is a powerful tool to combine information from multiple sources through valid, justifiable mathematical relationships, avoiding the human tendency to make subjective decisions based on known, unavoidable, irrational biases.

14.2 Fusion

The term "fusion" refers to "the process or result of joining two or more things together to form a single entity" [6]. Waltz and Llinas introduce the analogy of the human senses, which readily and automatically combine data from multiple perceptors (each with different measurement characteristics) to interpret the environment. They define the generic term "fusion" as "the process of combining or blending into a whole" [7]. Fusion is the process of disambiguating of two or more objects, variables, measurements, or entities that asserts—with a defined confidence value—that the two elements are the same. Simply put, the difference between correlation and fusion is that correlation says "these two elements are related." Fusion says "these two objects are the same."

Data fusion "combines data from multiple sensors and related information to achieve more specific inferences than could be achieved by using a single, independent sensor" [8]. Techniques for fusion come from signal processing, electrical engineering, statistics, control theory, artificial intelligence, decision theory, and psychology. The evolution of data fusion methods since the 1980s recognizes that fusion of information to improve decision making is a central process in many human endeavors, especially intelligence. Data fusion has been recognized as a mathematical discipline in its own right, and numerous conferences and textbooks have been dedicated to the subject.

Multisensor data fusion (a critical enabler for ABI) includes combining the results of multiple-single phenomenology sensors such as multiple radars tracking incoming missiles as well as multiplatform, multi-INT sensor data (e.g., the combination of imagery and electronic signal information) [9].

Steinberg and Bowman note that data fusion's relevance has often been limited to tracking and state estimation problems like multitarget tracking [10]; however, the mathematical techniques for data fusion can be applied to many problems in information theory such as intelligence analysis and ABI. They highlight the often confusing terminology used by multiple communities (see Figure 14.1) that rely on similar mathematical techniques with related objectives. Target tracking, for example, is a critical enabler for ABI but is only a small subset of the techniques in data fusion and information fusion.

14.2.1 A Taxonomy for Fusion Techniques

Recognizing that "developing cost-effective multi-source information systems requires a standard method for specifying data fusion processing and control functions, interfaces, and associated databases," the Joint Directors of Laboratories (JDL) proposed a general taxonomy for data fusion systems in the 1980s. This has become known as the JDL fusion model (Figure 14.2), and it is widely used to categorize data fusion related functions and technologies.

The model evolved since the 1980s when it originally contained three levels. Additional changes have been added by the JDL data fusion subpanel (DFS) and its alumni. Some discussions favor dividing the model into "low-level fusion" and "high-level fusion" to represent basic and advanced processing respectively. The fusion levels defined by the JDL are as follows:

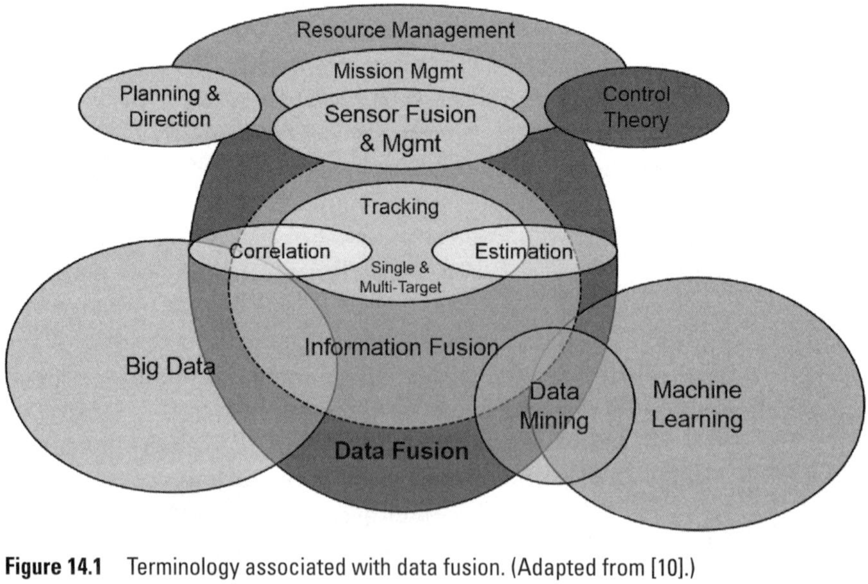

Figure 14.1 Terminology associated with data fusion. (Adapted from [10].)

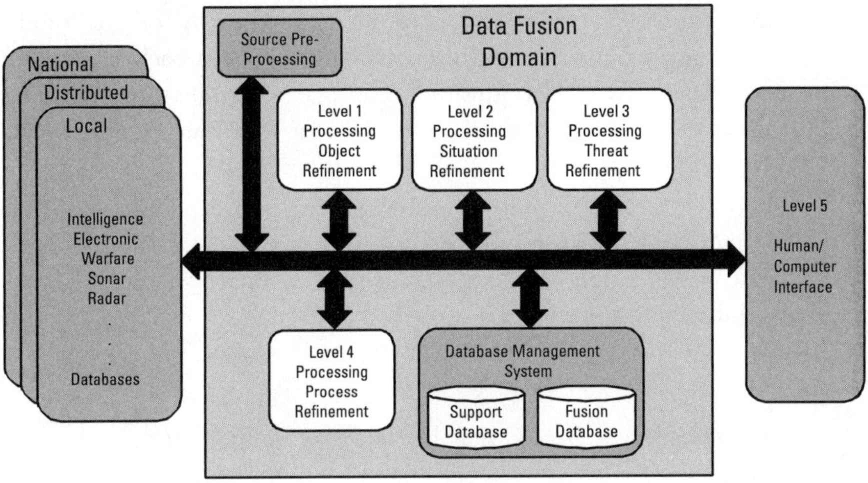

Figure 14.2 JDL fusion model. (Adapted from [8, 10, 11].)

- Source preprocessing, sometimes called level 0 processing, is data association and estimation below the object level. This step was added to the three-level model to reflect the need to combine elementary data (pixel level, signal level, character level) to determine an object's characteristics. Detections are often categorized as level 0.
- Level 1 processing, object refinement, combines sensor data to estimate the attributes or characteristics of an object to determine position, velocity, trajectory, or identity. This data may also be used to estimate the future state of the object. Hall and Llinas include sensor alignment, association, correlation, correlation/tracking, and classification in level 1 processing [11].
- Level 2 processing, situation refinement, "attempts to develop a description of current relationships among entities and events in the context of their environment" [8, p. 9]. Contextual information about the object (e.g., class of object and kinematic properties), the environment (e.g., the object is present at zero altitude in a body of water), or other related objects (e.g., the object was observed coming from a destroyer) refines state estimation of the object. Behaviors, patterns, and normalcy are included in level 2 fusion.
- Level 3 processing, threat refinement or significance estimation, is a high-level fusion process that characterizes the object and draws inferences in the future based on models, scenarios, state information, and constraints. Most advanced fusion research focuses on reliable level 3 techniques. This level includes prediction of future events and states.

- Level 4 processing, process refinement, augmented the original model by recognizing that continued observations can feed back into fusion and estimation processes to improve overall system performance. This can include multiobjective optimization or new techniques to fuse data when sensors operate on vastly different timescales [12, p. 12].
- Level 5 processing, cognitive refinement or human/computer interface, recognizes the role of the user in the fusion process. Level 5 includes approaches for fusion visualization, cognitive computing, scenario analysis, information sharing, and collaborative decision making. Level 5 is where the analyst performs correlation and fusion for ABI.

The designation as "levels" may be confusing to apprentices in the field as there is no direct correlation to the "levels of knowledge" associated with knowledge management. The JDL fusion levels are more accurately termed categories; a single piece of information does not have to traverse all five "levels" to be considered fused.

According to Hall and Llinas, "the most mature area of data fusion process is level 1 processing," and a majority of applications fall into or include this category [8]. Level 1 processing relies on estimation techniques such as Kalman filters, MHT, or joint probabilistic data association (JPDA, see Chapter 12) [12]. Data fusion applications for detection, identification, characterization, extraction, location, and tracking of individual objects fall in level 1. Additional higher level techniques that consider the behaviors of that object in the context of its surroundings and possible courses of action are techniques associated with levels 2 and 3. These higher level processing methods are more akin to analytic "sensemaking" performed by humans, but computational architectures that perform mathematical fusion calculations may be capable of operating with greatly reduced decision timelines. A major concern of course is turning what amounts to decision authority to silicon-based processors and mathematical algorithms, especially when those algorithms are difficult to calibrate. Another issue is how to assign blame when the multisensor fusion system makes a mistake—an ongoing debate in the evolution of driverless automobiles. Nevertheless, we have learned to accept a degree of automation in life-critical situations like certain regimes of commercial air travel (a well-researched multisensor state estimation problem).

While ongoing ABI research explores techniques across fusion levels 0-4. Many programs at DARPA (see Section 14.5) develop enablers for multisource tracking and fusion. The NRO's Sentient Enterprise sought solicitations for "cognitive processing for automated system processing and cognitive processing for automated data analysis"—what they call, "sensemaking" [13]. While silicon-based systems exhibit promise, the state of the art in information fusion for ABI still relies heavily on fusion level 5: a highly trained carbon-based neural fusion system.

14.2.2 Architectures for Data Fusion

The voluminous literature on data fusion includes several architectures for data fusion that follow the same pattern. Hall and Llinas propose three alternatives:

1. "Direct fusion of sensor data;
2. Representation of sensor data via feature vectors, with subsequent fusion of the feature vectors;
3. Processing of each sensor to achieve high-level inferences or decisions, which are subsequently combined [8]."

These three architectures are depicted in Figure 14.3. Architecture 1 represents upstream fusion where the results of individual sensors are mathematically associated and the results are fused early in the process. Feature extraction and entity resolution proceed on the fused result. Architecture 3 represents a

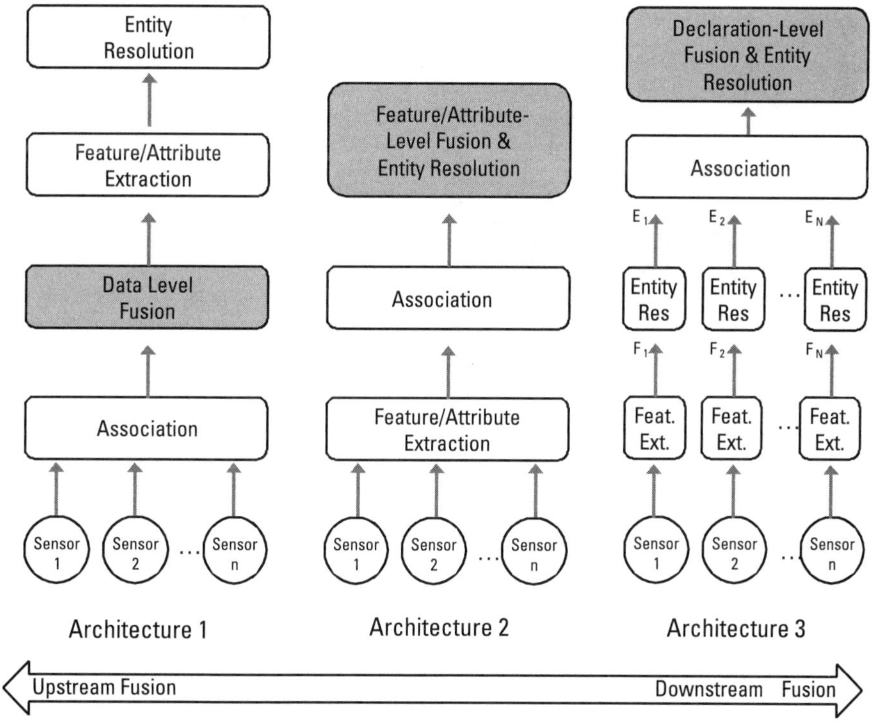

Figure 14.3 Architectures for upstream and downstream data fusion. (Adapted from [8].)

downstream fusion approach where sensor-specific processing on individual sensors extracts and resolves entities based on the input from a single sensor. The position and state information about the resultant entities are fused later in the process. Architecture 2 is a hybrid that extracts features and attributes from multiple sensors before association and then combines fusion and entity resolution as the final step. The advantages and disadvantages of upstream and downstream fusion are discussed in Section 14.2.3.

14.2.3 Upstream Versus Downstream Fusion

Bottomley et al. conducted an instrumented multi-INT experiment to quantify the benefits of upstream fusion using an architecture based on architecture 1 and architecture 3 in Figure 14.3 [14]. The upstream fusion case processes the raw streams together and fuses them using a maximum likelihood (ML) approach. In the downstream fusion case, the streams are processed separately to produce parameter estimates (e.g., position estimates), which are then fused using an optimal ML approach. Bottomley et al. advocate the upstream approach because it prevents information loss prior to fusion and results in better position estimates with fewer sensors. The procedure for comparison is shown in Figure 14.4. The experiment consisted of two emitting vehicles with unique IMINT

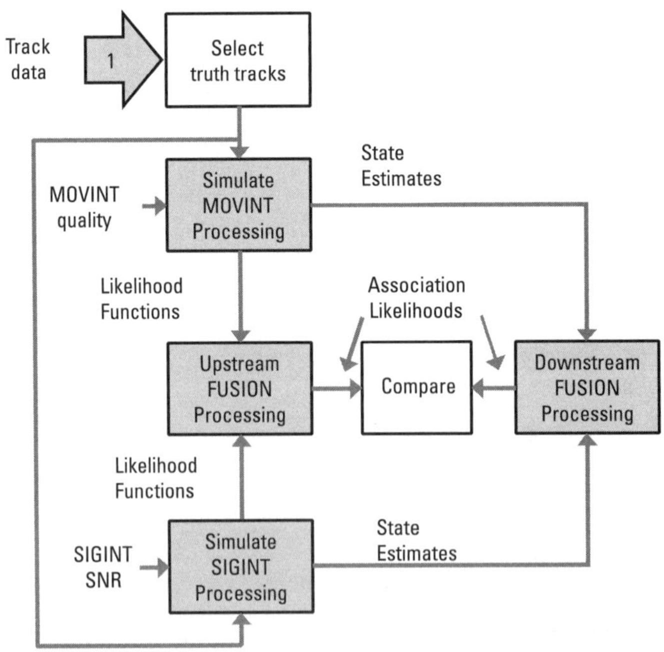

Figure 14.4 Experimental setup comparing upstream and downstream fusion. (Adapted from [14]. Reprinted with permission.)

and SIGINT signatures collected by two or three ground-based SIGINT sensors and one airborne WAMI sensor. Tracks were extracted from the WAMI data and "truthed" for the two instrumented vehicles, calculating "MOVINT" likelihoods and (position, velocity) state estimates for each object. Simulated SIGINT processing was used to obtain likelihoods and state estimates for each SIGINT emitter. The association estimation step in Figure 14.4 attempts to associate object A with emitter 1 versus object A with emitter 2. In the upstream processing case, the likelihoods are fused. In the downstream case, the state estimates are fused.

The results of the multisensor fusion experiment for two RF sensors and a single imagery-based MOVINT sensor are shown in Figure 14.5. The gray solid lines show the results of downstream fusion at four levels of SNR. The downstream fusion case does not "work" for this problem because the probability of correct association varies dramatically and even after eight measurements, it is not consistently lower than at the beginning of the experiment. In contrast, the upstream fusion results converge to a 0% probability of false association by the seventh measurement in the 0 SNR case. As SNR is increased, the upstream fusion case quickly resolves to the correct association. When upstream processing is used, the MOVINT sensor and SIGINT sensor behave as commensurate, meaning their measurements can be directly fused. They essentially behave as the same type of sensor. Figure 14.6 adds a third RF sensor, which

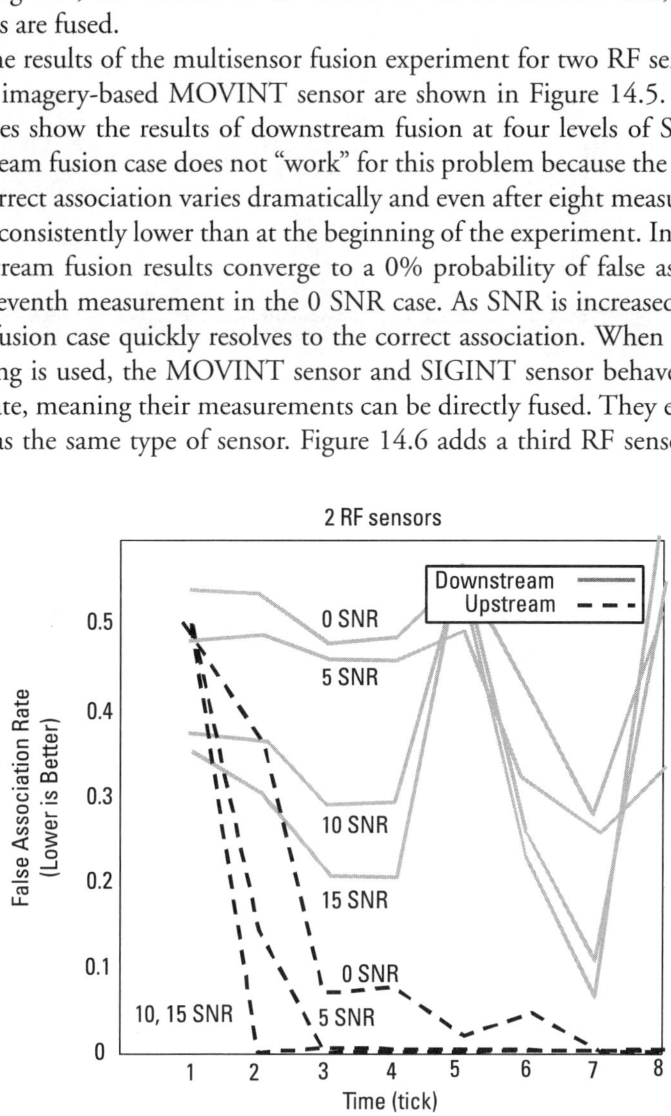

Figure 14.5 Comparison of upstream and downstream fusion using two RF sensors and a MOVINT sensor. (Adapted from [14]. Reprinted with permission.)

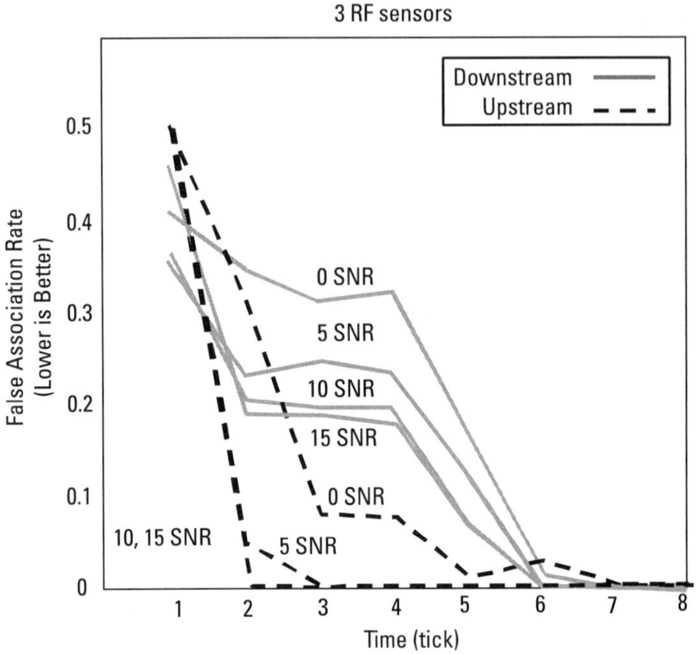

Figure 14.6 Comparison of upstream and downstream fusion using three RF sensors and a MOVINT sensor. (Adapted from [14]. Reprinted with permission.)

significantly improves performance and allows the fusion engine to resolve the entities in the downstream fusion case. The impact on the upstream fusion case is less pronounced, although the upstream processing case still outperforms the downstream processing case by a significant margin.

14.3 Mathematical Correlation and Fusion Techniques

Most architectures and applications for multi-INT fusion, at their cores, rely on various mathematical techniques for conditional probability assessment, hypothesis management, and uncertainty quantification/propagation. The most basic and widely used of these techniques, Bayes's theorem, Dempster-Shafer theory, and belief networks, are discussed in this section.

14.3.1 Bayesian Probability and Application of Bayes's Theorem

One of the most widely used techniques in information theory and data fusion is Bayes's theorem. Named after English clergyman Thomas Bayes who first documented it in 1763, the relation is statement of conditional probability and its dependence on prior information. Bayes's theorem calculates the probability of an event, A, given information about event B and information about the likelihood of one event given the other. The standard form of Bayes's theorem is given as:

$$P(A|B) = \frac{P(B|A)P(A)}{P(B)} \tag{14.1}$$

where

$P(A)$ is the prior probability, that is, the initial degree of belief in event A;

$P(A|B)$ is the conditional probability of A given that event B occurred (also called the posterior probability in Bayes's theorem);

$P(B|A)$ is the conditional probability of B, given that event A occurred, also called the likelihood;

$P(B)$ is the probability of event B.

This equation is sometimes generalized as:

$$posterior = \frac{likelihood \times prior}{marginal\ likelihood} \tag{14.2}$$

or, said as "the posterior is proportional to the likelihood times the prior" as:

$$P(A|B) \propto P(B|A)P(A) \tag{14.3}$$

Sometimes, Bayes's theorem is used to compare two competing statements or hypotheses, and given as the form:

$$P(A|B) = \frac{P(B|A)P(A)}{P(B|A)P(A) + P(B|\neg A)P(\neg A)} \tag{14.4}$$

where $P(\neg A)$ is the probability of the initial belief against event A, or $1-P(A)$, and $P(B|\neg A)$ is the conditional probability or likelihood of B given that event A is false.

Bayes's theorem is a compact, useful, simple, and mathematically pure technique for comparing probabilities. Taleb explains that this type of statistical thinking and inferential thinking is not intuitive to most humans due to evolution: "consider that in a primitive environment there is no consequential difference between the statements most killers are wild animals and most wild animals are killers" [4, p. 53]. In the world of prehistoric man, those who treated these statements as equivalent probably increased their probability of staying alive. In the world of statistics, these are two different statements that can be represented probabilistically. Bayes's theorem is useful in calculating quantitative probabilities of events based on observations of other events, using the property of transitivity and priors to calculate unknown knowledge from that which is known. In ABI, it is used to formulate a probability-based reasoning tree for observable intelligence events.

Application of Bayes's Theorem to Object Identification

Consider the intelligence problem of locating rare military equipment like a mobile missile launcher. Many countries employ decoys to confuse photographic interpreters. The fictional central Asian country Zazikistan has 10 mobile missile launchers, L. To confuse western analysts, the country employs 99 decoys, D, for every one real launcher. However, after years of analyzing Zazikistan, image analysts have become very adept at recognizing real and decoy launchers; consider the following summary of what they think they know:

- If the object is a decoy, D, there is a 90% chance they identify it as a decoy and a 10% chance of falsely identifying it as a real launcher, L. That is, $P(D|D) = 90\%$ and $P(L|D) = 10\%$.
- If the object is a launcher, L, there is a 90% chance they identify it as a launcher and a 10% chance of falsely identifying it as a decoy, D. That is, $P(L|L) = 90\%$ and $P(D|L) = 10\%$.

So, if you are an intelligence analyst working the Zazikistan mobile missile account, what is the probability that you correctly identify a launcher.[1] A decision tree is helpful in walking through the answer. Step 1 of shows the initial breakout of the 1,000 objects in Zazikistan: 10 real launches and 990 decoys. The imagery analyst goes to work in step 2, applying the conditional probability of identifying launchers given that the object is a launcher. Based on the conditional probabilities, he successfully identifies nine launchers and incorrectly identifies one real launcher as a decoy. Applying conditional probabilities to the right side of the decision tree, the analyst identifies 891 decoys correctly as decoys and misidentifies 99 decoys as launchers.

Based on this example, if an analyst identifies a launcher, what is the probability he or she is correct? A total of 108 launchers (99+9) were identified. Of these, nine were correctly identified. The probability of correctly identifying a launcher is 9/108 or 8.3%.

This is a shocking and nonintuitive result. Given the exceedingly high probability of correctly identifying a launcher (90%), most intelligence analysts

1. We recommend writing your answer down and continuing the exercise.

estimate their probability of correctly identifying the launcher as close to 90%. However, because the true objects are exceedingly rare (only 1% of the total sample), if an analyst observes a launcher it is 11 times more likely that he or she incorrectly identified the much more prevalent decoys.

The frequency or rarity of the objects in step 1 of Figure 14.7 is called the base rate. Numerous studies of probability theory and decision-making show that humans tend to overestimate the likelihood of events with low base rates. (This tends to explain why people gamble). Psychologists Amos Tversky and Daniel Kahneman refer to the tendency to overestimate salient, descriptive, and vivid information at the expense of contradictory statistical information as the representativeness heuristic [15].

The CIA examined Bayesian statistics in the 1960s and 1970s as an estimative technique in a series of articles in *Studies in Intelligence.* An advantage of the method noted by CIA researcher, Jack Zlotnick is that the analyst makes "sequence of explicit judgments on discrete units" of evidence rather than "a blend of deduction, insight, and inference from the body of evidence as a whole" [16]. He notes, "The research findings of some Bayesian psychologists seem to show that people are generally better at appraising a single item of evidence than at drawing inferences from the body of evidence considered in the aggregate" [17].

Application of Bayes's Theorem to Multisensor Fusion

The mathematical formulation of Bayes's Theorem is widely applied to probabilistic fusion for object identification and entity resolution in data fusion [8]. Consider a collection activity comprised of multiple (n) sensors that individually detect different characteristics of m objects. The joint probability of resolving each entity, O_j with multisensor information is given as:

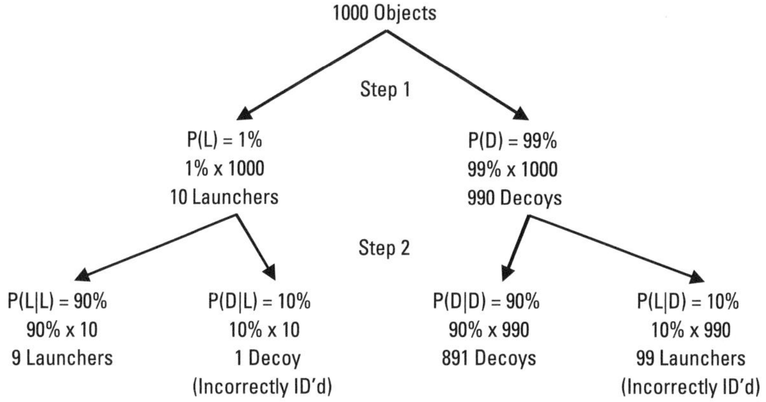

Figure 14.7 Decision tree for mobile missile and decoy differentiation.

$$P\left(O_j / D_1 \cap D_2 \cap \cdots \cap D_n\right), j = 1, \ldots, M \tag{14.5}$$

where

D_n is the declaration of evidence/identity from sensor n;

M is the total set of objects;

$P(O_j/D_1 \ldots D_n),j$ is the maximum a posteriori probability (MAP), the maximum joint probability for the identity of object, O_j.

The process for Bayesian combination of probability distributions from multiple sensors to produce a fused entity identification is shown in Figure 14.8. Each individual sensor produces a declaration matrix, which is that sensor's declarative view object's identity based on its attributes—sensed characteristics, behaviors, or movement properties. The individual probabilities are combined jointly using the Bayesian formula. Decision logic is applied to select the MAP that represents the highest probabilities of correct identity. Decision rules can also be applied to threshold the MAP based on constraints or to apply additional deductive logic from other fusion processes. The resolved entity is declared (with an associated probability). When used with a properly designed multisensor data management system, this declaration maintains provenance back to the original sensor data.

Bayes's formula provides a straightforward, easily programmed mathematical formulation for probabilistic combination of multiple sources of information; however, it does not provide a straightforward representation for a lack of information. A modification of Bayesian probability called Dempster-Shafer theory introduces additional factors to address this concern.

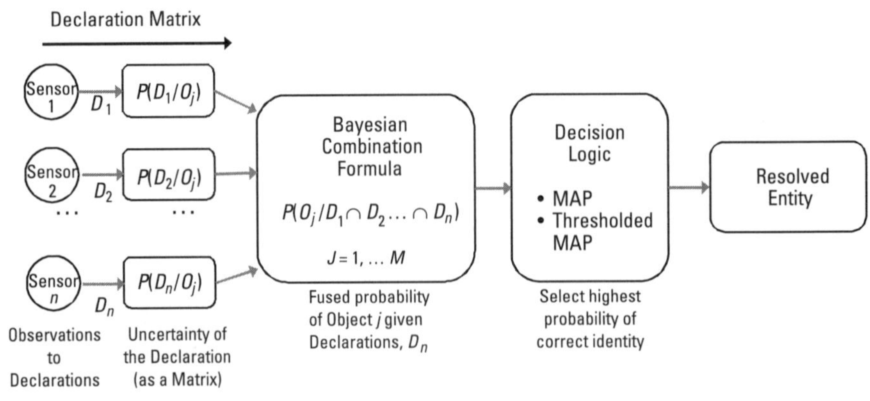

Figure 14.8 Application of Bayes's rule to fusion (Bayesian fusion). (Adapted from [7, p. 221]).

14.3.2 Dempster-Shafer Theory

Dempster-Shafer theory is a generalization of Bayesian probability based on the integration of two principles. The first is belief functions, which allow for the determination of belief from one question on the subjective probabilities for a related question. The degree to which the belief is transferrable depends on how related the two questions are and the reliability of the source [18]. The second principle is Dempster's composition rule, which allows independent beliefs to be combined into an overall belief about each hypothesis [19]. According to Shafer, "The theory came to the attention of artificial intelligence (AI) researchers in the early 1980s, when they were trying to adapt probability theory to expert systems" [20]. Dempster-Shafer theory differs from the Bayesian approach in that the belief in a fact and the opposite of that fact does not need to sum to 1; that is, the method accounts for the possibility of "I don't know." This is a useful property for multisource fusion especially in the intelligence domain.

A question that has a single outcome, Θ, from Θ_n possible outcomes in the set:

$$\Theta = \{\theta_1, \theta_2, \ldots \theta_n\} \tag{14.6}$$

has a "frame of discernment" for Θ represented by the power set of Θ, that is, all possible subsets. For $n = 3$, the frame of discernment of Θ is:

$$(\varnothing, \theta_1, \theta_2, \theta_2, \{\theta_1, \theta_2\}, \{\theta_1, \grave{e}_3\}, \{\theta_1, \theta_2, \theta_3\}) \tag{14.7}$$

Dempster-Shafer theory defines the concept of mass function, $m(A)$, which represents the proportion of all evidence that supports a given outcome, Θ_n, in the power set. (Imagine the concept of "massing" evidence for an outcome). The mass function represents the degree of belief in an outcome, which is represented as a set of probabilities. The probability, or belief, for each element of the power set is between 0 and 1. Also, $m(\varnothing) = 0$ (the probability of no outcome is zero), and the masses must sum to 1.

Dempster's rule of combination combines the joint mass from two masses m_1 and m_2 as follows [21]:

$$m_{1,2}(A) = (m_1 \oplus m_2)(A) = \frac{1}{1-K} \sum_{B \cap C = A \neq \varnothing} m_1(B) m_2(C) \tag{14.8}$$

where K represents the conflict between the two mass sets, that is, the probability that B and C provide conflicting evidence toward the outcome:

$$K = \sum_{B \cap C = A \neq \emptyset} m_1(B) m_2(C) \tag{14.9}$$

Normalizing by 1/(1-K) ensures that the viable belief functions sum to 1.0 after the combination of evidence.

Multisensor fusion approaches use Dempster-Shafer theory to discriminate objects by treating observations from multiple sensors as belief functions based on the object and properties of the sensor. Instead of combining conditional probabilities for object identification as shown in Figure 14.8, the process for fusion proposed by Waltz and Llinas is modified for the Dempster-Shafer approach in Figure 14.9. Mass functions replace conditional probabilities, and Dempster's combination rule accounts for the additional uncertainty when the sensor cannot resolve the target. The property of normalization by the null hypothesis is also important because it removes the incongruity associated with sensors that disagree.

Although this formulation adds more complexity, it is still easily programmed into a multisensor fusion system. The Dempster-Shafer technique can also be easily applied to quantify beliefs and uncertainty for multi-INT analysis including the beliefs of members of an integrated analytic working group.

Example: Using Dempster-Shafer Theory for Multisource Fusion

Dempster-Shafer theory is widely used in multisensor fusion for target tracking and recognition, but the method is widely applicable to problems of evidence combination from multiple sources. Consider the following example.

Following a period of unrest in the country of Zazikistan, analysts form a hypothesis, H, that the supreme leader is no longer in control of the country

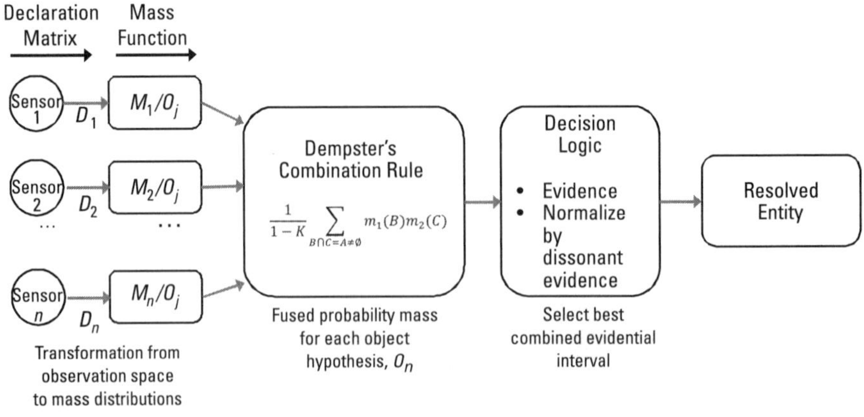

Figure 14.9 Application of Dempster-Shafer theory to fusion. (Adapted from [7, p. 223].)

and a coup de état is in progress. The converse of the hypothesis, that the supreme leader is still in control, is H_c. Uncertainty in the outcome is represented as u. Four sensors produce evidence E_1 through E_4 to confirm or deny the judgment, but each sensor is not completely reliable. Table 14.1 summarizes the belief values for the four sensors. The values of H, H_c, and u must sum to 1.0. For example, belief in the result of evidence E_1 from the GEOINT sensor produces a belief of 0.8 in H—that Zazikistan is under a coup, and a belief of 0.1 toward H_c that it is not. The uncertainty (the inability to distinguish between H and H_c), u, is 0.1. Note that the SIGINT sensor and HUMINT source have high values of u. The HUMINT source appears unlikely to validate the state of the coup, and the higher uncertainty value leads the analyst to believe he or she is not a particularly reliable source.

Dempster's rule applies to the combination of multiple independent beliefs. Consider the case where GEOINT and SIGINT are collected. The composite belief between the two, E_1E_2 is given by the equation:

$$(H \mid E_1 \cdot E_2) = (H|E_1)(H|E_2) + (H|E_1)(u|E_2) + (H \mid E_2)(u \mid E_1) \quad (14.10)$$

In plain English, (14.10) says, "The joint belief in hypothesis H given evidence E_1 and E_2 is the sum of 1) the belief in H given confirmatory evidence from both sensors, 2) the belief in H given confirmatory evidence from sensor 1 [GEOINT] but with uncertainty about the result from sensor 2, and 3) the belief in H given confirmatory evidence from sensor 2 [SIGINT] but with uncertainty about the result from sensor 1." Substituting yields:

$$(H \mid E_1 \cdot E_2) = 0.8 * 0.5 + 0.8 * 0.35 + 0.5 * 0.1 = 0.73 \quad (14.11)$$

Similarly, this process can be repeated for the next column to calculate the belief in the converse hypothesis:

$$(H_c \mid E_1 \cdot E_2) = (H_c|E_1)(H_c|E_2) + (H_c|E_1)(u|E_2) + (H_c \mid E_2)(u \mid E_1) \quad (14.12)$$

$$(H_c \mid E_1 \cdot E_2) = 0.1 * 0.15 + 0.1 * 0.35 + 0.15 * 0.1 = 0.065 \quad (14.13)$$

Table 14.1
Evidence and Belief Values for the Sample Problem

Sensor Type	Evidence	H	Hc	u
GEOINT	E1	0.8	0.1	0.1
SIGINT	E2	0.5	0.15	0.35
MASINT	E3	0.7	0.2	0.1
HUMINT	E4	0.2	0.5	0.3

And for the uncertainty, *u:*

$$(u \mid E_1 \cdot E_2) = (u|E_1)(u|E_2) = 0.35 * 0.1 = 0.035 \tag{14.14}$$

The values 0.73, 0.065, and 0.035 sum to 0.83, but the total belief in the system must sum to 1.0. There is a fourth term to the equation, *d,* which represents the dissonance in the belief system. This is when you believe *H* from E_1 and disbelieve *H* from E_2, or:

$$\begin{aligned}(d \mid E_1 \cdot E_2) &= (H|E_1)(H_c|E_2) \\ &+ (H_c|E_1)(H|E_2) = 0.8 * 0.15 + 0.1 * 0.5 = 0.17\end{aligned} \tag{14.15}$$

The final answer is normalized to remove dissonant values by dividing each belief by (1-*d*). The final beliefs are the following:

- Zazikistan is under a coup = 87.9%;
- Zazikistan is not under a coup = 7.8%;
- Unsure = 4.2%;
- Total = 100%.

Next, Table 14.2 demonstrates the inclusion of a third sensor.

Repeating the steps above, substituting $E_1 * E_2$ for the first belief and E_3 as the second belief, Dempster's rule can again be used to combine the beliefs for the three sensors:

- Zazikistan is under a coup = 95.3%;
- Zazikistan is not under a coup = 4.2%;
- Unsure = 0.5%;
- Total = 100%.

Note that the overall belief in the coup hypothesis increased slightly but that the uncertainty went down significantly by integrating information from

Table 14.2
Fusion of Multisensor Data

Sensor Type	Evidence	H	Hc	u
GEOINT & SIGINT	E1*E2	0.879	0.078	0.042
MASINT	E3	0.7	0.2	0.1

the MASINT sensor (which has an uncertainty of only 0.1). Finally, fuse the beliefs for the HUMINT sensor treating $E_1{}^*E_2{}^*E_3$ as the first belief and E_4 as the second. Note that in Dempster's rule, the order of combination does not matter. The resultant belief is:

- Zazikistan is under a coup = 92.7%;
- Zazikistan is not under a coup = 7.0%;
- Unsure = 0.3%;
- Total = 100%.

In this case, because the HUMINT source has only contributes 0.2 toward the belief in *H,* the probability that Zazikistan is under a coup actually decreases slightly. Also, because this source has a reasonably low value of *u,* the uncertainty was further reduced.

While the belief in the coup hypothesis is 92.7%, the plausibility is slightly higher because the analyst must consider the belief in hypothesis *H* as well as the uncertainty in the outcome. The plausibility of a coup is 93%. Similarly, the plausibility in H_c also requires addition of the uncertainty: 7.3%. These values sum to greater than 100% because the uncertainty between *H* and H_c makes either outcome equally likely in the rare case that all four sensors produce faulty evidence.

The Power of Negation

The previous example illustrated the utility of Dempster-Shafer theory for combining evidence from multiple sensors; however, the sensors mostly provided information that confirmed hypothesis H. Table 14.3 shows the table of evidence with a modification to the HUMINT source. In this case, the HUMINT source contributes nothing to the hypothesis that a coup is under way but provides perfect evidence that it is not, with an uncertainty of only 1%.

Applying Dempster's rule to this new case yields the following beliefs:

Table 14.3
Evidence and Belief Values for the Second Sample Problem

Sensor Type	Evidence	H	Hc	u
GEOINT	E1	0.8	0.1	0.1
SIGINT	E2	0.5	0.15	0.35
MASINT	E3	0.7	0.2	0.1
HUMINT	E4	0	1.0	0.01

- Zazikistan is under a coup = 16.87%;
- Zazikistan is not under a coup = 83%;
- Unsure = 0.1%.

With three sensors that contribute belief to *H,* the strongly dissenting belief from the new HUMINT source significantly influences the overall belief. (Note that for the boundary condition of $u_4 = 0$, $H_c = 1$). As u_4 increases, even slightly, the belief in H_c rapidly drops. For this reason, application of multisensor fusion techniques should rely on high-quality (low-uncertainty) sources but should also seek a balance of sensors that provide information about differing hypotheses.

14.3.3 Belief Networks

A belief network[2] is a "a probabilistic graphical model (a type of statistical model) that represents a set of random variables and their conditional dependencies via a directed acyclic graph (DAG)" [22]. This technique allows chaining of conditional probabilities—calculated either using Bayesian theory or Dempster-Shafer theory—for a sequence of possible events. In one application, Paul Sticha and Dennis Buede of HumRRO and Richard Rees of the CIA developed APOLLO, a computational tool for reasoning through a decision-making process by evaluating probabilities in Bayesian networks [23, 24]. Bayes's rule is used to multiply conditional probabilities across each edge of the graph to develop an overall probability for certain outcomes with the uncertainty for each explicitly quantified. Figure 14.10 illustrates a notional belief network for the key intelligence question "What will [Syrian President Bashir al] Assad do?" Analysts define relationships using the directed graph and assign probabilities to potential outcomes of answerable questions like "Does the leader support the military" and "What is U.S. president Barack Obama's posture toward Assad and the ISIS?"

14.4 Multi-INT Fusion For ABI

Correlation and fusion is central to the analytic tradecraft of ABI. One application of multi-INT correlation is to use the strengths of one data source to compensate for the weaknesses in another. SIGINT, for example, is exceptionally accurate in verifying identity through proxies because many signals have unique identifiers that are broadcast in the signal like the Maritime Mobile Service Identity (MMSI) in the ship-based navigation system, AIS. Signals also may

2. Also called a Bayesian network, Bayes net, or Bayesian Belief Network (BBN).

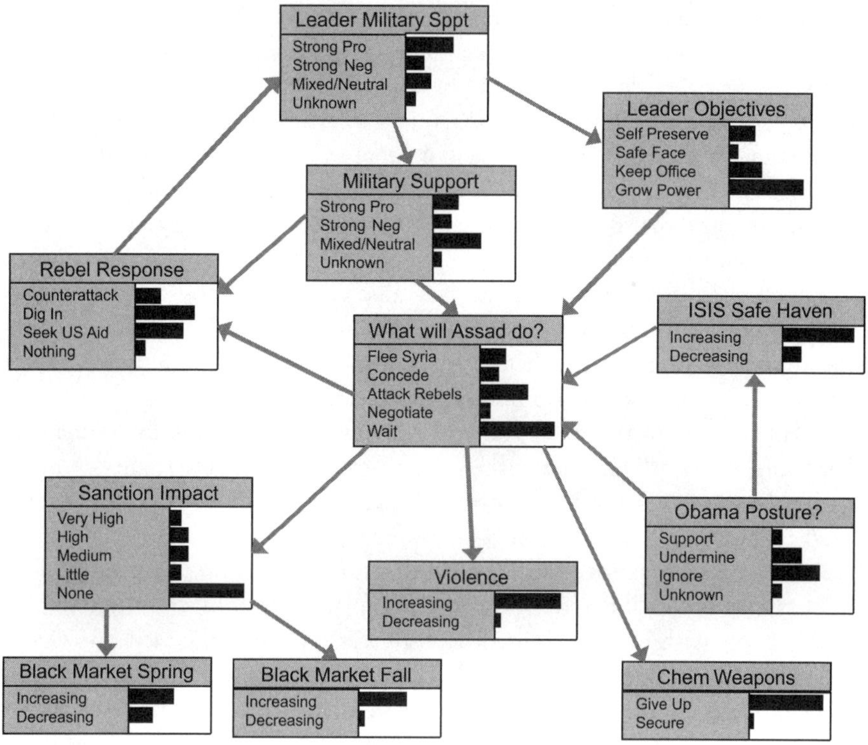

Figure 14.10 Example of a belief network (Bayes network) for an intelligence assessment. Unclassified example with notional values. (Adapted from [24].)

include temporal information, but SIGINT is accurate in the temporal domain because radio waves propagate at the speed of light—if sensors are equipped with precise timing capabilities, the exact time of the signal emanation is easily calculated. Unfortunately, because direction-finding and triangulation are usually required to locate the point of origin, SIGINT has measurable but significant errors in position (depending on the properties of the collection system). GEOINT on the other hand is exceptionally accurate in both space and time. A GEOINT collection platform knows when and where it was when it passively collected photons coming off a target. This error can be easily propagated to the ground using a sensor model. The WorldView-3 imaging platform has an accuracy of 3.5m (CE90) [25]. When imagery is coregistered to known ground control points, this accuracy can be further improved. Entity resolution and identity determination from GEOINT is extremely difficult if not impossible because the signature of individuals can rarely be determined from overhead photography or other collection means. Proxies in GEOINT are easily spoofed and confused. Correlation and fusion of SIGINT and GEOINT allows one INT to compensate for the weaknesses of the other.

Another approach to correlation and fusion for ABI is the integration of low-resolution wide-area sensors and high-resolution-point collection platforms. As described in Chapter 11, to obtain greater persistence and a larger area of coverage, a higher altitude is required. Because imaging resolution and signal strength are both inversely proportional to the distance from the target, high area coverage is generally inversely proportional to resolvability of the target. On the other hand, low-altitude platforms like a LEO imagery satellite or a medium-altitude UAV provide exceptionally exquisite resolution relative to their high-altitude counterparts for the same collection geometries and phenomenologies. The ability to correlate results of wide area collection with precise, entity resolving, narrow field-of-regard collection systems is an important use for ABI fusion and an area of ongoing research.

Finally, one of the most complex but promising areas is the domain of "hard/soft" fusion. "Hard" data is based on facts, including the results of highly validated remote sensing collection and processing capabilities. "Soft" data is based on empirical models, analyst knowledge, fuzzy logic, assumptions, and human-based observations [26, 27]. Hard/soft fusion is a promising area of research that enables validated correlation of information from structured remote sensing assets with human-focused data sources including the tacit knowledge of intelligence analysts. Gross et al. developed a framework for fusing hard and soft data under a university research initiative that included ground-based sensors, tips to law enforcement, and local news reports [28]. The University at Buffalo (UB) Center for Multisource Information Fusion (CMIF) is the leader of a multi-university research initiative (MURI) developing "a generalized framework, mathematical techniques, and test and evaluation methods to address the ingestion and harmonized fusion of hard and soft information in a distributed (networked) Level 1 and Level 2 data fusion environment" [29]. Team members include the Pennsylvania State University (PSU), Iona College (Iona), and Tennessee State University (TSU).

14.5 Examples of Multi-INT Fusion Programs

In addition to numerous university programs developing fusion techniques and frameworks, automated fusion of multiple sources is an area of ongoing research and technology development, especially at DARPA, federally funded research and development corporations (FFRDCs), and the national labs.

14.5.1 Example: A Multi-INT Fusion Architecture

A generic architecture for multisensor fusion, highlighting the JDL fusion levels, is shown in Figure 14.11. Level 0 and level 1 fusion extracts tracks from WAMI sensors and ground-based cameras. Electronic direction-finding and

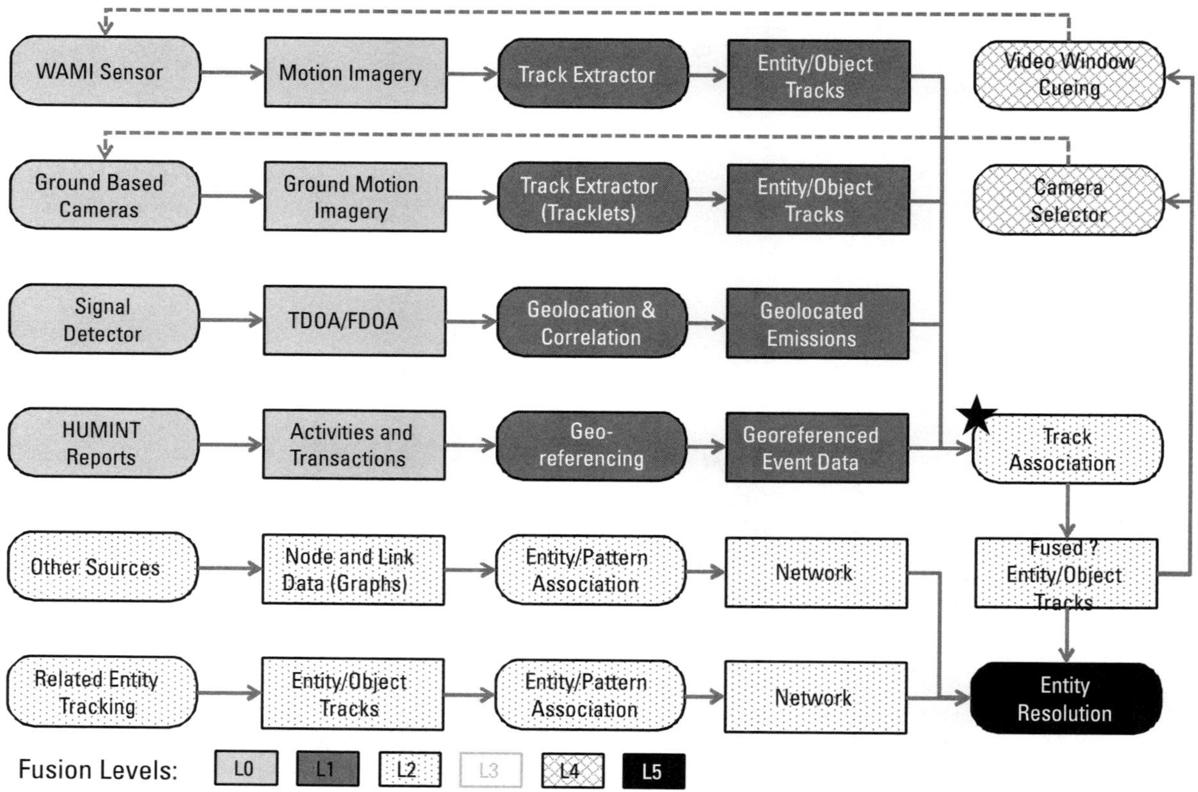

Figure 14.11 Example of a multisensor, multilevel information fusion problem for ABI.

HUMINT reports georeferenced using tools like LocateXT or Event Horizon produce geolocated detections of events of interest. Track association occurs at level 2.

Simultaneously, existing knowledge from other sources (in the form of node and link data) and tracking of related entities is combined through association analysis to produce network information. The network provides context to the fused multi-INT entity/object tracks to enhance entity resolution. Although entity resolution can be performed at level 2, this example highlights the role of human-computer interaction (level 5 fusion) in integration-before-exploitation to resolve entities. Finally, feedback from the fused entity/object tracks is used to retask GEOINT resources for collection and tracking in areas of interest.[3]

14.5.2 Example: The DARPA Insight Program

In 2010, DARPA instituted the Insight program to address a key shortfall for ISR systems: "the lack of a capability for automatic exploitation and cross-cueing of multi-intelligence (multi-INT) sources [30]. The program included simulation in a virtual sensor environment and a physical test bed at the National Training Center at Ft. Irwin, CA, where multiple sensor platforms cocollected multi-INT information and analysts applied advanced fusion tools to unravel the network and understand patterns of life.

The overall concept of the Insight program is highlighted in the process diagram in Figure 14.12. The process demonstrates how multiple sources fuse information including threat network information, tracks, and reports. A key aspect of the program is the ability to passively and actively update uncertainties about object state.

Methods like Bayesian fusion and Dempster-Shafer theory are used to combine new information inputs from steps 3, 4, 7, and 8. Steps 2 and 6 involve feedback to the collection system based on correlation and analysis to obtain additional sensor-derived information to update object states and uncertainties. The ambitious program seeks to "automatically exploit and cross-cue multi-INT sources" to improve decision timelines and automatically collect the next most important piece of information to improve object tracks, reduce uncertainty, or anticipate likely courses of action based on models of the threat and network.

BAE Systems and Science Applications International Corp. (SAIC) "developed technologies that combine intelligence information from imaging sensors, crowd-source and other social network or text-based sensors, and other

3. Level 3 fusion (threat prediction) is not shown in this model, although probabilistic forecasts or model-based behavior prediction could be added to anticipate likely future locations and to characterize possible threats. See Chapter 16 for more information.

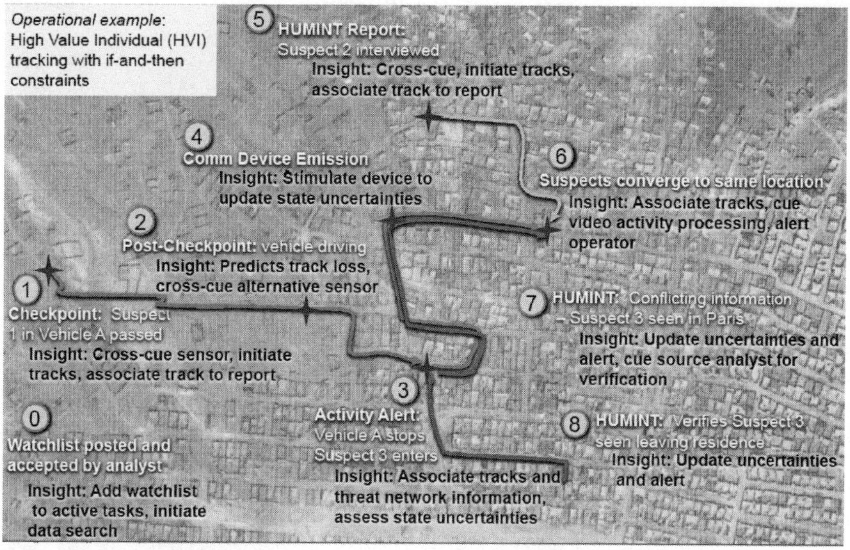

Figure 14.12 Concept of multi-INT fusion and association for DARPA's Insight program. (Source: Insight Industry Day Briefing, 21–22 September 2010. Approved for public release, distribution unlimited [31, p. 2].)

sources for further analysis, and cross-cue different intelligence sources automatically" under a \$79-million DARPA contract awarded in 2013 [32]. An example of the human-computer interface (an example of level 5 fusion) that integrates these tools is shown in Figure 14.13.

Insight also includes "detection and identification of threats through the use of behavioral discovery and prediction algorithms" including model-based correlation and network analysis tools that automatically fuse information and present it to the user [34]. The program developed a prototype graphical user interface and sought feedback from experienced multi-INT/ABI analysts on its application in an array of multisensor problems. They also performed unique, multisensor instrumented data collection exercises over the National Training Center (NTC) in Ft. Irwin, CA [31]. Insight is an ongoing development program that seeks to transition advanced multi-INT association and exploitation capabilities into existing programs of record, including the Army's Distributed Common Ground System (DCGS-A).

14.6 Summary

Analysts practice correlation and fusion in their workflows—the art of multi-INT. However, there are numerous mathematical techniques for combining information with quantifiable precision. Uncertainty can be propagated through

Figure 14.13 Example of the human-machine multisource fusion environment for DARPA's Insight program. (Image source: DARPA [33].)

multiple calculations, giving analysts a hard, mathematically rigorous measurement of multisource data. The art and science of correlation do not play well together, and the art often wins over the science. Most analysts prefer to correlate information they "feel" is related. Efforts to integrate structured mathematical techniques with the human-centric process of developing judgments must be developed. Hybrid techniques that quantify results with science but leave room for interpretation may advance the tradecraft but are not widely used in ABI today.

References

[1] Hume, D., *An Enquiry Concerning Human Understanding,* 1748.

[2] Silver, N., *The Signal and the Noise: Why So Many Predictions Fail—But Some Don't,* New York: Penguin, 2012.

[3] Kahneman, D., *Thinking, Fast and Slow,* London: Farrar, Straus, and Giroux, 2011.

[4] Taleb, N. N., *The Black Swan: The Impact of the Highly Improbable* (2nd ed.), New York: Random House, 2010.

[5] Tetlock, P. E., *Expert Political Judgment: How Good Is It? How Can We Know?* Princeton, New Jersey: Princeton University Press, 2006.

[6] "Definition: Fusion." Google.com.

[7] Waltz, E., and J. Llinas, *Multisensor Data Fusion,* Norwood, MA: Artech House, 1990.

[8] Hall, D. L., and J. Llinas, "Multisensor Data Fusion," in *Handbook of Multisensor Data Fusion, Theory and Practice* (eds. M. E. Liggins, D. L. Hall, and J. Llinas), Boca Raton, FL: CRC Press. 2008.

[9] Engle, M., S. Sarkani, and T. Mazzuchi, "Developing a Model for Simplified Higher Level Sensor Fusion," *Crosstalk,* February 2013, pp. 20–24,.

[10] Steinberg, A. N., and C. L. Bowman, "Revisions to the JDL Data Fusion Model," in *Handbook of Multisensor Data Fusion, Theory and Practice,* Boca Raton, FL: CRC Press, 2008.

[11] Hall, D. L., and J. Llinas, "An Introduction to Multisensor Data Fusion," *Proceedings of the IEEE,* Vol. 85, No. 1, January 1997,pp. 6–23.

[12] Hall, D. L., and A. N. Steinberg, "Dirty Secrets in Multisensor Data Fusion," DTIC ADA392879.

[13] "Sentient Enterprise Request for Information," National Reconnaissance Office, October 20, 2010, web. Available: https://www.fbo.gov/

[14] Bottomley, G. E., et al., "The Potential Gains of Upstream Fusion for SIGINT and MOVINT Data (Unclassified)," presented at the *Science of Multi-Intelligence (SOMI) Workshop,* Chantilly, VA, September 10, 2014.

[15] Kahneman, D., and A. Tversky, *Subjective Probability: A Judgment of Representativeness," in Judgment Under Uncertainty: Heuristics and Biases,* Cambridge, U.K.: Cambridge University Press, 1972.

[16] Zlotnick, J., "A Theorem for Prediction," *Studies in Intelligence,* Vol. 11, No. 4, 1967.

[17] Zlotnick, J., "Bayes' Theorem for Intelligence Analysis," *Studies in Intelligence,* Vol. 16, No. 2, 1972.

[18] Shafer, G., *A Mathematical Theory of Evidence,* Princeton, NJ: Princeton University Press, 1976.

[19] Dempster, A. P., "A Generalization of Bayesian Inference," *Journal of the Royal Statistical Society,* Series B, Vol. 30, 1968, pp. 205–247.

[20] Shafer, G., "Dempster-Shafer Theory," web. Available: http://www.corpsriskanalysisgateway.us/data/docs/dempster.pdf.

[21] "Dempster–Shafer Theory," Wikipedia.

[22] "Bayesian Network," Wikipedia.

[23] Pool, R., ed., *Field Evaluation in the Intelligence and Counterintelligence Context: Workshop Summary,* Washington, DC: National Academies Press, 2010.

[24] Sticha, P., D. Buede, and R. L. Rees, "APOLLO: An Analytical Tool for Predicting a Subject's Decision Making," presented at the *International Conference on Intelligence Analysis Methods and Tools,* McLean, VA, May 2, 2005.

[25] Satellite Imaging Corporation, "Worldview-3," web. Available: http://www.satimagingcorp.com/satellite-sensors/worldview-3/.

[26] "Unified Research on Network-Based Hard/Soft Information Fusion," Multidisciplinary University Research Initiative (MURI) Grant (Number W911NF-09-1-0392) by the US Army Research Office (ARO) to University at Buffalo (SUNY) and partner institutions.

[27] Llinas, J., R. Nagi, and J. Lavery, "A Multi-Disciplinary University Research Initiative in Hard and Soft Information Fusion: Overview, Research Strategies and Initial Results," presented at the *Proc. of 13th Conference on Information Fusion (FUSION),* Edinburgh, July 2010.

[28] Gross, G. A., et al., "Towards Hard+Soft Data Fusion: Processing Architecture and Implementation for the Joint Fusion and Analysis of Hard and Soft Intelligence Data," in *15th International Conference on Information Fusion,* FUSION 2012, Singapore, July 9–12, 2012, pp. 955–962.

[29] "MURI: Unified Research in Network-Based Hard/Soft Information Fusion," SUNY Buffalo, web.

[30] "Insight Broad Agency Announcement for Information Innovation Office," DARPA. BAA-10-94," September 20, 2010.

[31] Pagels, M., "Insight Industry Day." DARPA Information Innovation Office, September 22, 2010.

[32] "BAE Systems to Help DARPA Unify Imaging and Other Battlefield Intelligence Sensors," *Military & Aerospace Electronics,* web. Available: http://www.militaryaerospace.com/articles/2013/08/darpa-insight-bae.html.

[33] "Insight Program," DARPA, news release, web. Available: http://www.darpa.mil/uploadedImages/Content/NewsEvents/Releases/2013/InsightIllustrationv2.jpg.

[34] "DARPA Expanding Insight Program for Real-Time Analysis of ISR data," *Defense Systems,* December 3, 2012.

15

Knowledge Management

Knowledge is value-added information that is integrated, synthesized, and contextualized to make comparisons, assess outcomes, establish relationships, and engage decision-makers in discussion. Although some texts make a distinction between data, information, knowledge, wisdom, and intelligence, we define knowledge as the totality of understanding gained through repeated analysis and synthesis of multiple souces of information over time. Knowledge is the essence of an intelligence professional and is how he or she answers questions about key intelligence issues. This chapter introduces elements of the wide-ranging discipline of knowledge management in the context of ABI tradecraft and analytic methods. Concepts for capturing tacit knowledge, linking data using dynamic graphs, and sharing intelligence across a complex, interconnected enterprise are discussed.

15.1 The Need for Knowledge Management

Knowledge management is a term that first appeared in the early 1990s, recognizing that the intellectual capital of an organization provided competitive advantage and must be managed and protected. Knowledge management is a comprehensive strategy to get the right information to the right people at the right time so they can do something about it. So-called intelligence failures seldom stem from the inability to collect information, but rather the ability to integrate intelligence with sufficient confidence to make decisions that matter.

According to Dalkir, "Forty-five years ago, nearly half of all workers in industrialized countries were making or helping to make things." By 2000, that proportion had dropped to 20% [1, p. 2]. The intelligence community is a 17-member, \$52.7-billion enterprise [2], with over 10,000 locations involving as many as 854,000 people [3] almost entirely focused on the acquisition, processing, and management of knowledge.

Gartner's Duhon defines knowledge management (KM) as:

> a discipline that promotes an integrated approach to identifying, capturing, evaluating, retrieving, and sharing all of an enterprise's information assets. These assets may include databases, documents, policies, procedures, and previously un-captured expertise and experience in individual workers [4].

This definition frames the discussion in this chapter. The ABI approach treats data and knowledge as an asset—and the principle of data neutrality says that all these assets should be considered as equally "important" in the analysis and discovery process. Some knowledge management approaches are concerned with knowledge capture, that is, the institutional retention of intellectual capital possessed by retiring employees. Others are concerned with knowledge transfer, the direct conveyance of such a body of knowledge from older to younger workers through observation, mentoring, comingling, or formal apprenticeships. Much of the documentation in the knowledge management field focuses on methods for interviewing subject matter experts or eliciting knowledge through interviews. While these are important issues in the intelligence community, "increasingly, the spawning of knowledge involves a partnership between human cognition and machine-based intelligence" [5, p. 4]. This chapter focuses on the subset of knowledge management principles supporting ABI to include machine-readable constructs of knowledge, formal modeling of analytic assumptions, maintaining the provenance of knowledge, and collaborating to enhance the collective knowledge of a large-scale distributed organization.

15.1.1 Types of Knowledge

Knowledge is usually categorized into two types, explicit and tacit knowledge (Table 15.1). Explicit knowledge is that which is formalized and codified. This

Table 15.1

Comparison of the Properties of Explicit Versus Tacit Knowledge

Properties of Explicit Knowledge	Properties of Tacit Knowledge
Facts, know-what	Expertise, know-how and know-why
Schematized, documented, organized, and systematized	Abstract, difficult to describe but you can use the knowledge unconsciously when you need to
Only known when it is needed to be known	Intrinsic in everyday workflows
Conveyed through memorization, reporting, query, search, verbal communication, presentations, and lectures	Conveyed through experiential transfer or one-on-one coaching, observation, imitation, or practice

is sometimes called "know what" and is much easier to document, store, retrieve, and manage. Knowledge management systems that focus only on the storage and retrieval of explicit knowledge are more accurately termed information management systems, as most explicit knowledge is stored in databases, memos, documents, reports, notes, and digital data files.

Tacit knowledge is intuitive knowledge based on experience, sometimes called "know-how." Tacit knowledge is difficult to document, quantify, and communicate to another person. This type of knowledge is usually the most valuable in any organization. Lehaney notes that the only sustainable competitive advantage and "the locus of success in the new economy is not in the technology, but in the human mind and organizational memory" [6, p. 14]. Tacit knowledge is intensely contextual and personal. Most people are not aware of the tacit knowledge they inherently possess and have a difficult time quantifying what they "know" outside of explicit facts.

In the intelligence profession, explicit knowledge is easily documented in databases, but of greater concern is the ability to locate, integrate, and disseminate information held tacitly by experienced analysts. ABI requires analytic mastery of explicit knowledge about an adversary (and his or her activities and transactions) but also requires tacit knowledge of an analyst to understand why the activities and transactions are occurring.

While many of the methods in this textbook refer to the physical manipulation of explicit data, it is important to remember the need to focus on the "art" of multi-INT spatiotemporal analytics. Analysts exposed to repeated patterns of human activity develop a natural intuition to recognize anomalies and understand where to focus analytic effort. Sometimes, this knowledge is problem-, region-, or group-specific. More often than not, individual analysts have a particular knack or flair for understanding activities and transactions of certain types of entities. Often, tacit knowledge provides the turning point in a particularly difficult investigation or unravels the final clue in a particularly difficult intelligence mystery, but according to Meyer and Hutchinson, individuals tend to place more weight on concrete and vivid information over that which is intangible and ambiguous [7, p. 46]. Effectively translating ambiguous tacit knowledge like feelings and intuition into explicit information is critical in creating intelligence assessments. This is the primary paradox of tacit knowledge; it often has the greatest value but is the most difficult to elicit and apply.

Amassing facts rarely leads to innovative and creative breakthroughs. The most successful organizations are those that can leverage both types of knowledge for dissemination, reproduction, modification, access, and application throughout the organization.

15.2 Discovery of What We Know

Chapter 10 introduced the concept of "big data," and daily news reports remind us that voluminous digital detritus larger than the Library of Congress is created faster than you can read this sentence. However, the problem of information organization and discovery is not new, writes George Wright in a 1958 edition of *Studies in Intelligence:*

> The problem of storing an ever mounting accumulation of raw intelligence information and maintaining ready access to assorted needles in this haystack is one of the most baffling in the whole field of intelligence management. It is particularly difficult in CIA, where it is necessary to provide community-wide reference services and where no categories of data are excluded from the collection. The problem has been attacked manfully and partial solutions have been achieved; but these solutions have not kept pace with the growing mountain of documents and the sharpened requirements of intelligence analysts. CIA analysts still fall more or less frustrated between the impossibility of keeping adequate personal files and the deficiencies of the central reference service [8].

As the amount of information available continues to grow, knowledge workers spend an increasing amount of their day messaging, tagging, creating documents, searching for information, and performing queries and other information-focused activities [9, p. 114]. New concepts are needed to enhance discovery and reduce the entropy associated with knowledge management.

15.2.1 Recommendation Engines

Amazon.com, a multibillion-dollar web retailer of books, movies, electronics, clothing, and digital content popularized the concept of the algorithmic recommendation engine for online shopping. While the details of Amazon.com's algorithms are a closely guarded trade secret, a 2001 patent filing for "Personalized Recommendations of Items Represented Within a Database" discloses the foundational method of Amazon's techniques: content-based filtering and collaborative filtering [10].

Content-based filtering identifies items based on an analysis of the item's content as specified in metadata or description fields. Parsing algorithms extract common keywords to build a profile for each item (in our case, for each data element or knowledge object). Content-based filtering systems generally do not evaluate the quality, popularity, or utility of an item. "Content-based systems tend to be poorly suited for recommending movies, music titles, authors, restaurants, and other types of items that have little or no useful, parsable content" [10]. This may hold true for geospatially enabled data sets or other intelligence databases about entities that contain little parsable content about the intrinsic value of the data. Content-based filtering is a technique for identifying items with similar content.

In *collaborative filtering,* "items are recommended based on the interests of a community of users without any analysis of item content" [10]. Collaborative filtering ties interest in items to particular users that have rated those items. This technique is used to identify similar users: the set of users with similar interests. In the intelligence case, these would be analysts with an interest in similar data.

The Amazon.com recommendation system combines content-based filtering and collaborative filtering to increase the accuracy and lower the false alarm rate of its predictions. This process is similar to the way multiple ISR sensors are used to increase the probability of detection while lowering false alarm rate. A key to Amazon's technology is the ability to calculate the related item table offline, storing this mapping structure, and then efficiently using this table in real time for each user based on current browsing history. This process is described in Figure 15.1. Items the customer has previously purchased, favorably reviewed, or items currently in the shopping cart are treated with greater affinity than items browsed and discarded. The "gift" flag is used to identify anomalous items purchased for another person with different interests so these purchases do not skew the personalized recommendation scheme.

In ABI knowledge management, knowledge about entities, locations, and objects is available through object metadata. Content-based filtering identifies similar items based on location, proximity, speed, or activities in space and time. Collaborative filtering can be used to discover analysts working on similar problems based on their queries, downloads, and exchanges of related content. This is an internal application of the "who-where" tradecraft, adding additional metadata into "what" the analysts are discovering and "why" they might need it.

15.2.2 Data Finds Data

An extension of the recommendation engine concept to next-generation knowledge management is an emergent concept introduced by Jeff Jonas and Lisa Sokol called "data finds data." In contrast to traditional query-based information systems, Jonas and Sokol posit that if the system knew what the data meant, it would change the nature of data discovery by allowing systems to find related data and therefore interested users. They explain:

> With interest in a soon-to-be-released book, a user searches Amazon.com for the title, but to no avail. The user decides to check every month until the book is released. Unfortunately for the user, the next time he checks, he finds that the book is not only sold out but now on back order, awaiting a second printing. When the data finds the data, the moment this book

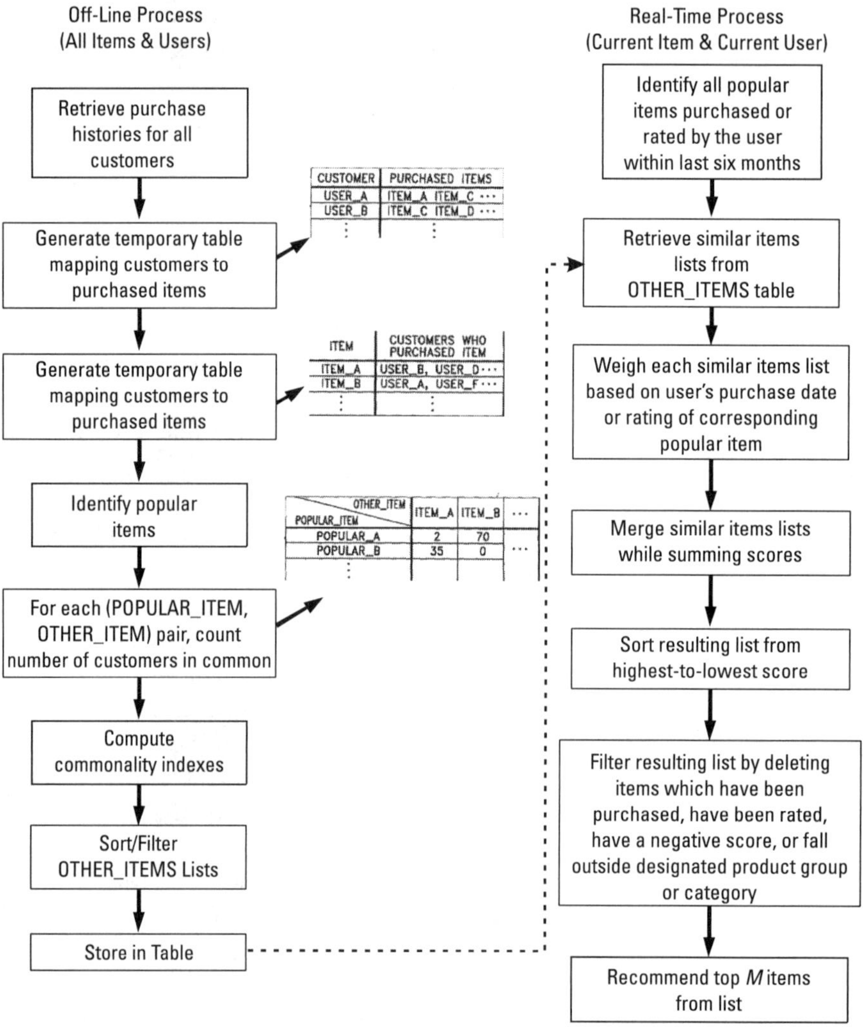

Figure 15.1 Process diagram for a recommendation engine. (Adapted from [10].)

is available, this data point will discover the user's original query and automatically email the user about the book's availability [11].

Jonas, now chief scientist at IBM's entity analytics group joined the firm after Big Blue acquired his company, SRD, in 2005. SRD developed data accumulation and alerting systems for Las Vegas casinos including non-obvious relationship analysis (NORA), famous for breaking the notorious MIT card counting ring in the best-selling book *Bringing Down the House* [12]. He postulates that knowledge-rich but discovery-poor organizations derive increasing wealth from connecting information across previously disconnected information silos using real-time "perpetual analytics" [13]. Instead of processing data using large bulk algorithms, each piece of data is examined and correlated on ingest for its relationship to all other accumulated content and knowledge in the system. Such a context-aware data ingest system is a computational embodiment of integrate before exploit, as each new piece of data is contextualized, sequence-neutrally of course, with the existing knowledge corpus across silos. Jonas says, "If a system does not assemble and persistent context as it comes to know it... the computational costs to reconstruct context after the facts are too high" [14].

Jonas elaborated on the implication of these concepts in a 2011 interview after the fact: "There aren't enough human beings on Earth to think of every smart question every day... every piece of data is the question. When a piece of data arrives, you want to take that piece of data and see how it relates to other pieces of data. It's like a puzzle piece finding a puzzle" [15, 16]. Treating every piece of data as the question means treating data as queries and queries as data.

15.2.3 Queries as Data

Information requests and queries are themselves a powerful source of data that can be used to optimize knowledge management systems or assist the user in discovering content. Former Google chief information officer (CIO) and vice-president of engineering Doug Merrill explained how Google uses queries to produce a massively scalable, self-correcting search engine in the following example adapted from a 2007 lecture [17]:

- An anonymous user types the word "inetligence" into Google's search engine and presses "return."
- The user views the returned results and does not click any links.
- The user then (within five seconds) types "intelligence" into the search box and presses "return."
- The user views the returned results and clicks the fifth suggestion on the page.

With this approach, Google does not need to know anything about the language or content, it simply mines the activities and transactions of the requestor and stores these results in a related data table that is continuously mined for correlations. Through this type of correlation, a knowledge management system can typify the requestor and continually improve the query results using the user actions resultant from "incorrect" queries. With enough users and a diverse base of queries, a contextually aware query system naturally evolves.

Similar techniques are becoming widespread as human-machine interfaces that understand the nature and context of user interactions become feasible.

15.3 The Semantic Web

The semantic web is a proposed evolution of the World Wide Web from a document-based structured designed to be read by humans to a network of hyperlinked, machine-readable web pages that contain metadata about the content and how multiple pages are related to each other. The semantic web is about relationships.

Although the original concept was proposed in the 1960s, the term "semantic web" and its application to an evolution of the Internet was popularized by Tim Berners-Lee in a 2001 article in *Scientific American.* He posits that the semantic web "will usher in significant new functionality as machines become much better able to process and 'understand' the data that they merely display at present" [18].

The semantic web is based on several underlying technologies, but the two basic and powerful ones are the extensible markup language (XML) and the resource description framework (RDF).

15.3.1 XML

XML is a World Wide Web Consortium (W3C) standard for encoding documents that is both human-readable and machine-readable [19]. Internet web pages are defined using the hypertext markup language (HTML), which describes page styles and layouts and how to handle hyperlinks to other documents. HTML is simple but the tag set cannot be extended. Also, it only describes how pages are rendered for a human user and does not provide metadata about the content expect for <keyword> tags. The syntax of XML is:

```
<tag attribute-name="attribute-value"></tag>
```

which essentially describes a document as a series of key-value pairs (attribute-name; attribute-value). Tags are case-sensitive. Attribute values must always appear in quotes. Content associated with each tag appears between the brackets. An XML description of this book follows:

```
<book category="engineering">
 <title lang="en">Activity-Based Intelligence: Principles and
Applications</title>
 <author>Patrick Biltgen</author>
 <author>Stephen Ryan</author>
 <year>2015</year>
 <publisher>Artech House</publisher>
</book>
```

XML can also be used to create relational structures. Consider the example shown below where there are two data sets consisting of locations (locationDetails) and entities (entityDetails) (adapted from [20]):

```
<locationDetails>
 <location ID="L1">
   <cityName>Annandale</cityName>
   <stateName>Virginia</stateName>
 </location ID>
 <location ID="L2">
   <cityName>Los Angeles</cityName>
   <stateName>California</stateName>
 </location ID>
</locationDetails>
<entityDetails>
 <entity locationRef="L1">
   <entityName>Patrick Biltgen</entityName>
 </entity>
 <entity locationRef="L2">
   <entityName>Stephen Ryan</entityName>
 </entity>
</entityDetails>
```

Instead of including the location of each entity as an attribute within entityDetails, the structure above links each entity to a location using the attribute locationRef. This is similar to how a foreign key works in a relational database. One advantage to using this structure is that the two entities can be linked to multiple locations, especially when their location is a function of time and activity.

XML is a flexible, adaptable resource for creating documents that are context-aware and can be machine parsed using discovery and analytic algorithms.

15.3.2 Resource Description Framework (RDF)

The RDF, a standard developed by the W3C, expresses meaning and defines relationships in the semantic web. RDF uses XML as an underlying representation, although it does not necessarily depend on XML syntax. RDF is often depicted as a graph of subject-predicate-object as shown in Figure 15.2.

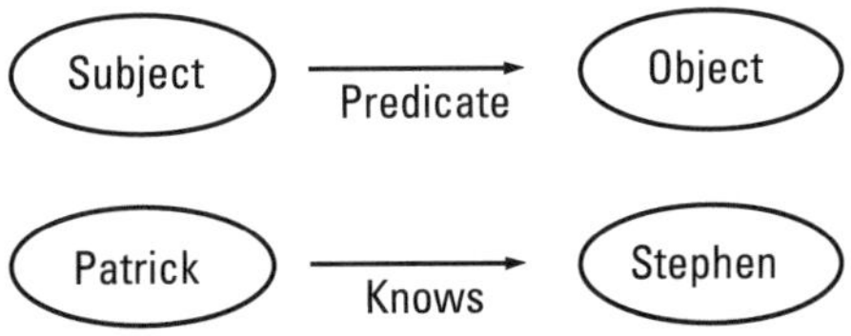

Figure 15.2 Example of a semantic triple.

Alternatively, the triple relationship shown at the bottom of Figure 15.2, “Patrick |knows| Stephen,” can be written in machine-readable triples as:

```
_:a <http://xmlns.com/foaf/0.1/name> "Patrick" .
_:a <http://xmlns.com/foaf/0.1/knows> _:b .
_:b <http://xmlns.com/foaf/0.1/name> "Stephen" .
```

or using the W3C Turtle nomenclature as:

```
@prefix foaf: <http://xmlns.com/foaf/0.1/>


[ foaf:name "Patrick" ] foaf:knows [
 foaf:name "Stephen" ];
 foaf:mbox <Steve@activitybasedintelligence.com> ] .
```

The FOAF vocabulary for RDF “began as the ‘RDFWeb’ project and established a widely adopted model for publishing simple factual data via a network of linked RDF documents” [21]. RDF provides a practical, standards-based construct for machine readable knowledge architectures. In practice, these enable advanced “knowledge agents,” which are human-machine assisted reasoning software tools that help humans more rapidly discover and correlate related information to improve the value of decision making.

15.4 Graphs for Knowledge and Discovery

Graphs are a mathematical construct consisting of nodes (vertices) and edges that connect them. Many real-world problems can be represented by graphs and analyzed using the discipline of graph theory. In information systems, graphs represent communications, information flows, library holdings, data models, or the relationships in a semantic web. In intelligence, graph models are used to represent processes, information flows, transactions, communications networks, order-of-battle, terrorist organizations, financial transactions, geospatial networks, and the pattern of movement of entities. Because of their widespread applicability and mathematical simplicity, graphs provide a powerful construct for ABI analytics.

Graphs are drawn with dots or circles representing each node and an arc or line between nodes to represent edges as shown in Figure 15.3. Directional graphs use arrows to depict the flow of information from one node to another. When graphs are used to represent semantic triplestores, the direction of the arrow indicates the direction of the relationship or how to read the simple sentence. Figure 5.8 introduced a three-column framework for documenting facts, assessments, and gaps. This information is depicted as a knowledge graph in Figure 15.3. Black nodes highlight known information, and gray nodes depict knowledge gaps. Arrow shading differentiates facts from assessments and

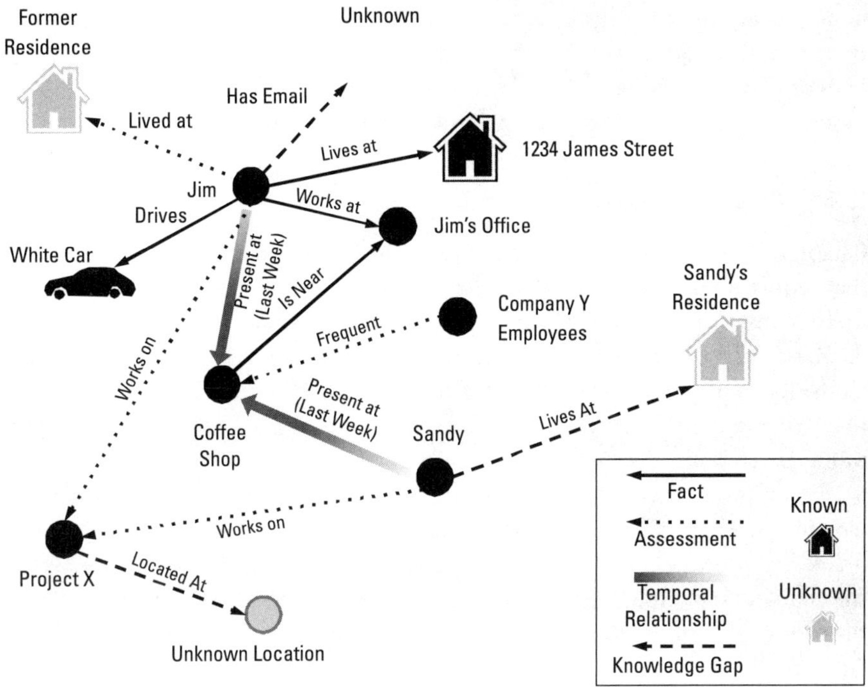

Figure 15.3 Example of a knowledge graph.

gaps. Shaded lines show information with a temporal dependence like the fact that Jim used to live somewhere else (a knowledge gap because we don't know where). Implicit relationships can also be documented using the knowledge graph: Figure 5.8 contains the fact "the coffee shop is two blocks away from Jim's office." The semantic triple "Jim | Works At | Jim's Office" is implicit from the existence of an entity called "Jim's Office."

The knowledge graph readily depicts knowns and unknowns. Because this construct can also be depicted using XML tags or RDF triples, it also serves as a machine-readable construct that can be passed to algorithms for processing. Graphs are useful as a construct for information discovery when the user doesn't necessarily know the starting point for the query. By starting at any node (a tip-off), an analyst can traverse the graph to find related information. This workflow is called "know-something-find-something." A number of heuristics for graph-based search assist in the navigation and exploration of large, multidimensional graphs that are difficult to visualize and navigate manually.

Deductive reasoning techniques integrate with graph analytics through manual and algorithmic filtering to quickly answer questions and convey knowledge from the graph to the human analyst. Analysts filter by relationship type, time, or complex queries about the intersection between edges and nodes to rapidly identify known information and highlight knowledge gaps. Increasingly, graph analytic capabilities have been integrated into commercial network analysis packages like BAE Systems NetReveal, Palantir, IBM Analyst Notebook, Semantica Professional, and dozens of other tools.

15.4.1 Graphs and Linked Data

Chapter 10 introduced graph databases as a NoSQL construct for storing data that requires a flexible, adaptable schema. Graphs—and graph databases—are a useful construct for indexing intelligence data that is often held across multiple databases without requiring complex table joins and tightly coupled databases. Consider two graph stores, G and S for GEOINT and SIGINT data respectively that contain intelligence "objects" organized as graphs γ_1 and γ_2 as shown in Figure 15.4.

γ_1 represents GEOINT information about an ongoing intelligence problem focused around object *C*. Information in the GEOINT graph store is linked to multiple raw data stores held at the agency level, including a report store and a map store. Object *C* is linked to both a report and a map. Examine the four-node graph, γ_2, which is a SIGINT analyst's graph object view of a related intelligence problem comprised of different objects D, E, F, and H. These objects also link to agency-level data holdings using graph relations (edges). A triple relationship is created between objects *C* and *D* reflecting that "*C* knows *D*." This relationship is stored in a relationship store at the enterprise-wide sharing level, but it documents the relationship between two objects in agency-level graph stores and therefore the linkages through the graph to the raw data sources in five agency-specific databases.

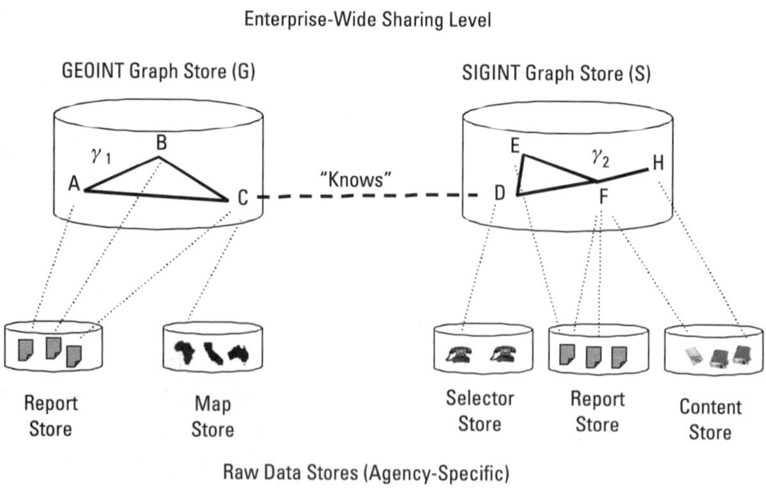

Figure 15.4 Using graphs for information sharing.

Using linked data, an analyst working issue "*C*" can quickly discover the map and report directly connected to *C*, as well as the additional reports linked to related objects. *C* can also be packaged as a "super object" that contains an instance of all linked data with some number of degrees of separation—calculated by the number of graph edges—away from the starting object. The super object is essentially a stack of relationships to the universal resource identifiers (URIs) for each of the related object, documented using RDF triples or XML tags.

15.4.2 Provenance

Provenance is the chronology of the creation, ownership, custody, change, location, and status of a data object. The term was originally used in relation to works of art to "provide contextual and circumstantial evidence for its original production or discovery, by establishing, as far as practicable, its later history, especially the sequences of its formal ownership, custody, and places of storage" [22]. In law, the concept of provenance refers to the "chain of custody" or the paper trail of evidence. This concept logically extends to the documentation of the history of change of data in a knowledge system.

The W3C implemented a standard for provenance in 2013, documenting it as "information about entities, activities, and people involved in producing a piece of data or thing, which can be used to form assessments about its quality, reliability or trustworthiness" [23]. The PROV-O standard is web ontology language 2.0 (OWL2) ontology that maps the PROV logical data model to RDF [24]. The ontology describes hundreds of classes, properties, and attributes.

Maintaining provenance across a knowledge graph is critical to assembling evidence against hypotheses. Each analytic conclusion must be traced back to each piece of data that contributed to the conclusion. Although ABI methods can be enhanced with automated analytic tools, analysts at the end of the decision chain need to understand how data was correlated, combined, and manipulated through the analysis and synthesis process. Ongoing efforts across the community seek to document a common standard for data interchange and provenance tracking.

15.4.3 Using Graphs for Multianalyst Collaboration

In the legacy, linear TCPED model, when two agencies wrote conflicting reports about the same object, both reports promulgated to the desk of the all-source analyst. He or she adjudicated the discrepancies based on past experience with both sources. Unfortunately, the incorrect report often persisted—to be discovered in the future by someone else—unless it was recalled. Using the graph construct to organize around objects makes it easier to discover discrepancies that can be more quickly and reliably resolved. When analyzing data spatially, these discrepancies are instantly visible to the all-source analyst because the same object simultaneously appears in two places or states. Everything happens somewhere, and everything happens in exactly one place.

15.5 Information and Knowledge Sharing

Intelligence Community Directive Number 501, Discovery and Dissemination or Retrieval of Information Within the Intelligence Community, was signed by the DNI on January 21, 2009. Designed to "foster an enduring culture of responsible sharing within an integrated IC," the document introduced the term "responsibility to provide" and created a complex relationship with the traditional mantra of "need to know" [25]. It directed that all authorized information be made available and discoverable "by automated means" and encouraged data tagging of mission-specific information. ICD 501 also defined "stewards" for collection and analytic production as:

> An appropriately cleared employee of an IC element, who is a senior official, designated by the head of that IC element to represent the [collection/analytic] activity that the IC element is authorized by law or executive order to conduct, and to make determinations regarding the dissemination to or the retrieval by authorized IC personnel of [information collected/analysis produced] by that activity [25].

With a focus on improving discovery and dissemination of information, rather than protecting or hoarding information from authorized users, data stewards gradually replace data owners in this new construct. The data doesn't belong to a person or agency. It belongs to the intelligence community. When applied, this change in perspective has a dramatic impact on the perspectives of information.

According to the 9/11 Commission, "the biggest impediment to all-source analysis—to a greater likelihood of connecting the dots—is the human or systemic resistance to sharing information" [26, p. 416]. Because the intelligence community is a federation of experts, many intelligence analysts see their knowledge as a source of power or job security, creating a natural reluctance to share. Kilawada and Holtshouse believe knowledge sharing needs a cultural component: "Companies need to create an environment that is conducive to sharing and then support it with strong technology and improved work processes" [27]. Prusak notes that knowledge "clumps" in groups and connectivity of individuals into groups and networks wins over knowledge capture [28, p. 154]. Organizations that promote the development of social networks and the free exchange of information witness the establishment of self-organizing knowledge groups. Bahra says that there are three main conditions to assist in knowledge sharing [29, p. 56]:

- Reciprocity: One helps a colleague, thinking that he or she will receive valuable knowledge in return (even in the future).
- Reputation: Reputation, or respect for one's work and expertise, is power, especially in the intelligence community.
- Altruism: Self-gratification and a passion or interest about a topic.

These three factors contribute to the simplest yet most powerful transformative sharing concepts in the intelligence community.

15.6 Wikis, Blogs, Chat, and Sharing

The historical compartmented nature of the intelligence community and its "need to know" policy is often cited as an impetus to information sharing. In 2005, CIA scientist D. Calvin Andrus published "The Wiki and the Blog: Toward a Complex Adaptive Intelligence Community," which postulated that "the intelligence community must be able to dynamically reinvent itself by continuously learning and adapting as the national security environment changes" [30, p. 12]. Andrus proposed information sharing and independent, self-organized action as a mechanism to enable this transformation, and introduced the intelligence community to "a new generation of Internet tools" including the Wiki and the blog.

Andrus's essay won the intelligence community's Galileo Award and was partially responsible for the start-up of a classified Wiki based on the platform and structure of Wikipedia called Intellipedia [31]. Shortly after its launch, the tool was used to write a high-level intelligence assessment on Nigeria. Thomas Fingar, the former deputy director of National Intelligence for Analysis (DDNI/A) cited Intellipedia's success in rapidly characterizing Iraqi insurgents' use of chlorine in improvised explosive devices highlighting the lack of bureaucracy inherent in the self-organized model [32]. Former director of national intelligence J. M. McConnell cited the tool during a congressional hearing in 2007, saying, "Analysts are also increasingly using interactive, classified blogs and wikis, much as the tech-savvy, collaboration minded user would outside the community…Such tools enable experts from different disciplines to pool their knowledge, form virtual teams, and quickly make complete intelligence assessments" [33]. As of January 2014, the U.S. government's top-secret Intellipedia has 113,379 content pages with 255,402 users, 290,355,786 page views, and 6,216,642 edits [34].

While Intellipedia is the primary source for collaborative, semiformalized information sharing on standing and emergent intelligence topics, most analysts collaborate informally using a combination of chat rooms, Microsoft

SharePoint sites, and person-to-person chat messages. In 2007, Denver-based Jabber won a $22-million contract to deploy a commercial instant messaging (IM) server platform for the Department of Defense. At the time, Jabber (now Cisco) said that "its Extensible Communications Platform is already widely deployed throughout the federal government, including the Defense and Homeland Security departments and the U.S. intelligence community" [35]. By 2009, intelligence agency employees were exchanging about five million instant messages per day [36]. Many analysts prefer chat to more formalized communication mechanisms because it is rapid, easy, and point-to-point. Knowledge transfer is fundamentally a social, human process that occurs most successful via peer-to-peer relationships [29, p. 152].

Because the ABI tradecraft reduces the focus on producing static intelligence products to fill a queue, ABI analysts tend to collaborate and share around in-work intelligence products. These include knowledge graphs on adversary patterns of life, shape file databases, and other in-work depictions that are often not suitable as finished intelligence products. In fact, the notion of "ABI products" is a source of continued consternation as standards bodies attempt to define what is new and different about ABI products, as well as how to depict the dynamics of human patterns of life on what is often a static Powerpoint chart.

Managers like reports and charts as a metric of analytic output because the total number of reports is easy to measure; however, management begins to question the utility of "snapping a chalk line" on an unfinished pattern-of-life analysis just to document a "product." Increasingly, interactive products that use dynamic maps and charts are used for spatial storytelling. Despite all the resources allocated to glitzy multimedia products and animated movies, these products are rarely used because they are time-consuming, expensive, and usually late to need

15.7 Crowdsourcing

Crowdsourcing is "the practice of obtaining needed services, ideas, or content by soliciting contributions from a large group of people and especially from the online community rather than from traditional employees or suppliers" [37]. Amazon.com launched an online crowdsourcing marketplace called Mechanical Turk, which it jokingly refers to as "artificial artificial intelligence." Named after the fake automated chess-playing machine from the 18th century, Amazon's mechanical turk farms Human Intelligence Tasks (HITs) to human workers who are paid to complete the task. Mechanical Turk gives "businesses and developers access to an on-demand, scalable workforce. Workers select from thousands of tasks and work whenever it's convenient" [38]. Participants in

Amazon's HITs receive some miniscule compensation each task completed, but most crowdsourcing platforms do not pay participants. Examples of some of the 208,000 HITs available to workers include the following:

- "Select the correct spelling for these search terms."
- "Is this web site suitable for a general audience?"
- "Find the item number for the product in this image."
- "Are these two products the same?"
- "Translate a paragraph from English to French."

Crowdsourcing large-scale, simple, menial tasks to a pool of workers supports ABI by farming out classification and identification tasks that confound automated algorithms. In 2012, the European Space Agency (ESA) installed NightPod, a motorized tripod that compensates for the motion of the space station to take automated pictures of the Earth. Currently, NASA's Johnson Space Center has over 1.2-million images taken from the International Space Station. While the motion-compensating NightPod produces clear imagery, the geolocation of the images is poor, limiting their usefulness for science [39]. A project called Image Detective automatically classifies and georeferences daytime images based on feature correlation, but the technique does not work on nighttime pictures (about 30% of the collection) because the algorithm is easily confused between cities, stars, and other objects like the moon [40]. "Anyone can help," says Alejandro Sanchez, a Ph.D. student at UCM. "Humans are much more efficient for complex image analysis" [39]. As of September 2014, the Cities at Night crowdsourced project completed 85,651 tasks (80%) with contributions from 14,387 volunteers [41]. For comparison, the Johnson Space Center employs a total of around 15,000 civil servants and contractors [42].

Because crowdsourced tasking generally relies on an inexperienced set of workers, the false alarm rate can be high. In practice, giving the same task to multiple workers and comparing the results easily remedies this. If multiple workers all agree on a judgment, the result is saved with high confidence. Conflicting results are identified and passed too more experienced experts for a final decision.

In addition to GEOINT exploitation tasks like identifying objects in imagery and georeferencing space-based photos, crowdsourcing is a useful concept for data conditioning. For example, a crowdsourcing platform could be used to clean up transaction data, stitch tracks, or geocode text files. Because some automated algorithms like entity extractors generate ambiguous results, crowdsourcing could be used to disambiguate entities and locations. Because many of the menial tasks do not require access to complete sources and methods—only snippets of information—in some cases, subsets of intelligence data could be distributed to the general public and the results reaggregated on a classified information system.

In 2013, Longmont, Colorado–based DigitalGlobe acquired Tomnod, a crowdsourced imagery analysis platform. Tomnod received significant publicity in 2014 during the crowdsourced satellite imagery search for mission Malaysian Airlines flight 370. This case study is discussed in Chapter 20. Crowdsourcing can also be used to aggregate the thoughts and opinions of a diverse user pool toward ideas and judgments on a range of topics. This application is discussed in Chapter 16.

15.8 Summary

Knowledge management is a crucial enabler for ABI because tacit and explicit knowledge about activities, patterns, and entities must be discovered and correlated across multiple disparate holdings to enable the principle of data neutrality. Increasingly, new technologies like graph data stores, recommendation engines, provenance tracing, wikis, and blogs, contribute to the advancement of ABI because they enhance knowledge discovery and understanding. Chapter 16 describes approaches that leverage these types of knowledge to formulate models to test alternative hypotheses and explore what might happen.

References

[1] Dalkir, K., *Knowledge Management in Theory and Practice,* Amsterdam: Butterworth Heinemann/Elsevier, 2005.

[2] "DNI Releases Budget Figure for the 2013 National Intelligence Program," Office of the Director of National Intelligence, October 30, 2013.

[3] Priest, D., and W. M. Arken, "A Hidden World, Growing Beyond Control," *The Washington Post,* July 19, 2010.

[4] Duhon, B., "It's All in Our Heads," *Inform,* Vol. 12, No. 8, September 1998, pp. 8–13,.

[5] Housel, T., and A. H. Bell, *Measuring and Managing Knowledge,* Boston, MA: McGraw-Hill/Irwin, 2001.

[6] Lehaney, B., S. Clark, E. Coakes, and J. Gillian, *Beyond Knowledge Management,* Hershey, PA: Idea Group Publishing, 2004.

[7] Meyer, R. J., and J. W. Hutchinson, "Wharton on Making Decisions," in *Bumbling Geniuses: The Power of Everyday Reasoning in Multistage Decision Making* (eds. Hoch, S. J., H. C. Kunreuther, and R. E. Gunther, Robert E.), New York: Wiley, 2001, p. 350.

[8] Wright, G. W., "Toward a Federal Intelligence Memory," *Studies in Intelligence,* Vol. 2, No. 3, 1958.

[9] Davenport, T. H., *Thinking for a Living: How to Get Better Performance and Results from Knowledge Workers,* Boston, MA: Harvard Business School Press, 2005.

[10] Jacobi, J. A., E. A. Benson, and G. D. Linden, "Personalized Recommendations of Items Represented Within a Database," U.S. Patent No. 7113917, 26 Sep 2006.

[11] Segaran, T., and J. Hammerbacher, *Beautiful Data: the Stories Behind Elegant Data Solutions,* Sebastapol, CA: O'Reilly Media, Inc., 2009.

[12] Malik, Om, "Jeff Jonas Video on How Data Makes Corporations Dumb," GigaOM. Web.

[13] Jonas, J., "What Do You Know? Introducing Perpetual Analytics," February 1, 2006, http/ jeffjonas.typepad.com/jeff_jonas/2006/02/what_do_you_kno.html.

[14] Jonas, J., "Accumulating Context: Now or Never," August 20, 2006.

[15] Jonas, J., "Jeff Jonas Interview Part 2: Data Finds Data," https://www.youtube.com/ watch?v=W3-JrG-_gcE, December 2011.

[16] Pacelli, M. "Jeff Jonas Talks Big Data," *Business Insider,* December 2011.

[17] Merrill, D., "Search 101," presentation at Technical University, Prague, Czech Republic, October 25, 2007, https://www.youtube.com/watch?v+syKY8CrHKcK#4=22033.

[18] Berners-Lee, T., J. Hendler, and O. Lassila, "The Semantic Web," *Scientific American,* May 17,2001.

[19] "XML," Wikipedia.

[20] Foley R., "XML," *Managing Data Exchange,* wikibooks.org.

[21] "FOAF Vocabulary Specification." Web, Available: http://xmlns.com/foaf/spec/.

[22] "Provenance," Wikipedia.

[23] "PROV-Overview." Web. Available: http://www.w3.org/TR/prov-overview/.

[24] "PROV-O: The PROV Ontology." Web. Available: http://www.w3.org/TR/2013/REC-prov-o-20130430.

[25] "Intelligence Community Directive (ICD) 501, Discovery and Dissemination or Retrieval of Information Within the Intelligence Community." Office of the Director of National Intelligence, 21 Jan 2009.

[26] "The 9/11 Commission report: Final Report of the National Commission on Terrorist Attacks upon the United States," Washington, D.C.: U.S. Government Printing Office, 2004.

[27] Kilawada, K., and D. Holtshouse, "The Knowledge Perspective in the Xerox Group," in *Managing Industrial Knowledge: Creation, Transfer, and Utilization* (eds. Nonaka, I., and D. J. Teece), London: Sage Publications, 2001.

[28] Prusak, L., "Practice and Knowledge Management," in *Knowledge Management in the Innovation Process* (eds. de la Mothe, J., and D. Foray), Boston, MA: Kluwer Academic Publishers, 2001, p. 272.

[29] Bahra, N., *Competitive Knowledge Management,* New York: Palgrave, 2001.

[30] Andrus, D. C., "The Wiki and the Blog: Toward a Complex Adaptive Intelligence Community," *Studies in Intelligence,* Vol. 49, No. 3, September 2005.

[31] Shrader, K., "Over 3,600 intelligence professionals tapping into Intellipedia, USATODAY.com," *USA Today,* November 2, 2006.

[32] Losey, S., "U.S. Intel Agencies Modernize Info Sharing," *Defense News,* May 7, 2007.

[33] McConnell, J. M., Director of National Intelligence, Confronting the Terrorist Threat to the Homeland: Six Years after 9/11, 2007.

[34] Smathers, J., "Intellipedia Usage Statistics, FOIA Request. Approved for Release by NSA on 02-05-2014. FOIA Case #76167." .

[35] "Jabber Gets DOD IM Win | Techrockies.com." [Online]. Available: http://www.techrockies.com/jabber-gets-dod-im-win/s-0009850.html. [Accessed: 03-Sep-2014].

[36] Hoover, J. N., "CIA, NSA Adopting Web 2.0 Strategies," *InformationWeek,* March 10, 2009.

[37] "Crowdsourcing— Definition and More from the Free Merriam-Webster Dictionary." [Online]. Available: http://www.merriam-webster.com/dictionary/crowdsourcing. [Accessed: 07-Sep-2014].

[38] Amazon.Com, "Amazon Mechanical Turk - Welcome." [Online]. Available: https://www.mturk.com/mturk/welcome. [Accessed: 07-Sep-2014].

[39] National Aeronautics and Space Administration, "Space Station Sharper Images of Earth at Night Crowdsourced For Science," 14-Aug-2014. [Online]. Available: http://www.nasa.gov/mission_pages/station/research/news/crowdsourcing_night_images/#.VAxkVWRdX9N. [Accessed: 07-Sep-2014].

[40] "crowdcrafting · Project: Dark Skies ISS." [Online]. Available: http://crowdcrafting.org/app/darkskies/. [Accessed: 07-Sep-2014].

[41] Alejandro Sánchez de Miguel, J. G. C., et al., "Atlas of Astronaut Photos, of Earth at Night," *News and Reviews in Astronomy & Geophysics,* Vol. 55, No. 4, August 2014.

[42] National Aeronautics and Space Administration, "About Johnson Space Center: People." [Online]. Available: http://www.nasa.gov/centers/johnson/about/people/index.html. [Accessed: 06-Sep-2014] .

16

Anticipatory Intelligence

After reading chapters on persistent surveillance, big data processing, automated extraction of activities, analysis, and knowledge management, you might be thinking that if we could just automate the steps of the workflow, intelligence problems would solve themselves. Nothing could be further from the truth. In some circles, ABI has been conflated with predictive analytics and automated sensor cross-cueing, but the real power of the ABI method is in producing anticipatory intelligence. Anticipation is about considering alternative futures and what might happen…not what will happen. This chapter describes technologies and methods for capturing knowledge to facilitate exploratory modeling, "what-if" trades, and evaluation of alternative hypotheses.

16.1 Introduction to Anticipatory Intelligence

Anticipatory intelligence is a systemic way of thinking about the future that focuses our long range foveal and peripheral vision on emerging conditions, trends, and threats to national security. Anticipation is not about prediction or clairvoyance. It is about considering a space of potential alternatives and informing decision-makers on their likelihood and consequence. Modeling and simulation approaches aggregate knowledge on topics, indicators, trends, drivers, and outcomes into a theoretically sound, analytically valid framework for exploring alternatives and driving decision advantage. This chapter provides a survey of the voluminous approaches for translating data and knowledge into models, as well as various approaches for executing those models in a data-driven environment to produce traceable, valid, supportable assessments based on analytic relationships, validated models, and real-world data.

Figure 16.1 highlights the hierarchy of analytic functions and their respective level of effort. Chapter 14 introduced this figure and focused on visualization, query/drill down, and statistical analysis. This chapter describes techniques for forecasting and anticipatory analysis that focus on what might happen.

16.1.1 Prediction, Forecasting, and Anticipation

In a quote usually attributed to physicist Neils Bohr or baseball player Yogi Berra, "Prediction is hard, especially about the future." The terms "prediction," "forecasting," and "anticipation" are often used interchangeably but represent significantly different perspectives, especially when applied to the domain of intelligence analysis.

A *prediction* is a statement of what will or is likely to happen in the future. Usually, predictions are given as a statement of fact: "in the future, we will all have flying cars." This statement lacks any estimate of likelihood, timing, confidence, or other factors that would justify the prediction.

Forecasts, though usually used synonymously with predictions, are often accompanied by quantification and justification. Meteorologists generate forecasts: "There is an 80% chance of rain in your area tomorrow."

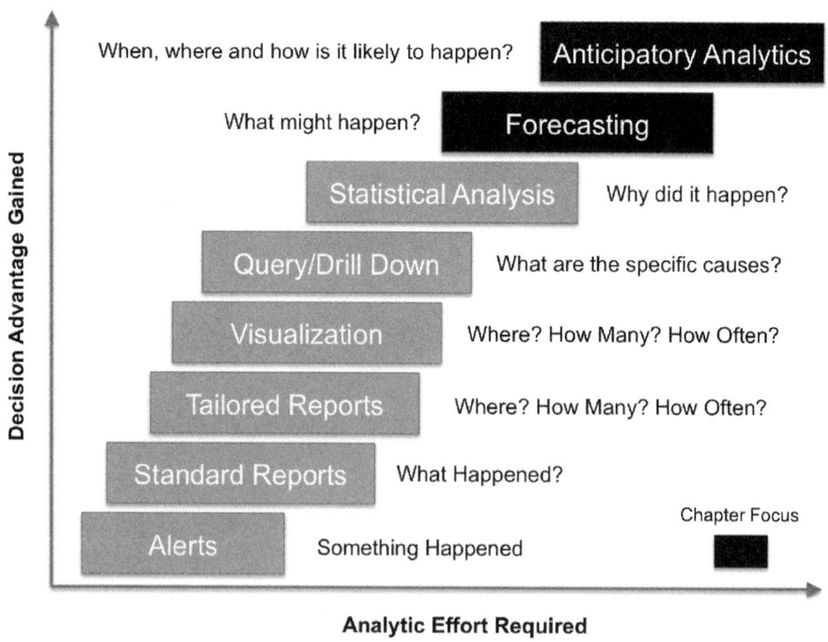

Figure 16.1 Hierarchy of analysis functions and outputs.

Forecasts of the distant future are usually inaccurate because underling models fail to account for disruptive and nonlinear effects. In the 1989 film *Back to the Future: Part II,* protagonist Marty McFly travels 30 years in the future to the year 2015 where he sees flying cars, hoverboards, and three-dimensional phone booths. In the real year 2015, the prediction of flying everything failed to materialize while the writers failed to prognosticate that the suitcase-sized telecommunications devices of the 1980s would become ubiquitous and phone booths would become obsolete.

While predictions are generated by pundits and crystal ball-waving fortune tellers, forecasts are generated analytically based on models, assumptions, observations, and other data.

Anticipation is the act of expecting or foreseeing something, usually with presentiment or foreknowledge. While predictions postulate the outcome with stated or implied certainty and forecasts provide a mathematical estimate of a given outcome, anticipation refers to the broad ability to consider alternative outcomes. Anticipatory analysis combines forecasts, institutional knowledge (see Chapter 15), and other modeling approaches to generate a series of "what if" scenarios. The important distinction between prediction/forecasting and anticipation is that anticipation identifies what may happen. Anticipatory analysis sometimes allows analysis and quantification of possible causes. Sections 16.2–16.6 in this chapter will describe modeling approaches and their advantages and disadvantages for anticipatory intelligence analysts.

16.2 Modeling for Anticipatory Intelligence

Anticipatory intelligence is based on models. Models, sometimes called "analytic models," provide a simplified explanation of how some aspect of the real world works to yield insight. Models may be tacit or explicit. Tacit models are based on knowledge and experience. They exist in the analyst's head and are executed routinely during decision processes whether the analyst is aware of it or not. Explicit models are documented using a modeling language, diagram, description, or other relationship.

16.2.1 Models and Modeling

Waltz describes several approaches for anticipatory modeling in Figure 16.2.

The most basic approach is to construct a model based on relevant context and use the model to understand or visualize a result. Another approach, comparative modeling (2), uses multiple models with the same contextual input data to provide a common output. This approach is useful for exploring alternative hypotheses or examining multiple perspectives to anticipate what may happen and why. A third approach called model aggregation combines

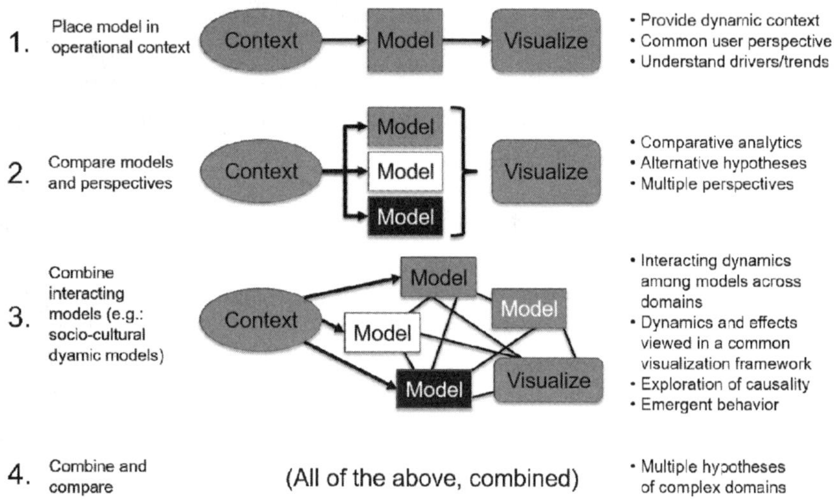

Figure 16.2 Combination and comparison of models in a framework. (Image courtesy of Edward Waltz. Reprinted with permission.)

multiple models to allow for complex interactions. The third approach has been applied to human socio-cultural behavior (HSCB) modeling and human domain analytics on multiple programs over the past 20 years with mixed results (see Section 16.6). Human activities and behaviors and their ensuing complexity, nonlinearity, and unpredictability, represent the most significant modeling challenge facing the community today.

16.2.2 Descriptive Versus Anticipatory/Predictive Models

Descriptive models present the salient features of data, relationships, or processes. They may be as simple as a diagram on a white board or as complex as a wiring diagram for a distributed computer networks. Analysts often use descriptive models to identify the key attributes of a process (or a series of activities). An example of a famous descriptive model, the launch and processing of the space shuttle as described by NASA, is shown in Figure 16.3.

Figure 16.3 describes the process by which the space shuttle is launched, conducts on-orbit operations, re-enters, lands, and is processed for subsequent missions. Each of the steps in the process is associated with observables, objects, and facilities. This type of process is standard for traditional GEOINT analysis as it describes large fixed facilities like refurbishment hangars, launch pads, and the assembly building. Objects like the shuttle, booster recovery ships, and the deployed payload may also be studied. Inductive reasoning allows an analyst to observe any step of the process and infer that precursors must have been present and that future events in the process are likely to occur.

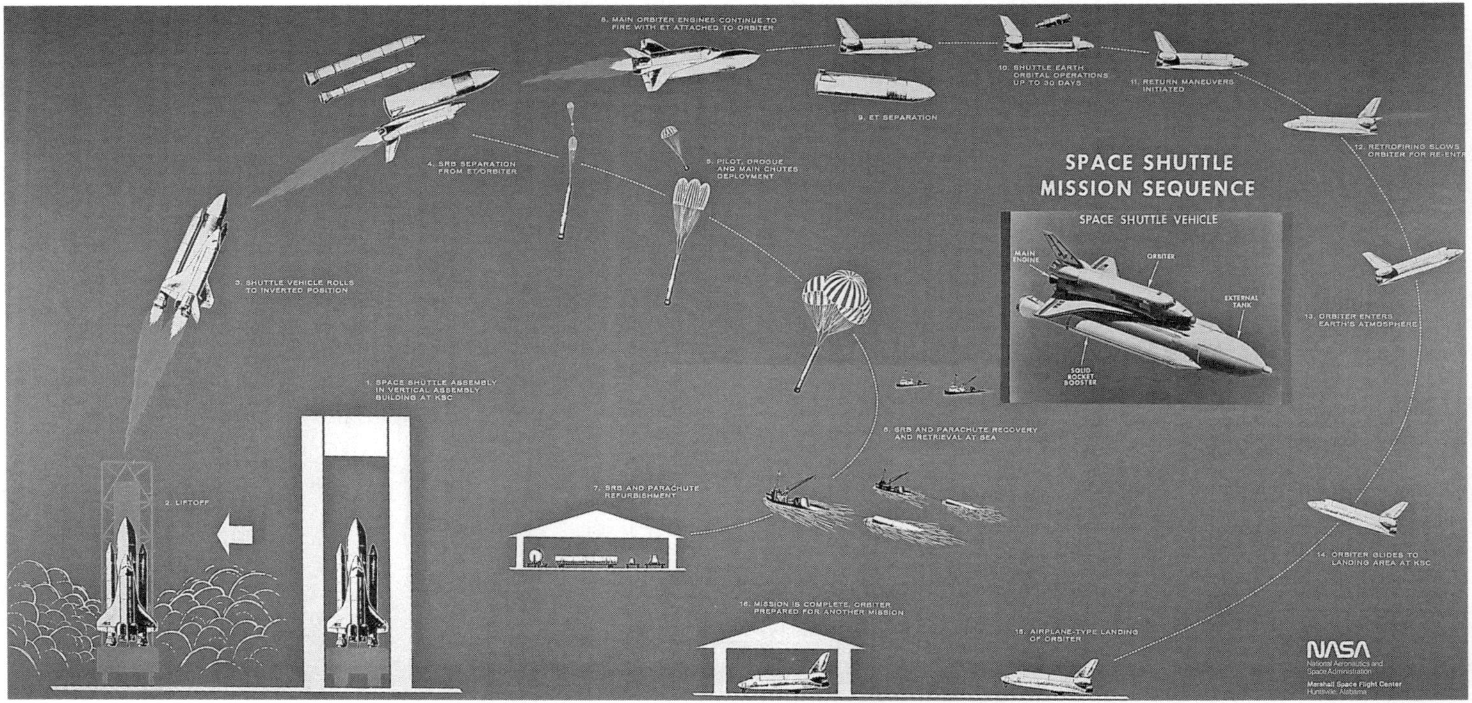

Figure 16.3 Space shuttle mission profile. (*Source:* NASA.)

Predictive modeling says that future events will occur, but anticipatory modeling allows the analyst to formulate hypotheses and explore what might happen. The shuttle is likely to land on the runway at Kennedy Space Center, but could it land anywhere else? If the booster recovery ships have departed, is a launch certain? The descriptive model represents the analyst's explicit understanding of what is likely to happen and baselines understanding of the complex process across the analytic community.

Descriptive models cannot be "executed" or "simulated," although the descriptive model may be used as a basis for the construction of such models. When the term "modeling" is used in the analysis community, it usually means "translate a tacit or explicit descriptive model into an analytic model suitable for computer simulation." This machine-readable code is executed to provide insight.

16.3 Machine Learning, Data Mining, and Statistical Models

Machine learning traces its origins to the 17th century where German mathematician Leibnitz began postulating formal mathematical relationships to represent human logic. In the 19th century, George Boole developed a series of relations for deductive processes (now called Boolean logic). By the mid 20th century, English mathematician Alan Turing and John McCarthy of MIT began experimenting with "intelligent machines," and the term "artificial intelligence" was coined. Machine learning is a subfield of artificial intelligence concerned with the development of algorithms, models, and techniques that allow machines to "learn."

Natural intelligence, often thought of as the ability to reason, is a representation of logic, rules, and models. Humans are adept pattern-matchers. Memory is a type of historical cognitive recall. Although the exact mechanisms for "learning" in the human brain are not completely understood, in many cases it is possible to develop algorithms that mimic human thought and reasoning processes. Many machine-learning techniques including rule-based learning, case-based learning, and unsupervised learning are based on our understanding of these cognitive processes.

16.3.1 Rule-Based Learning

In rule-based learning, a series of known logical rules are encoded directly as an algorithm. This technique is best suited for directly translating a descriptive model into executable code. For example, an analyst encoding the space shuttle processing cycle from Figure 16.3 into a rule-based model might specify that the orbiter can only land at certain predefined runways. If a subsequent analyst queried the model to find out where the space shuttle can land, the "learned" model would look only in the defined locations without considering other runways with similar properties that might be used in an emergency.

Rule-based learning is the most straightforward way to encode knowledge into an executable model, but it is also the most brittle for obvious reasons. The model can only represent the phenomena for which rules have been encoded. This approach significantly reinforces traditional inductive-based analytic approaches and is highly susceptible to surprise.

16.3.2 Case-Based Learning

Another popular approach is called case-based learning. This technique presents positive and negative situations for which a model is learned. The learning process is called "training"; the model and the data used are referred to as the "training set." In image classification, a model may be presented with many representations of a cat. During the training process, a human operator would preselect instances of a cat (positive examples). The learning process would identify key features associated with the classifier. Simultaneously, the human operator would also present instances of not-cat. Chairs, dogs, space shuttles, and elephants would comprise part of the training set so the model could identify key attributes that are the same and different across positive and negative examples.

This learning approach is useful when the cases—and their corresponding observables, signatures, and proxies—can be identified a priori. In the space shuttle example, a wide area search algorithm might be presented with images that contain launch towers and large processing facilities. Key features like fuel tanks, railroad tracks, and launch pads might be exemplars used to train the model. Case-based learning suffers from a high false alarm rate when there are weak distinguishing characteristics between positive and negative exemplars in the training set. For rare objects like the space shuttle, the geospatial characteristics are represented by strong signatures. It is easy to distinguish the space shuttle's unique vehicle assembly building from other large buildings.

In the case of counterterrorism, many terrorists participate in normal activities and look like any other normal individual in that culture. The distinguishing characteristics that describe "a terrorist" are few, making it very difficult to train automatic detection and classification algorithms. Furthermore, when adversaries practice denial and deception, a common technique is to mimic the distinguishing characteristics of the negative examples so as to hide in the noise [1]. This approach is also brittle because the model can only interpret cases for which it has positive and negative examples.

16.3.3 Unsupervised Learning

Another popular and widely employed approach is that of unsupervised learning where a model is generated based upon a data set with little or no "tuning" from a human operator. This technique is also sometimes called data mining because the algorithm literally identifies "nuggets of gold" in an otherwise unseemly heap of slag.

Several unsupervised learning techniques including response surface modeling, radial basis functions, and Kriging are used for data mining but one of the most widely used techniques is artificial neural networks (ANNs). An ANN is "an interconnected group of artificial neurons that uses a mathematical or computational model for information processing based on a connectionist approach to computation" [2]. The technique traces its origin to a 1943 article by neurophysiologist Warren McCulloch and mathematician Walter Pitts [3]. This approach is based on the premise that the computational elements themselves are very simple, like the neurons in the human brain. Complex behaviors arise from connections between the neurons that are modeled as an entangled web of relationships that represent signals and patterns.

Although there are many types of neural networks, the most common technique is the feedforward neural network, also referred to as a multilevel perceptron. These are typically comprised of three layers of interconnected neurons: the input layer, the hidden layer, and the output layer. A single response, R_k, is defined by the equation:

$$R_k = c_k + d_k\left[e_k + \sum_{j=1}^{N_H}\left(f_{jk}\left(\frac{1}{1+e^{-\left(a_j+\sum_{i=1}^{N}(b_{ij}X_i)\right)}}\right)\right)\right] \tag{16.1}$$

where

X_i is the value of the ith input variable;

a_j is the intercept term for the jth hidden node;

b_{ij} is the coefficient for the ith design variable;

c_k is the response scaling intercept term for the kth response;

d_k is the response scaling coefficient for the kth response;

e_k is the intercept term for the kth response;

f_{jk} is the coefficient for the jth hidden node and kth response;

N_H is the number of hidden nodes.

The process for training a neural network—that is, determining the unknown coefficients of the neural network equation—is shown in Figure 16.4. Coefficient values are postulated with some initial condition. For a model with a user-specified number of hidden nodes, H, the response, R_k, is evaluated and compared to the actual response for each value of the training set to calculate a model fit error. During the training process (typically constrained by a

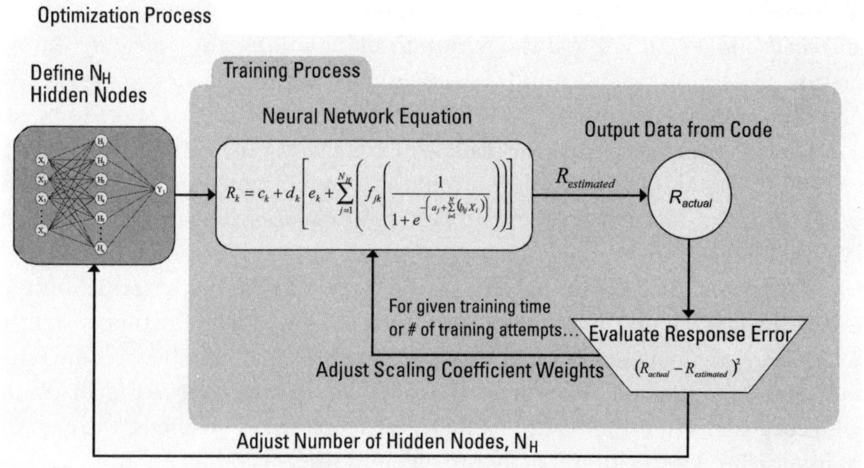

Figure 16.4 Training process for artificial neural networks. (©2007 Patrick Biltgen [2]. Reprinted with permission.)

user-specified number of attempts), the model coefficients are altered to minimize the error across the training set. Sometimes, as shown in Figure 16.4, the training process is encapsulated within an optimization process that changes the number of hidden nodes, N_H, to test different model topologies. The optimization of the topology and coefficients leads to the most accurate neural network that matches the training data. Recently, as computational power and data set sizes simultaneously increase, multilayer neural networks have become more popular. This approach is called "deep learning."

Although their application across many domains was limited before about 2000 due to the complexity of model training and optimization, automated modeling tools, distributed training, and extremely high-speed computers have improved the penetration of the technique across hundreds of disciplines. Many machine learning and classification approaches have a variant of the ANN embedded in their logic processing.

Unsupervised learning is a popular approach to autonomously and incrementally develop models using multiple machine-learning algorithms. Many of these techniques are based on ANNs or related techniques. Learned patterns are stored as executable models that are compared to real-time data to classify activities, associate observed patterns with activities of interest, generate forecasts of future activities, and detect anomalies.

16.3.4 Sensemaking

Chapter 3 introduced the term sensemaking, popularized in a model postulated by Pirolli and Card in the 1990s [4]. Moore defines sensemaking as "a set of philosophical assumptions, substantive propositions, methodological framings, and methods" [5]. It is a holistic, critical thinking process that integrates information to develop meaning under uncertainty. This holistic view contrasts the decompositional view of analysis, engineering, and science that seeks to break complex phenomenon into the smallest discrete elements. All of the modeling techniques in this chapter use the atomic approach; algorithms can be broken down to individual machine instructions and equations can be decomposed into individual terms and relations.

While many of the formal cognitive processes for human sensemaking are not easily documented, sensemaking is the process by which humans constantly weigh evidence, match patterns, postulate outcomes, and infer between missing information. Although the term analysis is widely used to refer to the job of intelligence analysts, many sensemaking tasks are a form of synthesis: the process of integrating information together to enhance understanding.

Some researchers feel that sensemaking is a cognitive process consisting of nothing more than posits, assumptions, logic, and rules. It follows that "automated sensemaking" can be implemented in some computational or algorithmic form with sufficiently powerful computers. Essentially, "automated sensemaking" refers to a model of the intelligence analyst and his/her ability to produce judgments himself/herself.

Naturally, there has been significant backlash from the analytic community against automated sensemaking. Analysts abhor the idea of machine created intelligence driving decisions. A famous example of models-gone-wrong occurred on May 6, 2010, when a single erroneous trade by a large mutual fund firm upset automatic trading models and caused the Dow Jones Industrial Average to plummet over 1,000 points in minutes, an incident known as the "flash crash" [6].

On the other hand, human-generated models are limited to what we know and what we can conceive. Advanced knowledge-processing, model-generation, and machine-learning approaches like IBM's WATSON may make nonintuitive suggestions that defy human reasoning. In 2014, demonstrating "cognitive cooking" technology, a specially trained version of WATSON created "Bengali Butternut BBQ Sauce," a delicious combination of butternut squash, white wine, dates, Thai chilies, tamarind, and more [7].

British statistician George Box famously said "all models are wrong, some models are useful." Artificial intelligence, data mining, and statistically created models are generally good for describing known phenomenon and forecasting outcomes (calculated responses) within a trained model space but are unsuitable for extrapolating outside the training set. Models must be used where appropriate, and while computational techniques for automated sensemaking have been proposed, many contemporary methods are limited to the processing, evaluation, and subsequent reaction to increasingly complex rule sets.

16.4 Rule Sets and Event-Driven Architectures

An emerging software design paradigm, event-driven architectures, are "a methodology for designing and implementing applications and systems in which events transmit between loosely coupled software components and services" [8]. These software architectures produce, detect, and react to events. Events are defined as a change in state, which could represent a change in the state of an object, a data element, or an entire system. An event-driven architecture applies to distributed, loosely coupled systems that require asynchronous processing (the data arrives at different times and is needed at different times). Three types of event processing are typically considered:

- Simple event processing (SEP): The system response to a change in condition and a downstream action is initiated (e.g., when new data arrives in the database, process it to extract coordinates).
- Event stream processing (ESP): A stream of events is filtered to recognize notable events that match a filter and initiate an action (e.g., when this type of signature is detected, alert the commander).
- Complex event processing (CEP): Predefined rule sets recognize a combination of simple events, occurring in different ways and different times, and cause a downstream action to occur (e.g., if the Federal Reserve chairman makes a negative comment and the price of oil is above $150/barrel and there is conflict in the Middle East, then liquidate my stock portfolio.

Event processing engines of all three types have become increasingly popular as components in information monitoring systems, especially for financial transaction processing (e.g., automated stock trading).

16.4.1 Event Processing Engines

An event processing engine receives events generated from other sources, compares them to rules, and generates events for other downstream processes [9]. A popular event-driven architecture is the iPhone and Android application "if this then that" (IFTTT) that allows users to create "recipes" for event handling and response. The "if" part of IFTTT includes input channels and event generators like Facebook, Evernote, and e-mail. The "this" part of the recipe is the trigger on the input channel.

IFTTT engines have proliferated across multiple fields including intelligence for automating rule-based workflows. One example is the joint enterprise modeling and analytics (JEMA) tool. JEMA allows users to record and save workflows (e.g., "keystrokes" for accessing, preparing, conditioning, and manipulating data) as executables that can be reused and repeated [10]. According to developer KEYW, JEMA is widely accepted across the intelligence community as "a visual analytic model authoring technology, which provides drag-and-drop authoring of multi-INT, multi-discipline analytics in an online collaborative space" [11]. Because JEMA automates data gathering, filtering, and processing, analysts shift the focus of their time from search to analysis.

Many companies use simple rule processing for anomaly detection, notably credit card companies whose fraud detections combine simple event processing and event stream detection. Alerts are triggered on anomalous behaviors. Although credit card companies are often cited as examples of firms that practice "ABI," the distinction is a bit of a misnomer. Although the targets and precise tactics are unknown, the channel (a financial transaction with a credit card) and the intent (fraudulent purchases) are always the same.

16.4.2 Simple Event Processing: Geofencing, Watchboxes, and Tripwires

Another type of "simple" event processing highly relevant to spatiotemporal analysis is a technique known as geofencing. A geofence is a geometric area projected onto the surface of the Earth. When an event happens within the area (a watchbox), an alert is triggered. A variation on the technique limits events to those that cross a spatial boundary (a tripwire). The ARGUS-IS persistent surveillance sensor developed by DARPA featured tripwires that generated automatic tracking events when vehicles or dismounts left a particular area.

Effective watchboxes are often limited to small geographic regions and specific signatures; creating a country-sized watchbox for "all activities" results in many false alarms. This technique is most effective when used to monitor a known area after an analyst uses ABI techniques to identify and characterize discrete locations.

16.4.3 CEP

A widely-used, Java-based, open source solution for CEP is Esper and the Esper Query Language (EQL). These tools "provide a highly scalable, memory-efficient, in-memory computing, SQL-standard, minimal latency, real-time streaming-capable Big Data processing engine for historical data, or medium to high-velocity data and high-variety data" [12]. Because Esper is written in Java (and a sister application, NEsper is written for Microsoft .NET), the event processing engine can be easily integrated with other Java or C# processes. Esper has been deployed for business process automation, network and application monitoring, financial fraud detection, sensor network monitoring, and many other applications [12].

Another product, LUX, by the Illumina Consulting Group (ICG), is a configurable real-time analytics solution built to address high-volume, high-rate scenarios. According to ICG, the LUX user interface "lets analysts interrogate all available information streams directly and adjust their problems on the fly: from highly granular, when very specific items of interest—like individual persons—can be observed, to very broad, when multiple, diverse events or signals must be correlated across time or geography" [13]. LUX features a web-based, interactive rule-set generation capability. Rules can be defined from scratch, but LUX features templates where prevalidated rules can be modified for different purposes. Simple rules can be combined and correlated into rule sets representing increasingly complex real-world behaviors.

In October 2014, as hysteria related to a potential outbreak of Ebola transmitted from west African countries heightened, LUX demonstrated a capability for real-time event stream monitoring and complex event processing. A user created rule sets based on information from the AIS (see Chapter 11) and the GDELT project, which "monitors the world's broadcast, print, and web news…in over 100 languages and identifies people, locations, organizations, counts, themes, sources, and events" [14]. An example of the LUX user interface and rules browser is shown in Figure 16.5.

In this use case, a LUX user defines multiple watchboxes. Figure 16.5 shows three watchboxes. An inner one monitors activities around the port areas. The middle one triggers when ships are off the coast near African ports. An outer one provides alerts on arrivals and departures from the west African coast as they go to sea. Rule sets produce alerts on any inbound or outbound ships crossing the boundaries of the watchboxes. Ships in each watchbox are identified and tagged using AIS metadata (from open sources) on the ship's name, unique identifier, course, speed, flag, declared port, and cargo. LUX monitors the ships in each watchbox for future activities, including tracking the objects to their next port.

A dynamic area of interest is a watchbox that moves with an object. In the Ebola tracking example, the dynamic areas of interest are centered on each tagged ship with a user-defined radius. This allows the user to identify when two ships come within close proximity or when the ship passes near a geographic feature like a shoreline or port, providing warning of potential docking activities.

To facilitate the monitoring of thousands of objects, rules can be visualized on a watchboard that uses colors, shapes, and other indicators to highlight rule activation and other triggers. A unique feature of LUX is the timeline view, which provides an interactive visualization of patterns across individual rules or sets of rules as shown in Figure 16.6 and how rules and triggers change over time.

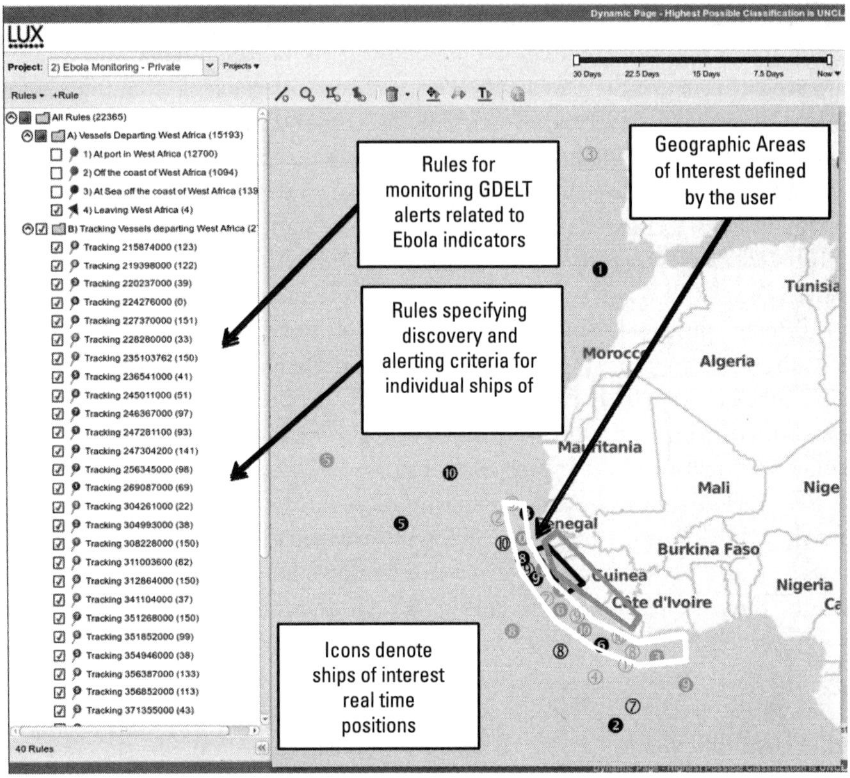

Figure 16.5 Example of the LUX user interface for Ebola event monitoring.

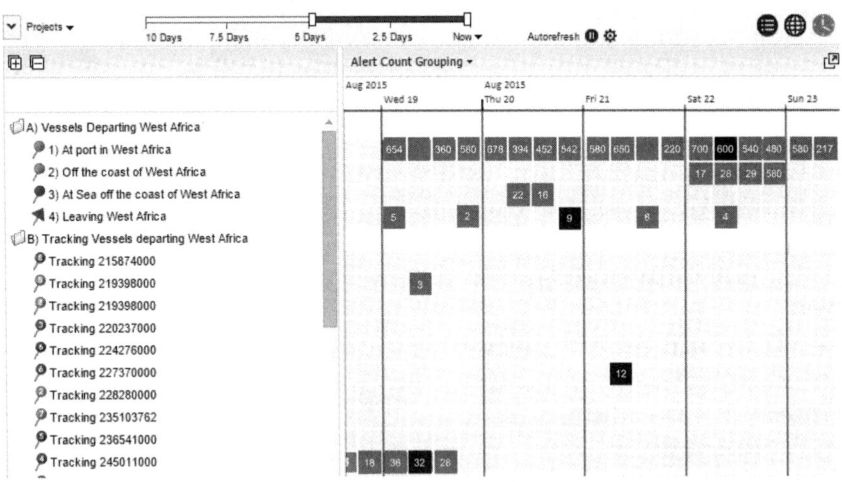

Figure 16.6 Example of the LUX rules timeline viewer. (©2014 Illumina Consulting Group. Reprinted with permission.)

The top row provides an event count of all vessels currently at port in west Africa. The next two rows provide counts of vessels in the other two areas of interest off the coast and at sea. The fourth row contains critical event triggers that show vessels leaving west Africa. This is a complex event that keeps track of vessels moving from the state of "in port" to "off the coast" to "at sea" within a defined time range. By fusing information from GDELT, analysts correlate ship activities with reports of Ebola cases in the region, quarantines, illnesses among crew members, or other factors. LUX also creates alerts that can be sent to a user, another software process, or encapsulated as a KML file for geospatial analysis and visualization.

Rule sets and alerts also flag abnormal behaviors (for example, when a ship deviates from its stated heading or moves toward a port other than the declared destination in the AIS feed). A rule set may also be used to notify when the AIS reporting has stopped, an indicator that the ship does not want to be monitored. This type of event can trigger a search task to relocate the missing object. When alerts are linked to subsequent collection, this is often referred to as tipping and cueing.

16.4.4 Tipping and Cueing

The original USD(I) definition for ABI referred to "analysis and subsequent collection" and many models for ABI describe the need for "nonlinear TCPED" where the intelligence cycle is dynamic to respond to changing intelligence needs. This desire has often been restated as the need for automated collection in response to detected activities, or "automated tipping and cueing."

Although the terms are usually used synonymously, a tip is the generation of an actionable report or notification of an event of interest. When tips are sent to human operators/analysts, they are usually called alerts. A cue is a more related and specific message sent to a collection system as the result of a tip. Automated tipping and cueing systems rely on tip/cue rules that map generated tips to the subsequent collection that requires cueing.

Numerous community leaders have highlighted the importance of tipping and cueing to reduce operational timelines and optimize multi-INT collection. A 2008 report by the Joint Defense/Intelligence Science board recommended that the community "develop techniques for closed loop dynamic tasking to take advantage of operational sensor integration through tipping and cueing" [15]. DNI Clapper, speaking at the 2012 GEOINT Symposium, referred to the integration of SIGINT and GEOINT in the same time domain, to perform activity-based change detection on a region over time "to better forecast events and quickly alert analysts to where the action is." Clapper said that automated change detection across intelligence systems will lead to greater "cross-agency tipping and cueing of our collection whenever things or activities of interest are detected" [16].

Several existing or proposed collection and mission management systems implement this capability. Defense officials noted that the air force's multisensor *Blue Devil 1* aircraft "allows an operator to use one sensor in real time to tip off another for target validation" [17]. Lewis, Messinger, and Gartley proposed identifying unusual objects in a scene using a polarimetric sensor and cueing an FMV sensor to subsequently track the object [18]. In a rare public appearance at the 2014 GEOINT Symposium, national reconnaissance office (NRO) director Betty Sapp highlighted the operational successes of the Sentient program, saying, "We've demonstrated that we can not only be responsive but predictive in where we aim our space assets" [19]. Sapp noted that Sentient develops "machine-speed tasking, collection, and processing" [20].

Although many intelligence community programs conflate "ABI" with tipping and cueing, the latter is an inductive process that is more appropriately paired with monitoring and warning for known signatures after the ABI methods have been used to identify new behaviors from an otherwise innocuous set of data. In the case of modeling, remember that models only respond to the rules for which they are programmed; therefore tipping and cueing solutions may improve efficiency but may inhibit discovery by reinforcing the need to monitor known places for known signatures instead of seeking the unknown unknowns.

16.5 Exploratory Models

Data mining and statistical learning approaches create models of behaviors and phenomenon, but how are these models executed to gain insight. Exploratory modeling is a modeling technique used to gain a broad understanding of a problem domain, key drivers, and uncertainties before going into details [21]. Section 16.5.1 highlights some of the key techniques applied to exploratory modeling across multiple disciplines.

16.5.1 Basic Exploratory Modeling Techniques

There are many techniques for exploratory modeling. Some of the most popular include Bayes nets, Markov chains, Petri nets, and discrete event simulation.

A Bayes net is a simple modeling technique based on the concepts of Bayesian probability introduced in Chapter 15. Bayes nets are represented a directed acyclic graph where the edges of the graph represent conditional dependencies between variables, attributes, events, or outcomes. A descriptive model like the one in Figure 16.3 can be translated into a probability-based simulation model by translating each process event in the depiction to a node in a graph and by establishing conditional transition probabilities between the events.

Markov chains, a related technique, are stochastic state transition models where the state of $x^{(i+1)}$ is conditionally independent of all other points given the state of $x^{(i)}$:

$$P(x^{(i+1)} \mid x^{(i)}, x^{(i-1)}, \ldots, x^{(1)}) = P(x^{(i+1)} \mid x^{(i)}) \tag{16.2}$$

In other words, the probability distribution of the next state depends only on the previous state [22]. Markov chains are typically depicted as a directed graph as shown in Figure 16.7, but calculations are performed using a state transition matrix, *T,* as shown in Figure 16.7.

Because the transition probability is not dependent on how the process arrived at a certain state (called the Markov property or "memorylessness"), it is suitable for some intelligence problems but not for others. Processes such as industrial production and political election cycles that have definable transition probabilities that can be explicitly stated in a transition matrix are appropriate. Many more complex problems may have dependencies to other states of the system and require a different modeling technique.

Petri nets are a modeling technique used for concurrent modeling and state transitions of distributed systems. The net is as a set of places, transitions, and arcs. Arcs travel between places and transitions (and vice versa), but no two places or transitions may be directly connected. This property differentiates Petri nets from Markov chains (the latter may be defined as the set of only places and arcs). The transition property acts as a gate that enables or inhibits a transition between states. Markers or tokens are used to describe the state of one or more particles or masses in the system. In the space shuttle processing cycle of Figure 16.3, transitions act as gates between steps and define that certain conditions must be reached. For example, the tacit knowledge that "two boosters and an external fuel tank are required before shuttle assembly begins" could be encoded as a transition inhibited until components (objects) or their

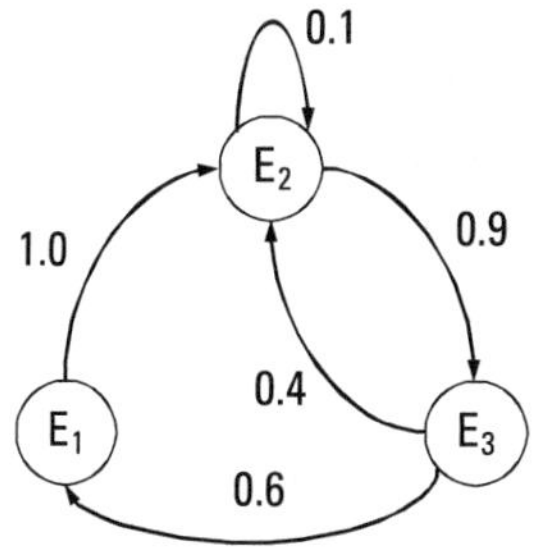

$$T = \begin{bmatrix} 0.0 & 1.0 & 0.0 \\ 0.0 & 0.1 & 0.9 \\ 0.6 & 0.4 & 0.0 \end{bmatrix}$$

Figure 16.7 Example of a Markov chain.

movements (transactions) have been observed. This technique is useful in anticipatory analysis for intelligence applications because different collection and analysis methods may focus on places, transitions (and their properties), or the pathway between them—he last of which may be considered a transaction in ABI parlance.

Discrete event simulation (DES) is another state transition and process modeling technique that models a system as a series of discrete events in time. In contrast to continuously executing simulations (see agent-based modeling and system dynamics in Sections 16.5.3 and 16.5.4), the system state is determined by activities that happen over user-defined time slices. Because events can cross multiple time slices, every time slice does not have to be simulated. This property means that many discrete event simulations can run much faster than their continuous-time counterparts. DES is widely used in industrial engineering and other process modeling, specifically in queuing analysis for the design of facilities (e.g., airports and hospitals). Manufacturing engineers use DES for process modeling where events represent manufacturing and assembly operations in a factory.

Bayes nets, Markov chains, Petri nets, and DESs are widely used by the modeling community. They are straightforward and easy to validate, and a number of commercial and open-source software packages exist for composing models using one or many of these techniques. While these methods can be used for exploratory analysis and forecasting, they are extremely limited in their ability to model the unknown as each rule, transition probability, event time, or other factors must be explicitly stated by the model builder. They are useful for identifying critical steps in a related process and may focus analyst attention to these areas, but they are usually inappropriate for discovering unknown behaviors and relationships.

16.5.2 Advanced Exploratory Modeling Techniques

There is also a class of modeling techniques for studying emergent behaviors and modeling of complex systems with a focus on discovery emerged due to shortfalls in other modeling techniques. Two of these, agent-based modeling (ABM) and system dynamics, are significantly more complicated in model construction, validation, execution, and integration; however, they provide a powerful capability to discover unknown behaviors and relationships due to the unforeseen dynamic behaviors that result from purposefully complex interactions of otherwise simple rules and objectives.

16.5.3 ABM

ABM is an approach that develops complex behaviors by aggregating the actions and interactions of relatively simple "agents." According to ABM pioneer

Andrew Ilachinski, "agent-based simulations of complex adaptive systems are predicated on the idea that the global behavior of a complex system derives entirely from the low-level interactions among its constituent agents" [23]. Human operators define the goals of agents. In simulation, agents make decisions to optimize their goals based on perceptions of the environment. The dynamics of multiple, interacting agents often lead to interesting and complicated emergent behaviors. See Figure 16.8.

ABM is a valid technique when analysts can specify some simple rules and objectives they believe an adversary would employ but cannot define macro-level behaviors. Simulations reveal the likely decision pathways based on how multiple agents interact and influence a changing environment.

One of the primary drawbacks of the method is that it is nearly impossible to validate. When the simulation produces counterintuitive results, it is difficult to determine whether this is a valid discovery or a computational error. For this reason, ABM is not useful for forecasting precise outcomes but may be valid for anticipating possible courses of action in complex situations.

16.5.4 System Dynamics Model

System dynamics is another popular approach to complex systems modeling that defines relationships between variables in terms of stocks and flows. Developed by MIT professor Jay Forrester in the 1950s, system dynamics was concerned with studying complexities in industrial and business processes [24]. The technique was employed in the 1970s by the Club of Rome to study Earth's natural resources in the face of an exploding global population. In their Limits to Growth study, Forrester's system dynamics model incorrectly forecast the demise of society by the early 21st century because it failed to account for a number of factors, notably technological developments in agriculture and transportation [25].

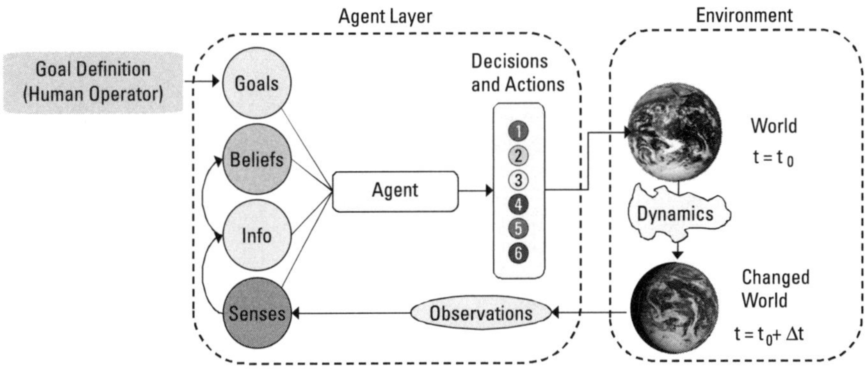

Figure 16.8 Process for ABM. (©2007 Patrick Biltgen [2]. Reprinted with permission.)

By the early 2000s, system dynamics emerged as a popular technique to model the human domain and its related complexities. Between 2007 and 2009, researchers from MIT and other firms worked with IARPA on the Pro-Active Intelligence (PAINT) program "to develop computational social science models to study and understand the dynamics of complex intelligence targets for nefarious activity" [26]. Researchers used system dynamics to examine possible drivers of nefarious technology development (e.g., weapons of mass destruction) and critical pathways and flows including natural resources, precursor processes, and intellectual talent. An example of the system dynamics model developed for PAINT is shown in Figure 16.9 where lines between nodes represent the direction and impact (positive or negative). For example, an increase in the economy produces an increase in popular support (after a time delay). This in turn produces an increase in government influence.

Another aspect of the PAINT program was the design of probes. Since many of the indicators of complex processes are not directly observable, PAINT examined input activities like sanctions that may prompt the adversary to do something that is observable. This application of the system dynamics modeling technique is appropriate for anticipatory analytics because it allows analysts to test multiple hypotheses rapidly in a surrogate environment. In one of the examples cited by MIT researchers, analysts examined a probe targeted at human resources where the simulators examined potential impacts of hiring away key personnel resources with specialized skills. This type of interactive, anticipatory analysis lets teams of analysts examine potential impacts of different courses of action.

System dynamics models have the additional property that the descriptive model of the system also serves as the executable model when time constants

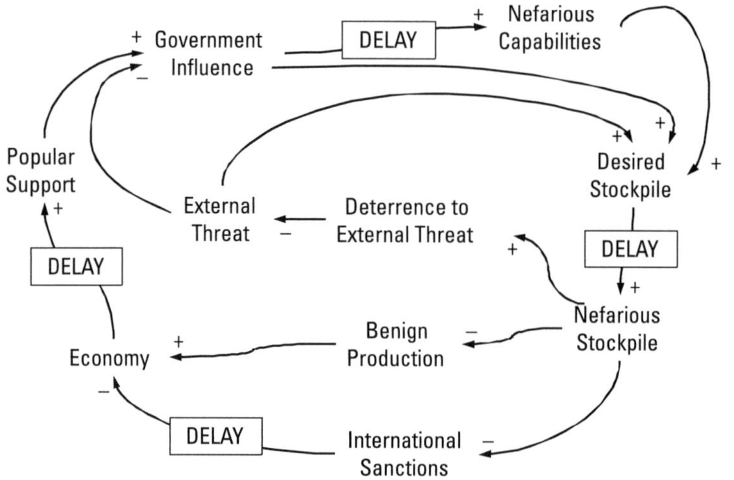

Figure 16.9 Example of system dynamics model for IARPA PAINT. (Adapted from [26].)

and influence factors are added to the representation. The technique suffers from several shortcomings including the difficulty in establishing transition coefficients, the impossibility of model validation, and the inability to reliably account for known and unknown external influences on each factor.

16.6 Model Aggregation

In January of 2015, monster storm Juno blanketed New York City with over two feet of snow—except that it didn't. This event happened inside a model called North American Mesoscale (NAM). Three days before the event, the National Weather Service forecast an 80% chance NYC would receive at least 12" of snow and a 62% chance for at least 18" of snow [27]. The NAM model agreed with a second model from the European Centre for Medium-Range Weather Forecasts (ECMWF), which itself became a media darling when years earlier it precisely forecast 2012's "superstorm" Sandy. However, an alternative model, the global forecast system (GFS) predicted between six and twelve inches [28]. Meteorologists ignored one possible outcome and threw their weight behind the most severe prediction. Although they erred on the side of caution, the "snowperbole" was excoriated by the public as an overreaction, yielding a waste of taxpayer dollars and loss in revenue and productivity as schools were shuttered, 7,700 flights were cancelled, and the NYC subway was closed for the first time in history. The forecasters admitted they got it wrong.

Statistician and political blogger Nate Silver gained notoriety during the 2012 presidential election. While most media outlets prognosticated a 50/50 split, Silver predicted an increasingly probable win for Barack Obama in the weeks leading up to the election earning a vitriolic response from right-leaning pundits. Silver's probabilistic forecasting model was based on aggregation of multiple models from national polls. Silver accounted for the statistical bias traditionally inherent in each poll, calibrating each model for observed or perceived errors. The role of anticipatory analytics should be to make these possible realities plausible in the mind's eye of the analyst.

Analysts can improve the fidelity of anticipatory modeling by combining the results from multiple models, using the second and third techniques advocated by Waltz in Figure 16.2. One framework for composing multiple models is the multiple information model synthesis architecture (MIMOSA), developed by Information Innovators. MIMOSA "aided one intelligence center to increase their target detection rate by 500% using just 30% of the resources previously tasked with detection freeing up personnel to focus more on analysis" [29]. MIMOSA uses target sets (positive examples of target geospatial regions) to calibrate models for geospatial search criteria like proximity to geographic features, foundation GEOINT, and other spatial relationships.

Merging multiple models, the software aggregates the salient features of each model to reduce false alarm rate and improve the predictive power of the combined model.

The concept of model aggregation is also referred to as multiresoluton modeling (MRM), where a multiple models are used to describe the same phenomenon at different levels of resolution. In MRM like in multi-INT analysis, the positive characteristics of one class of models are used to compensate for the weaknesses of others. Although "high-resolution" models are often preferred, the choice of resolution level is one of economy; low-resolution models are often easier to develop, calibrate, and execute. Since the intelligence discipline is dominated by tight timelines and high levels of uncertainty, the creation of high-fidelity models is often impractical. According to Davis and Bigelow, "low-resolution models are needed for analytic agility" [30].

An approach for multiresolution modeling of sociocultural dynamics was developed by DARPA for the COMPOEX program in 2007. COMPOEX provided multiple types of agent-based, system dynamics, and other models in a variable resolution framework that allowed military planners to swap different models to test multiple courses of action across a range of problems. A summary of the complex modeling environment is shown in Figure 16.10. COMPOEX includes modeling paradigms such as concept maps, social networks, influence diagrams, differential equations, causal models, Bayes networks, Petri nets, dynamic system models, event-based simulation, and agent-based models [31].

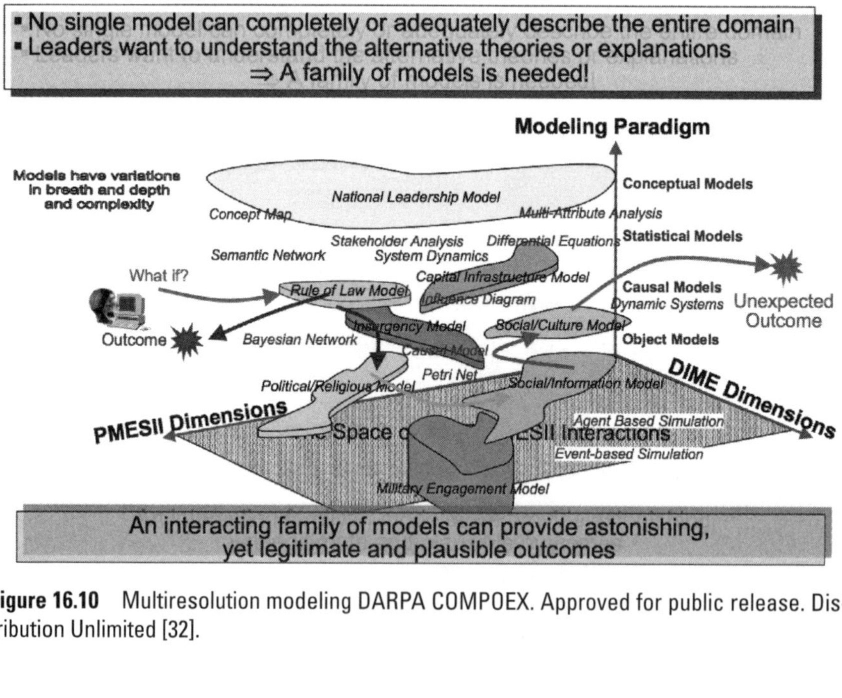

Figure 16.10 Multiresolution modeling DARPA COMPOEX. Approved for public release. Distribution Unlimited [32].

Another feature of COMPOEX was a graphical scenario planning tool that allowed analysts to postulate possible courses of action, as shown in Figure 16.11.

Each of the courses of action in Figure 16.11 was linked to one or more of the models across the sociocultural behavior analysis hierarchy, abstracting the complexity of models and their interactions away from analysts, planners, and decision makers. The tool forced models at various resolutions to interact (Figure 16.10) to stimulate emergent dynamics so planners could explore plausible alternatives and resultant courses of action.

Multiresolution modeling approaches allow integration of different models that analyze behaviors, events, and transactions at different levels. For example, some modeling approaches focus on the class of object or the instance of the object—the equipment—because that is what can be sensed. Objects can usually be modeled using physics-based or process models. However, an important tenet of ABI is that these objects are operated by someone (who). Knowing something about the "who" provides important insights into the anticipated behavior of those objects.

In Tom Clancy's *The Hunt for Red October*, a new class of advanced Soviet submarine is missing and presumed to be racing across the North Atlantic toward the United States. The type of object: a submarine. The class of object: a new Typhoon-class submarine equipped with a nearly silent magnetohydrodynamic, "caterpillar," drive. The specific instance: the Red October. In ABI analytics, the what and where of the single submarine provides some information, but the who and why enables anticipation of what might happen. The sub is piloted by Captain Ramius (who) and he intends to defect (why). Clancy's Jack Ryan, a CIA Directorate of Intelligence analyst, has a mental model of Ramius based deep historical analysis of his biographical, relational, and other information. In a confrontation with U.S. Navy Captain Mancuso, Ryan predicts Ramius will make his next "Crazy Ivan" starboard because "he always

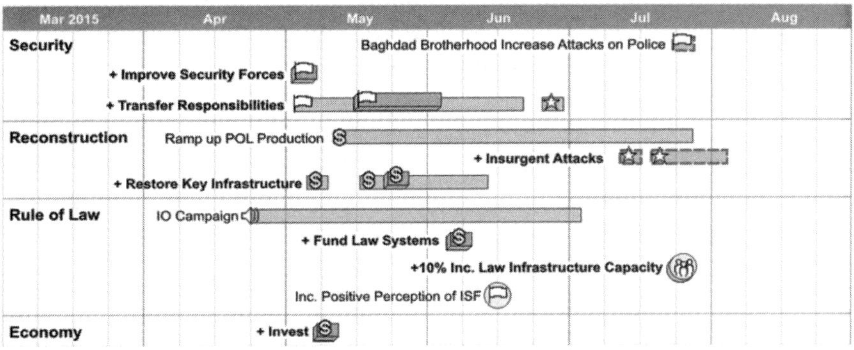

Figure 16.11 COMPOEX campaign planning tool. (Approved for public release. Distribution unlimited [32].)

goes to starboard in the bottom half of the hour." Although Ryan bluffed, the premise is that by understanding *who,* anticipatory models of object behavior can be improved.

16.7 The Wisdom of Crowds

Most of the anticipatory analytic techniques in this chapter refer to analytic, algorithmic, or simulation-based models that exist as computational processes; however, it is important to mention a final and increasingly popular type of modeling approach based on human input and subjective judgment.

James Surowiecki, author of *The Wisdom of Crowds,* popularized the concept of information aggregation that surprisingly leads to better decisions than those made by any single member of the group. It offers anecdotes to illustrate the argument, which essentially acts as a counterpoint to the maligned concept of "groupthink." Surowiecki differentiates crowd wisdom from group think by identifying four criteria for a "wise crowd" [33]:

- Diversity of opinion: Each person should have private information even if it's just an eccentric interpretation of known facts.
- Independence: People's opinions aren't determined by the opinions of those around them.
- Decentralization: People are able to specialize and draw on local knowledge
- Aggregation: Some mechanism exists for turning private judgments into a collective decision

IARPA sponsored the aggregate contingent estimation (ACE) program, developing a "powerful prediction engine that combines the opinions of many experts to make superior forecasts on world events" [34]. It has long been known that simply averaging a large set of independent judgments creates an estimate that is usually more accurate than every individual judgment in the group," IARPA ACE program manager Jason Matheny said [35]. A related DARPA program called FutureMAP was canceled in 2003 amidst congressional criticism regarding "terrorism betting parlors"; however, the innovative idea was reviewed in depth by Yeh in *Studies in Intelligence* in 2006 [36]. Yeh found that prediction markets could be used to quantify uncertainty and eliminate ambiguity around certain types of judgments. George Mason University launched IARPA-funded SciCast, which forecasts scientific and technical advancements [37]. This combinatorial market "combines different answers and other bits of information together in real time to provide a more accurate, moving picture of how all of these answers and areas are interacting with one another" [38]. This type of modeling approach is useful for producing forecasts of future events, with likelihoods determined by many individuals who tacitly aggregate information from different sources, weigh evidence, and produce judgments. Though decidedly low-tech, prediction markets provide a powerful tool for improving the quality of long-range forecasts and anticipating future possibilities

16.8 Shortcomings of Model-Based Anticipatory Analytics

By now, you may be experiencing frustration that none of the modeling techniques in this chapter are the silver bullet for all anticipatory analytic problems. The challenges and shortcomings for anticipatory modeling are voluminous.

The major shortcoming of all models is that they can't do what they aren't told. Rule-based models are limited to user-defined rules, and statistically generated models are limited to the provided data. As we have noted on multiple occasions, intelligence data is undersampled, incomplete, intermittent, error-prone, cluttered, and deceptive. All of these are ill-suited for turnkey modeling.

A combination of many types of modeling approaches is needed to perform accurate, justifiable, broad based anticipatory analysis. Each of these needs validation, but model validation, especially in the field of intelligence, is a major challenge. We seldom have "truth" data. The intelligence problem and its underlying assumptions are constantly evolving as are attempts to solve it, a primary criterion for what Rittel and Weber call "wicked problems" [39].

Handcrafting models is slow, and a high level of skill is required to use many modeling tools. Furthermore, most of these tools do not allow easy sharing across other tools or across modeling approaches, complicating the ability to share and compare models. This challenge is exacerbated by the distributed nature of knowledge in the intelligence community.

When models exist, analysts depend heavily on "the model." Sometimes it has been right in the past. Perhaps it was created by a legendary peer. Maybe there's no suitable alternative. Overdependence on models and extrapolation of models into regions where they have not been validated leads to improper conclusions. Although models promise to provide a physics-based representation of the world, Silver notes that, "We can never make perfectly objective predictions. They will always be tainted by our subjective point of view" [40]. This subjective view applies to model construction, validation, execution, and the interpretation of results.

A final note: weather forecasting relies on physics-based models with thousands of real-time data feeds, decades of forensic data, ground truth, validated physics-based models, one-of-a-kind supercomputers, and a highly trained cadre of scientists, networked to share information and collaborate. It is perhaps the most modeled problem in the world. Yet weather "predictions" are often wrong, or at minimum imprecise. What hope is there for predicting human behaviors based on a few spurious observations?

16.9 Modeling in ABI

In the early days of ABI, analysts in Iraq and Afghanistan lacked the tools to formally model activities. As analysts assembled data in an area, they developed a tacit mental model of what was normal. Their geodatabases representing a pattern of life constituted a type of model of what was known. The gaps in those databases represented the unknown. Their internal rules for how to correlate data, separating the possibly relevant from the certainly irrelevant composed part of a workflow model as did their specific method for data conditioning and georeferencing.

However, relying entirely on human analysts to understand increasingly complex problem sets also presents challenges. Studies have shown that experts (including intelligence analysts) are subject to biases due to a number of factors like perception, evaluation, omission, availability, anchoring, groupthink, and others. In the 1970s, Kahneman and Tversky collaborated on a series of articles about cognitive fallacies in decision-making, culminating in their 2002 Nobel Prize in Economics for their work on prospect theory [41]. Heuer examined the impact of cognitive biases on intelligence analysis [42]. Analytic models that treat facts and relationships explicitly provide a counterbalance to inherent biases in decision-making. Models can also quickly process large amounts of data and multiple scenarios without getting tired, bored, or discounting information [43]. These factors can improve the quality of ABI analysis by illuminating alternatives that an analyst may not otherwise consider.

This chapter focuses on executable, algorithmic-based models but also describes approaches for capturing knowledge including crowdsourced knowledge of many analysts. Combining models with intuition improves decision-making capacity. Statistics from the National Weather Service found that "humans improved the accuracy of precipitation forecasts by about 25% over the computer guidance alone" [40]. The synergy between humans and computers, so-called human-machine teaming, is gaining momentum across the intelligence community as a solution that provides the best of both worlds.

Current efforts to scale ABI across the community focus heavily on activity, process, and object modeling as this standardization is believed to enhance information sharing and collaboration. Algorithmic approaches like JEMA, MIMOSA, PAINT, and LUX have been introduced to worldwide users. A recent handout from NGA describes an approach to "driving anticipatory collection from analytic models" [44]. These efforts seek to integrate automated processing, rules engines, event-driven architectures, and other techniques to enable anticipatory analytics and improve the efficiency of subsequent collection.

16.10 Summary

Models provide a mechanism for integrating information and exploring alternatives, improving an analyst's ability to discover the unknown. However, if models can't be validated, executed on sparse data, or trusted to solve intelligence problems, can any of them be trusted? If "all models are wrong," in the high-stakes business of intelligence analysis, are any of them useful?

Model creation requires a multidisciplinary, multifaceted, multi-intelligence approach to data management, analysis, visualization, statistics, correlation, and knowledge management. The best model builders and analysts discover that it's not the model itself that enables anticipation. The exercise in data gathering, hypothesis testing, relationship construction, code generation, assumption definition, and exploration trained the analyst. To build a good model, the analyst had to consider multiple ways something might happen—to consider the probability and consequence of different outcomes. The data landscape, adversary courses of action, complex relationships, and possible causes are all discovered in the act of developing a valid model. Surprisingly, when many analysts set out to create a model they end by realizing they became one.

References

[1] Bennett, M., and E. Waltz, *Counterdeception Principles and Applications for National Security,* Norwood, MA: Artech House, 2007.

[2] Biltgen, P., "A Methodology for Capability-Based Technology Evaluation for Systems-of-Systems," Georgia Institute of Technology, Atlanta, GA, 2007.

[3] McCulloch, W. S., and W. H. Pitts, "A Logical Calculus of the Ideas Immanent in Nervous Activity," *Bulletin of Mathematical Biophysics,* Vol. 5, 1943, pp. 115–133.

[4] Pirolli, P., and S. Card, "Information Foraging," *Psychological Review,* Vol. 106, No. 4, 1999, pp. 643–675.

[5] Moore, D. T., *Sensemaking: A Structure for an Intelligence Revolution,* Washington, D.C.: National Defense Intelligence College Press, 2011.

[6] "Findings Regarding the Market Events of May 6, 2010." U.S. Securities and Exchange Commission (SEC) and the Commodity Futures Trading Commission (CFTC), 30 Sep 2010.

[7] Fingas, J., "IMB's Watson computer makes a delicious BBQ sauce," *Engadget,* May 27, 2015. http://www.engadget.com/2014/05/27/watson-bbq-sauce.

[8] "Event-Driven Services in SOA," *JavaWorld,* web.

[9] Michaelson, B. M., "Event-Driven Architecture Overview," Object Management Group, February 2, 2006.

[10] Porche, I. R., et al., "Data Flood: Helping the Navy Address the Rising Tide of Sensor Information, Santa Monica, CA: RAND, 2014.

[11] "KEYW at GEOINT 2013*" Web. Available: http://www.keywcorp.com/geoint.

[12] "Esper—Complex Event Processing." Web. Available: http://esper.codehaus.org/.

[13] "LUX: Finding Connections in a World of Data," Illumina Consulting Group, 2014.

[14] "The GDELT Project." Web. Available: http://gdeltproject.org/.

[15] "Report of the Joint Defense Science Board and Intelligence Science Board Task Force on Integrating Sensor-Collected Intelligence," Office of the Undersecretary of Defense for Acquisition, Technology, and Logistics, November 2008.

[16] "Opening Keynote Address by James R. Clapper Jr., Director of National Intelligence," presented at the *USGIF GEOINT Symposium 2012,* Orlando, FL, October 9, 2012.

[17] Butler, A., "Air Force Mulls Continued Blue Devil 1 Ops," *Aviation Week and Space Technology,* March 18, 2013.

[18] Lewis, C. M., D. Messinger, and M. G. Gartley, "Activity-Based Intelligence Tipping and Cueing Using Polarimetric Sensors," *Proc. SPIE 9099, Polarization: Measurement, Analysis, and Remote Sensing XI,* 2014, p. 90990C.

[19] Sapp, B., "Keynote Presentation at the 2013* GEOINT Symposium," Tampa, FL, April 2014.

[20] Alderton, M., "From Airborne to Spaceborne, NRO Director Shares Recipe for the Next Generation in Space Innovation," *Trajectory,* 2014.

[21] Davis, P. K., "Exploratory Analysis Enabled by Multiresolution, Multiperspective Modeling," in *Proceedings of the Winter Simulation Conference,* 2000, Vol. 1, pp. 293–302.

[22] Acar, A. C., "BIN504—Lecture XI, Bayesian Inference and Markov Chains."

[23] Ilachinski, A., "Irreducible Semi-Autonomous Adaptive Combat (ISAAC): An Artificial-Life Approach to Land Warfare (U)," *Technical Report for the Center for Naval Analyses,* August 1997.

[24] Forrester, J., "Counterintuitive Behavior of Social Systems," *Technology Review,* Vol. 73, No. 3, pp. 52–68, 1971.

[25] Medows, D. H., D. L. Medows, R. Randers, and W. W. Behrens III, *Club of Rome, The Limits to Growth; a Report for the Club of Rome's Project on the Predicament of Mankind,* New York: Universe Books, 1972.

[26] Anderson, E., et al., "System Dynamics Modeling for Pro-Active Intelligence (PAINT)," MIT Sloan School of Management, Working Paper CISL#2009-17, November 2009.

[27] "Winter Storm Juno: A Pummeling for the History Books," The Daily Beast," web, January 26, 2015.

[28] "Snowmageddon or Snowperbole? Juno was both." SciTech Now.

[29] "Information Innovators Inc. - Providing IT Services and Solutions," Information Innovators. Web. Available: http://www.iiinfo.com/services.

[30] Davis, P. K., "Experiments in Multiresolution Modeling (MRM)," Santa Monica, CA: RAND, 1998.

[31] Kott, A., and P. S. Corpac, "COMPOEX Technology to Assist Leaders in Planning and Executing Campaigns in Complex Operational Environments," in *12th International Command and Control Research and Technology Symposium,* "Adapting C2 to the 21st Century," 2007.

[32] Kott, A., and P. S. Corpac, "Technology to Assist Leaders in Planning and Executing Campaigns in Complex Operational Environments, Conflict Modeling, Planning, and Outcomes Experimentation Program (COMPOEX)," DARPA, June 19, 2007.

[33] Surowiecki, J., *The Wisdom of Crowds: Why the Many Are Smarter Than the Few and How Collective Wisdom Shapes Business, Economies, Societies and Nations,* London: Little Brown, 2004.

[34] "Forecasting Ace." Web. Available: http://www.crowdsourcing.org/site/forecasting-ace/wwwforecastingacecom/9916.

[35] "ARA Boosts Forecasting Methods to Improve Predictions, Intelligence Gathering," November 2011, http://www.ara.com/Newsroom_Whatsnews/press_releases/forecasting-methods.htm.

[36] Yeh, P. F, "Using Prediction Markets to Enhance U.S. Intelligence Capabilities," *Studies in Intelligence,* Vol. 50, No. 4, 2006.

[37] "Main—SciCast Predict," web. Available: https://scicast.org/

[38] Tucker, P., "This Is How America's Spies Could Find the Next National Security Threat," *Defense One,* web, February 20, 2014.

[39] Rittel H. W. J., and M. M. Webber, "Dilemmas in a General Theory of Planning," *Policy Sciences,* Vol. 4, No. 2, June 1973, pp. 155–169.

[40] Silver, N., *The Signal and the Noise: Why So Many Predictions Fail—But Some Don't,* New York: Penguin, 2012.

[41] Tversky, A., and D. Kahneman, "Judgment Under Uncertainty: Heuristics and Biases," *Science,* Vol. 185, September 1974, pp. 1124–1131.

[42] Heuer, R. J., *The Psychology of Intelligence Analysis,* Center for the Study of Intelligence, 1999.

[43] Hoch, S. J., "Wharton on Making Decisions," in *Combining Models with Intuition to Improve Decisions* (eds. Hoch, S. J., H. C. Kunreuther, and R. E. Gunther), New York: Wiley, 2001, p. 350.

[44] "Next-Generation Collection Portfolio Overview," National Geospatial-Intelligence Agency, Handout at the 2013* *GEOINT Symposium.* Approved for public release, April 2014.

17

ABI in Policing

Patrick Biltgen and Sarah Hank

Law enforcement and policing shares many common techniques with intelligence analysis. Since 9/11, police departments have implemented a number of tools and methods from the discipline of intelligence to enhance the depth and breadth of analysis. This chapter demonstrates concepts like persistent surveillance, georeference to discover, data integration, and anticipatory analysis using law enforcement problems.

17.1 The Future of Policing

The 2002 film *Minority Report* depicts a utopian society in the not-too-distant future where all crimes are prevented before they happen. Relying on three networked clairvoyant "pre-cogs," Tom Cruise's chief John Anderton analyzes clips of predictive video to locate future victims of crimes and rapidly deploys a team of helicopter fast-roping police commandos to arrest the perpetrators before the events even happen [1]. This fictionalized scenario is often referenced as the "holy grail" of anticipatory analysis.

Although precise prediction of future events is impossible, there is a growing movement among police departments worldwide to leverage the power of spatiotemporal analytics and persistent surveillance to resolve entities committing crimes, understand patterns and trends, adapt to changing criminal tactics, and better allocate resources to the areas of greatest need. This chapter describes the integration of intelligence and policing—popularly termed "intelligence-led policing"—and its evolution over the past 35 years.

17.2 Intelligence-Led Policing: An Introduction

The term "intelligence-led policing" traces its origins to the 1980s at the Kent Constabulatory in Great Britain. Faced with a sharp increase in property crimes and vehicle thefts, the department struggled with how to allocate officers amidst declining budgets [2, p. 144]. The department developed a two-pronged approach to address this constraint. First, it freed up resources so detectives had more time for analysis by prioritizing service calls to the most serious offenses and referring lower priority calls to other agencies. Second, through data analysis it discovered that "small numbers of chronic offenders were responsible for many incidents and that patterns also include repeat victims and target locations" [3, p. 33]. Focusing analysis and intelligence gathering to resolve these unique entities and understand their patterns of life, officers could narrow the patrol area to the geographic regions most likely to contain the suspect entities. The Kent Policing Model is credited with a 24% drop in these types of crimes over a three-year period [4].

A generalized model for intelligence-led policing is shown in Figure 17.1. The focus of analysis and problem solving is to analyze and understand the influencers of crime using techniques like statistical analysis, crime mapping, and network analysis. Police presence is optimized to deter and control these influencers while simultaneously collecting additional information to enhance analysis and problem solving. A technique for optimizing police presence is described in Section 17.5.

Intelligence-led policing applies analysis and problem solving techniques to optimize resource allocation in the form of focused presence and patrols. Accurate dissemination of intelligence, continuous improvement, and focused synchronized deployment against crime are other key elements of the method. It is generally believed that police presence influences the environment to deter criminal activity in the targeted geographic region. Therefore, by analyzing the geotemporal distribution of crimes, directed patrols can be more efficiently

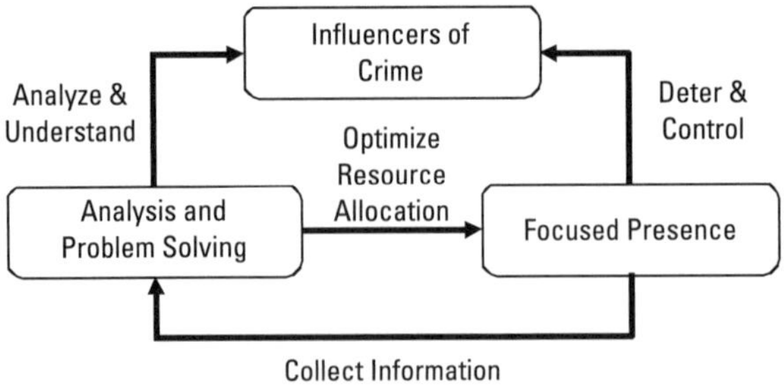

Figure 17.1 Generalized model for intelligence-led policing.

allocated. Placing officers near likely crime locations also has the secondary effect of reducing response times, perhaps lessening the severity of crimes or leading to an increase in apprehension rate due to proximity.

17.2.1 Statistical Analysis and CompStat

The concept of ILP was implemented in the New York City police department in the 1980s by police commissioner William Bratton and Jack Maple. Using a computational statistics approach called CompStat, "crime statistics are collected, computerized, mapped and disseminated quickly" [5]. Wall-sized "charts of the future" mapped every element of the New York transit system. Crimes were mapped against the spatial nodes and trends were examined. Maple says, "You have to get the crimes every day, the day they happen, and you've got to map them. You've got to map them in every precinct, every district, every squad room, and then everybody has to know about it" [5].

Though its methods are controversial, CompStat is widely credited with a significant reduction in crime in New York. The method has since been implemented at other major cities in the United States with a similar result, and the methods and techniques for statistical analysis of crimes is standard in criminology curricula.

17.2.2 Routine Activities Theory

A central tenet of ILP is based on Cohen and Felson's routine activities theory, which is the general principle that human activities tend to follow predictable patterns in time and space. In the case of crime, the location for these events is defined by the influencers of crime(Figure 17.2). Koper provides exposition of these influencers: "crime does not occur randomly but rather is produced by the convergence in time and space of motivated offenders, suitable targets, and the absence of capable guardians."

Since these three forces must converge in time and space for crime to occur, the endemic conditions for a crime tend to fall in the same place. These areas of highly concentrated crimes are typically referred to as "hot spots." Understanding the prevalence of hot spots enables the development of policing strategies to combat crime by focusing resources in a geographic area.

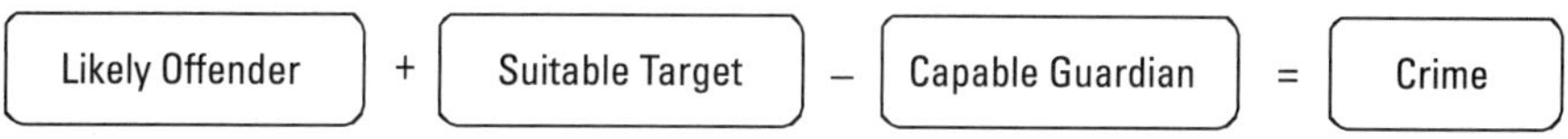

Figure 17.2 Summary of the influencing factors in routine activities theory.

17.3 Crime Mapping

Crime mapping is a geospatial analysis technique that geolocates and categorizes crimes to detect hot spots, understand the underlying trends and patterns, and develop courses of action. Crime hot spots are a type of spatial anomaly that may be characterized at the address, block, block cluster, ward, county, geographic region, or state level—the precision of geolocation and the aggregation depends on the area of interest and the question being asked.

17.3.1 Standardized Reporting Enables Crime Mapping

In 1930, Congress enacted Title 28, Section 534 of the U.S. code, authorizing the Attorney General and subsequently the FBI to standardize and gather crime information [6]. The FBI implemented the Uniform Crime Reporting Handbook, standardizing and normalizing the methods, procedures, and data formats for documenting criminal activity. This type of data conditioning enables information sharing and pattern analysis by ensuring consistent reporting standards across jurisdictions.

The Metropolitan District of Columbia Police Department provides a web-accessible database of Part I crimes in the District of Columbia [7]. This data is mapped in Figure 17.3. The clusters of high-crime activity indicated by the dark areas are calculated based on hotspots of crime. These correspond with the location of D.C. metro stops. The darkest area in the upper left is the Friendship Heights metro station and shopping district. The pattern of crime along Connecticut Avenue and Wisconsin Avenue illustrates a distracting trend in crime mapping. These data points represent reported crimes. There is a tendency to report crimes near the block of the nearest major street, creating the illusion that Connecticut Avenue is extremely crime-ridden while the next adjacent streets are almost completely peaceful. In practice, this geolocation error due to spatial reporting bias makes it difficult to build an activity model at the atomic level.

17.3.2 Spatial and Temporal Analysis of Patterns

Visualizing each observation as a dot at the city or regional level is rarely informative. For example, in the map in Figure 17.3, discerning a meaningful trend requires extensive data filtering by time of day, type of crime, and other criteria. One technique that is useful to understand trends and patterns is the aggregation of individual crimes into spatial regions. The D.C. metro crime data is reported according to census reporting area. By conflating the individual crime statistics with the polygons representing those areas as shown in Figures 17.4 and 17.5, it is possible to visualize and understand how crime changes over time.

Figure 17.3 Hotspot map of Washington, D.C., Ward 3. (Data Source: Metropolitan Washington, D.C., Police Department [7]. Background Map Source: OpenStreetMap.)

In the 1990s, Washington, D.C., had an unfortunate reputation as the murder capital of the United States [8]. Since then, it has passed that distinction on to other cities and undergone a major revival, though some neighborhoods have benefited from this more than others. Mental geographic boundaries, a part of the perception of which areas of the city are safe and which areas are unsafe, play a major role in perceived crime rates. The northwest (NW) quadrant of city is generally considered the safest, while the southeast (SE) quadrant has a notoriously bad reputation [9]. The Green line of the metro (shown as a dotted line in the reference map in Figure 17.3) has served as a common reference point for dangerous areas of the city in the recent past [10]. When visualizing the Metropolitan Police Department's open data set of reported crime using mapping techniques, many expect to see trends that align with their own mental map. Smart use of crime data can help realign incorrect expectations and focus crime prevention efforts in places where they are truly needed.

For the public and for law enforcement officials, the main concern when measuring neighborhood safety is often violent crime, including homicide, sexual abuse, robbery, and assault with a deadly weapon. Nonviolent crimes such as theft are an important but usually a secondary priority for residents and police.

Theft is the most commonly occurring crime in the Metropolitan District of Columbia Police Department data set and thus skews the visualization to make commercial areas seem very dangerous (Figure 17.3). Eliminating nonviolent crime from the data set and controlling for population density provides a more accurate representation of personal safety levels in a designated area. In the map in Figure 17.4, the total number of violent crimes per 100 people is aggregated per census tract for the year 2013.

Several trends are evident from the map, some which align with popular expectation, and some which do not. Natural geographic boundaries are a strong divider between areas of high and low crime. For example, the only areas of the city with zero violent crimes are west of Rock Creek (and more accurately, Rock Creek Park). This falls into place nicely with the expectation of the northwest quadrant being safer, yet many census tracts in the northwest have rates of violent crime on par with some tracts in southeast DC (≤ 2 violent crimes per 100 residents). Although these more dangerous tracts are categories as being in the northwest quadrant, they border but do not cross the border of the park.

On the opposite end of the spectrum, tracts to the south and east of the Anacostia River consistently fall into the higher ranges of violent criminal activity. "East of the River" is a common reference used to describe a unique area of D.C. that has historically been neglected relative to other areas of the city and thus suffers from high rates of poverty, and—as this map shows—crime. Less well-versed residents of D.C. refer to this entire part of the city as "Anacostia" and conflate the higher rates of crime with the entire southeast quadrant. Mapping aggregated crime data by census tract reveals that the rate of violent crime does not necessarily relate to quadrants, but rather to natural geographic barriers such as parks and rivers. Other geographic markers like landmarks, streets, and historical places may also act as historical anchors for citizens' perspectives on crime.

Another advantage of aggregating data by area using a GIS is the ability to visualize change over time. Figure 17.5 shows the overall average change in violent crime per 100 people over the interval from 2007 to 2013.

Tracts with increasing and decreasing crime are scattered across the entire city with little geographic pattern. The patchwork of varying rates of crime is present in wealthier neighborhoods west of Rock Creek Park as well as in gentrified areas such as the 14th street corridor and parts of Capitol Hill. The only area that shows consistency is, again, "east of the river". A majority of tracts in

Figure 17.4 Total violent crimes per 100 people per census tract in 2013, (Data source: Metropolitan Washington, D.C., Police Department [7].)

this area show slight increases in crime over this time period, and the contiguous nature of those tracts gives this trend geographic significance. It is interesting to note, however, that the highest increases are found in isolated areas of the northwest quadrant.

Publically available online crime records from the Metropolitan Police Department are only available to 2007. While the District of Columbia has changed significantly since the peak of violence from the early 1990s, it is usually difficult to determine causality for aggregated trend data over only a six-year

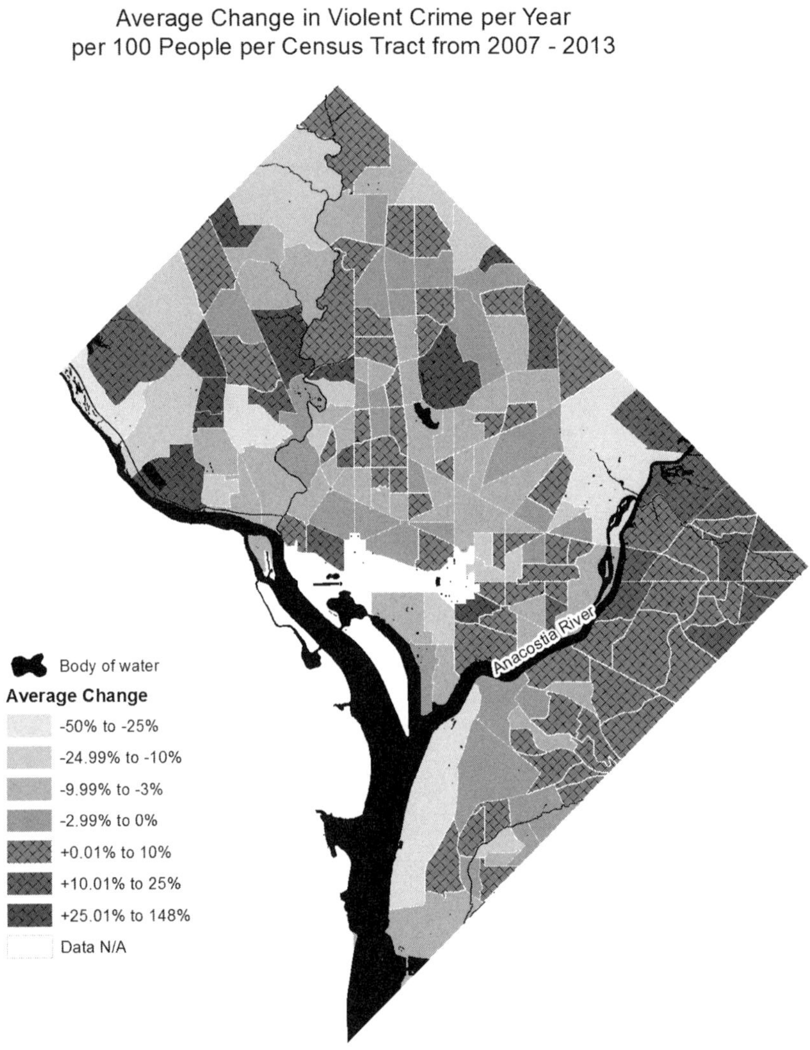

Figure 17.5 Average change in violent crimes per 100 people per census tract from 2007 to 2013. (Data source: Metropolitan Washington, D.C., Police Department [8].)

period. Short duration trends of this nature are sometimes called microtrends. Analysts should take note that change maps can serve as a starting point for analysis but should not be used to generalize or infer trends about broader populations or socioeconomic causes without the underlying data.

A deeper investigation into other geographic factors including infrastructure, demographic shifts, and investment would be necessary to clarify the trends in the majority of the city and determine whether a focused police effort would be effective in lowering violent crime rates.

Intelligence-led policing and crime mapping are two techniques to identify the spatial prevalence of crimes and to attempt to discern a pattern of activity over a given area. But who are the offenders? Why do the crimes take place? What patterns do the criminal entities follow? The identification of the criminal actors involved requires an approach based on other elements of ABI.

17.4 Unraveling the Network

Understanding hot spots and localizing the places where crimes tend to occur is only part of the story, and reducing crimes around hot spots is only a treatment of the symptoms rather than the cause of the problem. Crime mapping and intelligence-led policing focus on the ABI principles of collecting, characterizing, and locating activities and transactions. Unfortunately, these techniques alone are insufficient to provide entity resolution, identify and locate the actors and entities conducting activities and transactions, and identify and locate networks of actors. These techniques are generally a reactive, sustaining approach to managing crime. The next level of analysis gets to the root cause of crime to go after the heart of the network to resolve entities, understand their relationships, and proactively attack the seams of the network.

The Los Angeles Police Department's Real-Time Analysis and Critical Response (RACR) division is a state-of-the-art, network enabled analysis cell that uses big data to solve crimes. Police vehicles equipped with roof-mounted license plate readers provide roving wide-area persistent surveillance by vacuuming up geotagged vehicle location data as they patrol the streets.

One of the tools used by analysts in the RACR is made by Palo Alto-based Palantir Technologies. Named after the all-seeing stones in J. R. R. Tolkien's *Lord of the Rings,* Palantir is a data fusion platform that provides a clean, coherent abstraction on top of different types of data that all describe the same real world problem" [11]. Palantir enables "data integration, search and discovery, knowledge management, secure collaboration, and algorithmic analysis across a wide variety of data sources" [12]. Using advanced artificial intelligence algorithms—coupled with an easy-to-use graphical interface—Palantir helps trained investigators identify connections between disparate databases to rapidly discover links between people.

Before Palantir was implemented, analysts missed these connections because field interview (FI) data, department of motor vehicles data, and automated license plate reader data was all held in separate databases. The department also lacked situational awareness about where their patrol cars were and how they were responding to requests for help. Palantir integrated analytic capabilities like "geospatial search, trend analysis, link charts, timelines, and histograms" to help officers find, visualize, and share data in near-real time [13].

Figure 17.6 shows an example of a social network analysis mapping to look for correlations across multiple data sources.

The focus of the investigation begins in the lower right. A police report was filed for a crime scene (known spatial location). A red Camaro with license plate number QX 5104 was reported at the scene. An unknown entity must have been driving the red Camaro and may have either participated in a criminal activity or may have witnessed the crime. License plate reader data indicates that the red Camaro (QX 5104) was also seen at 487 Evergreen Trail at some time in the past (sequence neutrality). James Phillips, previously has an arrest record, lives at this residence. The red Camaro is not registered at this address, rather a black Mustang (NCC 1701) is registered at the address according to DMV records. James Phillips owns telephone 555-1345.

This network map combined information from police reports, arrest records, financial records, telephone records, DMV records, real estate, and public news. Investigators now have several courses of actions. First, they would almost certainly put out an alert for the red Camaro. They may choose to question Mr. Phillips about the red Camaro seen at his residence. Does he know the owner? Where might he/she be now? They may also seek a warrant to review Mr. Phillips' call logs. With whom did he speak around the time the red Camaro was seen at his house? Who owns that phone? Police may also place surveillance on 487 Evergreen Trail. If they have reasonable suspicion that Mr. Phillips was involved—perhaps based on his previous arrest records or nefarious financial activities—they may seek a warrant to tap his cell phone.

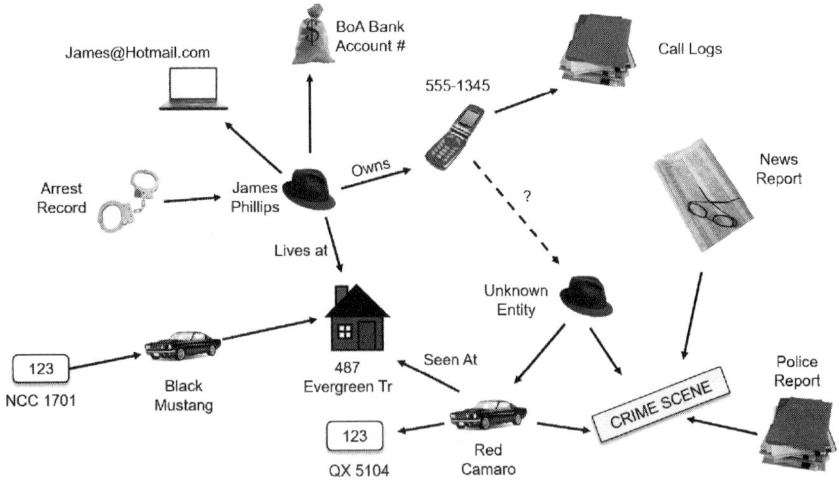

Figure 17.6 Example of social network analysis to correlate network information to identify an unknown entity.

Network analysis correlates data across multiple databases so previously undiscovered relationships are more readily apparent to the analyst. This technique has become increasingly important as adversaries exploit the advantageous properties of a distributed network. William Wechsler, deputy assistant secretary of defense for counternarcotics and global threats noted, "a network of adversaries requires a network to defeat it" [14]. Data integration, information sharing, and application of intelligence methodologies become important tools in combating these diverse, network-based threats.

17.5 Predictive Policing

Techniques like crime mapping, intelligence-led policing, and network analysis, when used together, enable all five principles of ABI and move toward the Minority Report nirvana described at the beginning of the chapter. This approach has been popularized as "predictive policing."

Jeff Brantingham, a professor of anthropology at UCLA, developed a series of predictive algorithms commercialized as PredPol. The software is based on the premise that "human behavior, especially when in search of resources, follows very predictable patterns" [15]. Once thieves (or foragers) have success in an area, they tend to return to that area and commit the same crime. Unlike *Minority Report,* PredPol does not predict who will commit the crime, but integrates spatiotemporal information to predict "where and when the crime is most likely to occur" [16]. The software produces 500 x 500 ft colored boxes on a map that dynamically move based on patrol data, time of data, and prevalence of other crimes. This information is fed to patrol cars to provide situational awareness-quality information that redirects patrols to high-threat areas. Although some critics have questioned the validity of PredPol's predictions, "during a four-month trial in Kent [UK], 8.5% of all street crime occurred within PredPol's pink boxes...predictions from police analysts scored only 5%" [17]. An example of the PredPol interface is shown in Figure 17.7. The darker colored boxes indicate a higher probability of crime.

PredPol calculates the predicted change in crime, $\lambda(x,y,t)$, as:

$$\lambda(x,y,t) = \mu(x,y) + \sum_{\{k:t_k<t\}} g(x - x_k, y - y_k, t - t_k; M_k) \tag{17.1}$$

where $\mu(x,y)$ is the background rate of the crime type independent of time and g is a kernel density function based on a marked point process. This is a mathematical representation of the notion that crime tends to happen in a similar place, and decays according to a predefined rate. Brantingham asserts that this type of decay is similar to modeling aftershocks in earthquakes: The risk of

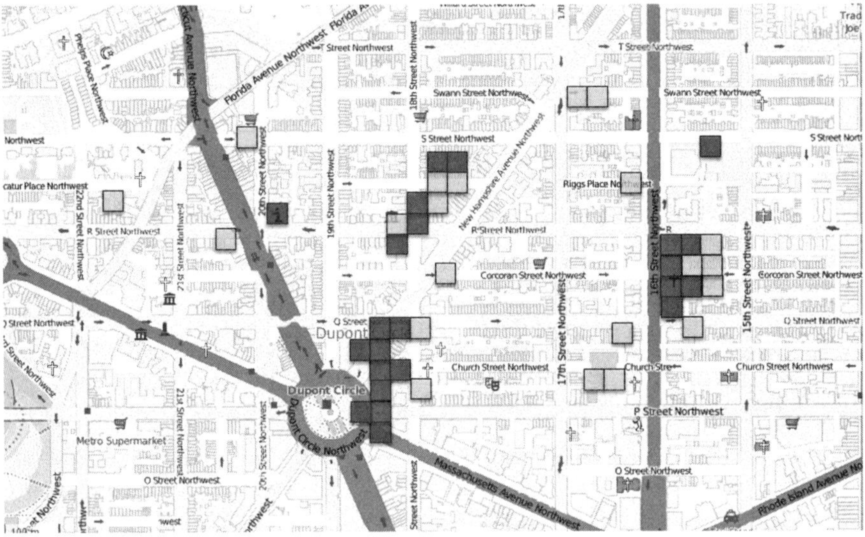

Figure 17.7 Example of PredPol prediction boxes. (Adapted from [18, p. 1]. Map data source: OpenStreetMap.)

influence increases with the magnitude of the event and decreases in space and time away from each event. Predpol uses the isotropic kernel [19, p. 101].

$$g(x,y,t;M)=\frac{K_0}{(t+c)^p}\cdot\frac{e^{\alpha(M-M_0)}}{\left(x^2+y^2+d\right)^q} \tag{17.2}$$

where K_0, M_0, and α control the number of aftershocks; (c,d) control the kernel distribution function, and (p, q) define the exponential rate of decay [20]. The coefficients of these equations are tuned to calibrate the model for the geographic area and type of crime.

Variations on this approach are used in other predictive policing programs, including CrimeStat, a program distributed by the National Institute of Justice. Gerber demonstrated how the kernel density estimation technique can be used to predict crime from Twitter data [21]. The technique is also used in spatial modeling of human behaviors including the settling patterns and housing voucher distribution [22] and the spatiotemporal changes in food retailing [23].

In Section 17.2.2, we introduced routine activities theory, which posits that crimes tend to happen near the same place and same time because humans are creatures of habit. Once we find something that works, we keep doing it. CompStat and crime mapping introduced the capability to model spatial distributions of crime to look for trends and patterns—but implementation of policing strategies based on these techniques alone relies on a general "hunch" based on a visual examination of spatial data. Predictive policing, as implemented by PredPol, adds a closed-form analytic solution to "predict" where crime will happen based on aggregation of spatial and temporal information. Essentially, it is an algorithmic implementation of routine activities theory.

17.6 Summary

Law enforcement and intelligence analysis share many similar techniques. Both disciplines require an understanding of trends and patterns. They focus heavily on resolving the identity of unique high-priority entities. Anticipating what may happen and applying courses of action to influence outcomes is a key desired effect. Increasingly, both law enforcement and intelligence analysis are applying spatiotemporal analytic techniques to characterize activities and transactions, understand patterns of life, and resolve unknowns.

17.7 Further Reading

Voluminous literature is available on intelligence-led policing. A number of federal and state departments of justice provide free handbooks online. Interested readers are encouraged to review *Predictive Policing: The Role of Crime Forecasting in Law Enforcement Operations,* an incredibly detailed and visually stunning account of the methods and techniques published in 2013 by RAND for the National Institute of Justice (ISBN: 978-0-8330-8148-3).

17.8 Chapter Author Biography

Sarah Hank is a GIS analyst in the energy consulting industry in Washington, D.C., where she has lived since 2005. She creates mapping projects for work and fun with ArcGIS, QGIS, MapBox, and TileMill. In 2011, she provided GIS training to volunteers at a nongovernment organization working to enhance awareness of sexual harassment in Egypt through crowd mapping. Her interests and expertise include crime mapping, gentrification, water security, and political geography of the Middle East. Ms. Hank holds a B.A. in geography and international affairs with a minor in GIS from George Washington University. A summary of her work in the field is available online at www.sarahkhank.com.

References

[1] *Minority Report.* Dir. Ronald Shusett, DreamWorks Pictures, 2002. Film.

[2] McGarrell, E. F., J. D. Freilich, and S. Chermak, "Intelligence- Led Policing As a Framework for Responding to Terrorism," *Journal of Contemporary Criminal Justice,* Vol. 23, No. 2, May 2007, pp. 142–158.

[3] Treverton, G., et al., "Moving Toward the Future of Policing," RAND Corporation, RAND MG1102, 2011.

[4] Anderson, R., *Intelligence Led Policing: International Perspectives on Policing in the 21st Century,* produced by the International Association of Law Enforcement Intelligence Analysts, Inc., 1997, https://members.ialeia.org/files/other/KP%20intl%20perspectives.pdf.

[5] Dussault, R., "Jack Maple: Betting on Intelligence," March 31, 1999.

[6] "About the Uniform Crime Reporting (UCR) Program," FBI, 2011

[7] "Metropolitan Police Department Statistics and Data." Web. Available: http://mpdc.dc.gov/page/statistics-and-data.

[8] Vulliamy, E., "Drugs: Redemption in Crack," *The Observer,* October 23, 1994.

[9] Layton, L., "Metrobuses Face Rock Attacks on Streets of Southeast D.C.," *The Washington Post,* August 3, 2003.

[10] Larimer, S., "Is the Green Line Dangerous? Or Just Misunderstood?" TBD.com, April 28, 2011.

[11] "Palantir Technologies: Our Platforms," web. Available: http://www.palantir.com/platforms/.

[12] "Palantir: An Open Source Development Success Story," *Directions Magazine,* February 12, 2013.

[13] "Palantir Impact Study: Responding to Crime in Real Time at the LAPD," Palantir Technologies, web, March 2014.

[14] Miles, D., "Drug Trafficking Threatens National Security, Official Says," American Forces Press Service, May 17, 2012.

[15] Kahn, C., "At LAPD, Predicting Crimes Before They Happen," NPR, November 26, 2001.

[16] Berg, N., "Predicting crime, LAPD-style," *The Guardian,* June 25, 2014.

[17] "Predictive Policing: Don't Even Think About It," The Economist, July 20, 2013.

[18] "PredPol Predicts Gun Violence with Open Government Data," PredPol, web.

[19] Mohler, G. O., et al., "Self-Exciting Point Process Modeling of Crime," *Journal of the American Statistical Association,* Vol. 106, No. 493, March 2011, pp. 100–108.

[20] Ogata, Y., "Space-Time Point Process Models for Earthquake Occurrences," *Annals of the Institute of Statistical Mathematics,* Vol. 50, No. 2, 1998, pp. 379–402.

[21] Gerber, M. S., "Predicting Crime Using Twitter and Kernel Density Estimation," *Decision Support Systems,* Vol. 61, May 2014, pp. 115–125.

[22] Wilson, R., "Using Dual Kernel Density Estimation to Examine Changes in Voucher Density Over Time," *Cityscape: A Journal of Policy Development and Research,* Vol. 13, No. 3, 2012, pp. 225–234.

[23] Jansenberger, E. M. and P. Staufer-Steinnocher, "Dual Kernel Density Estimation as a Method for Describing Spatio-Temporal Changes in the Upper Austrian Food Retailing Market," presented at the *7th AGILE Conference on Geographic Information Science,* Heraklion, Greece, 2004.

18

ABI and the D.C. Beltway Sniper

Mark Phillips

On the morning of October 2, 2002, James Martin was shot and killed by a sniper outside of a grocery store in Silver Spring Maryland, a suburb of the nation's capitol. So began a 23-day reign of terror, punctuated by random killings across two states and the District of Columbia. The killers were arrested successfully through the dedicated efforts of hundreds of law enforcement personnel. This chapter takes a closer look at the events that took place through the lens of the four pillars of ABI applied to entity resolution in this law enforcement and homeland security context.

18.1 Introduction

From October 2, 2002, until their eventual capture on October 24, 2002, Lee Boyd Malvo and John Muhammad randomly killed 10 people and critically wounded three in the environs around Washington, D.C., (Table 18.1). Furthermore, using forensic techniques based on the evidence gathered, Malvo and Muhammad were eventually connected to seven additional murders as well as seven other nonfatal shootings stretching from Washington State, to Louisiana, to Florida as shown in the network diagram in Figure 18.1 [1, 2]. Investigations into the "D.C. Sniper" shootings were led by the Montgomery County (MD) police department, with assistance from the FBI and the Bureau of Alcohol, Tobacco, and Firearms (ATF) as well as other state and local law enforcement. The FBI alone had over 400 agents assigned to the investigation [3].

Law enforcement personnel handled thousands of tips in the hopes that they would lead to the apprehension of the shooters. "During the course of the investigation, the telephone tip lines received more than 100,000 calls

Table 18.1
List of D.C. Beltway Sniper Victims and Associated Events [4]

Victim	Place	Date	Time
1	Wheaton, MD	October 2, 2002	6:04 AM
2	Rockville, MD	October 3, 2002	7:41 AM
3	Aspen Hill, MD	October 3, 2002	8:12 AM
4	Silver Spring, MD	October 3, 2002	8:37 AM
5	Kensington, MD	October 3, 2002	9:58 AM
6	Washington, D.C.	October 3, 2002	9:20 PM
7	Fredericksburg, VA	October 4, 2002	2:30 PM
8	Bowie, MD	October 7, 2002	8:09 AM
9	Manassas, VA	October 9, 2002	8:18 PM
10	Fredericksburg, VA	October 11, 2002	9:40 AM
11	Falls Church, VA	October 14, 2002	9:19 PM
12	Ashland, VA	October 19, 2002	8:00 PM
13	Aspen Hill, MD	October 22, 2002	5:55 AM

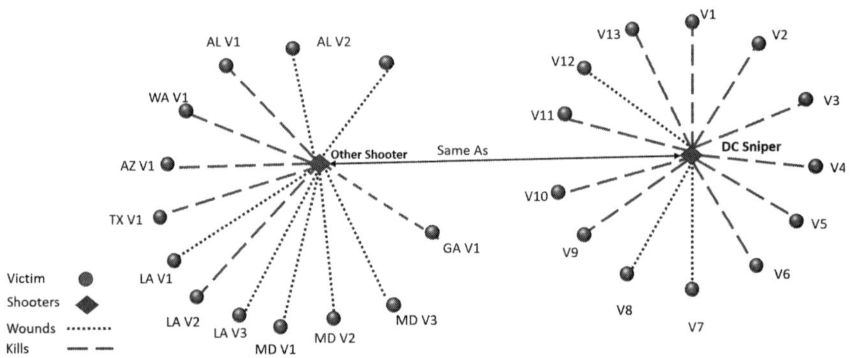

Figure 18.1 Malvo and Muhammad–related shootings, as determined by forensic investigation of multiple crimes. (Data Source: [1, 4].)

generating some 16,000 investigative leads" [5, p. 62]. The amount of information was daunting. Investigators noted, "While the amount of potentially valuable information may increase substantially with each agency that joins the investigation, there is a commensurate increase in the demand for efficient analysis. The sheer amount of material can overwhelm investigative personnel" [5, p. 17].

Although the ABI methods were not formalized at the time of this incident, this chapter retrospectively shows how ABI could be applied to problems of this type. We examine the problem from the standpoint of the four pillars: georeference-to-discover, integrate before exploit, sequence neutrality, and data neutrality. Given the benefit of a historical review, how might the pillars support the investigation? Are there situations where applying the pillars would have changed the results? If the pillars would have yielded greater efficiency in analysis, did law enforcement have the tools to apply the techniques?

18.2 Georeference to Discover

As the events of October 2002 unfolded, the public became acutely aware of the ongoing terrorist threat and fear griped the region. Authorities reached out to the public to provide tips and the amount of information provided to law enforcement went up exponentially. Some of it was useful but a lot of it turned out to be "noise" (i.e., information that was either wrong or had no bearing on the case). As mentioned in the opening paragraphs of this chapter, 16,000 of the tips provided to police resulted in investigative leads. Add to that number the vast amount of information collected by law enforcement in the performance of their investigation. In this particular case, the investigators had a large amount of "sparse information"; there was seldom enough evidence at a crime scene alone, or tips alone, to get a picture of the criminals. This is an ideal situation for the application of georeference to discover. Figure 18.2 shows the georeferenced locations of the shootings across Virginia, Maryland, and the District of Columbia. Despite efforts to discern a predictable pattern, georeferencing the individual shootings yields little information because of the sparsity and random distribution of the events.

Early in the investigation, the identities of Mohammad and Malvo were unknown, so georeferencing discrete locations associated with them or their relatives in the region is not possible until at least October 18. Literature searches do not indicate whether the tips called into the Joint Operations Center for the Montgomery County, MD, police or the accounts available from witnesses at the crime sites were georeferenced. Sightings of cars associated with the various crime scenes appear not to have been cataloged and georeferenced. In an eerie episode of "incidental collection," on October 8, Baltimore Police stopped a dark Caprice—the type of vehicle identified as potentially at the shooting in D.C. five days earlier—for erratic driving. Muhammad is the driver but since there is no warrant for him at this time in Maryland, the car is released [6].

In order to apply georeference to discover with the data available, there would need to be a common, structured database that was easily searchable and could be tied to some geospatial intelligence system. Such a system did not exist in 2002. Additionally, the criminal databases of Washington, Louisiana, Maryland, and Virginia were not tied together in a common searchable way. Federal law enforcement assistance was needed to coordinate across states. Technology is available today to create a system that allows georeferencing and integration

Figure 18.2 Georeferenced data for the Mohammad and Malvo victims. (Map data source: OpenStreetMap.)

of data across multiple crime scenes and databases. This type of system may help police investigators analyze data in a new and more efficient manner.

18.3 Integration Before Exploitation

Integration before exploitation is a closely tied corollary to georeference to discover, and its value is starkly evident in the review of the investigation. Early in the investigation police became interested in a white cargo truck or van [7]. Rather than georeference all information available (including the sighting of the gray Caprice at the shooting on the evening of October 3) to provide strength and add context to the data, police appeared to preferentially focus on one piece of data over others. The next several days were filled with tracking and stopping white cargo vans, none of which proved to be related to the shootings [8].

The issue of police following up on a hot lead is not in question. However, to use ABI terms, the entity (van) was not resolved to the degree that it could be identified. For example, the Ford E-series (also known as the Econoline) utility van is one of the top 20 automobiles sold in the United States [9]. Its default color is white. There are thousands of white cargo vans in the D.C. metro area alone. This is a case where the law enforcement officials were under such incredible pressure to gain traction in the case they exploited a clue and pursued that one particular clue, especially since it fit some of their preconceived notions of how the shootings occurred [8].

A better approach, which may not have been possible at the time due to information available or the means to integrate it, would have been to georeference all crime scene and surrounding area data. Then, using these data points add context to the crime scene information. In fact, the FBI was asked to make digital maps of the crime scenes; these maps, when taken together with tips and other information, might have proved a good basis for the investigators' georeferencing. Doing this would have allowed police to note that although a white van was present at a crime scene, a dark sedan (once described as appearing to be an old police car) was also present, perhaps leading in another investigative direction. Although neither is definitive nor a resolved entity, once additional information is available and georeferenced there may be additional identifiers for the cars and their owners, especially as potential discrete locations for Mohammad and Malvo were identified later in the investigation.

18.4 Sequence Neutrality

The FBI's historical review of the case notes a big break in the case came from the snipers themselves. On October 17, police received a phone call from one of the snipers. During this conversation, the caller bragged that "he was responsible for the murder of two women (actually, only one was killed) during the robbery of a liquor store in Montgomery, Alabama, a month earlier" [10]. Given this link, police in Alabama were able to link the caller to a particular crime in Montgomery. Fortunately for law enforcement, the shooter left his fingerprints on a weapon magazine at the site of the shooting. The police also had in their possession ballistic information regarding the weapons used in this shooting. Federal authorities we able to positively identify the shooter as Lee Boyd Malvo by matching the fingerprints on the magazine to an open arrest warrant in Washington state.

Next, federal agents found a link to a known associate of Malvo: John Muhammad. The identification of the association between Malvo and Muhammad was provided through one of the many tips received for a previous shooting in Tacoma, Washington. This association provided the opportunity of obtaining two additional vital clues: 1) Muhammad owned a Bushmaster rifle (the weapon suspected in the shootings), and 2) Muhammad owned a Blue Chevrolet Caprice with a known license plate—a proxy for the entities. These pieces of information are the ones that actually led to the arrest of the shooters on October 24, 2002.

Figure 18.3 illustrates the associations between the shooters, the victims, and other entities involved in the case. This type of figure is known as a "graph analysis" or "link analysis" chart. Graph analysis allows the analyst or investigator to pull away from geographic space and see all the relationships contained within the sea of data. Many times this is a helpful complement to georeference to discover and frequently leads to unknown relationships of significance. In this case, from the graph chart alone, an analyst can understand the significance of the caller's linking of two crime scenes; the caller becomes a "critical node" to the investigation in graph parlance.

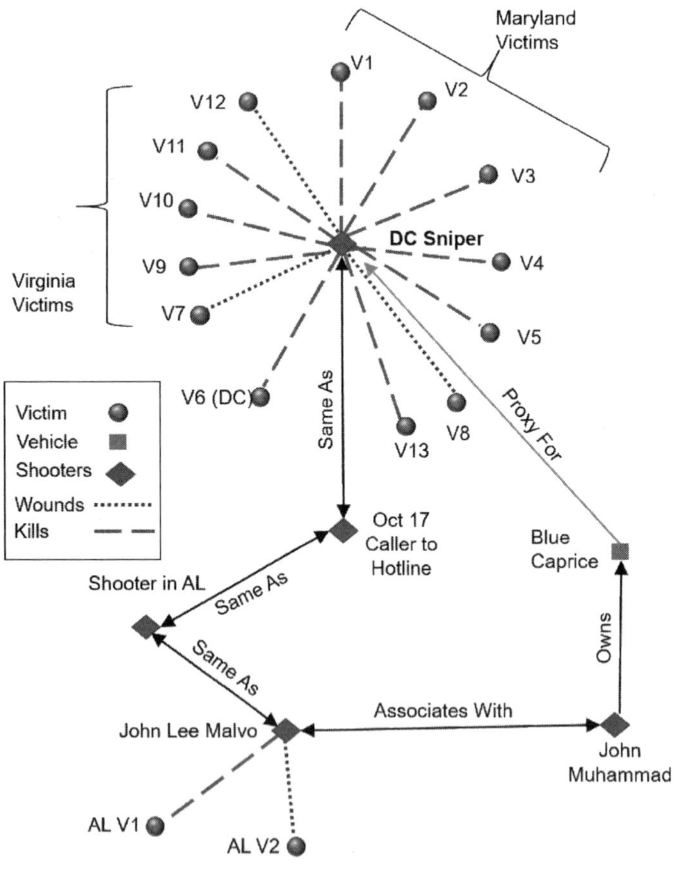

Figure 18.3 Relating evidence using sequence neutrality.

The analysis that was initiated by the shooter's call is a classic example of sequence neutrality. Recall that sequence neutrality allows for the fact that data collected in the past may answer a question today or that data collected today may give meaning to data collected in the past. In this case, Malvo's call to police provided a vital link to information collected in the past; namely tying the D.C. snipers to an unrelated shooting in Alabama. Too often, analysts and investigators seek new data instead of searching information collected in the past. Here the past crime provided not only the identity of the shooter, but also a known associate and the vehicle he drove. However, the linkage also ties information from the past to the present providing the whereabouts of the shooters from the Montgomery, Alabama, and Tacoma, Washington, cases.

18.5 Data Neutrality

Any piece of evidence may solve a crime. This is a well-known maxim within criminal cases and is another way of stating the ABI pillar of data neutrality. Investigators rarely judge that one piece of evidence is more important to a case than another with equal pedigree. Evidence is evidence. Coupled with the concept of data neutrality, crime scene processing is essentially a processes of incidental collection. When a crime scene is processed, investigators might know what they are looking for (a spent casing from a rifle) but may discover objects they were not looking for (an extortion note from a killer). Crime scene specialists enter a crime scene with an open mind and collect everything available. They generally make no value judgment on the findings during collection nor do they discard any evidence, for who knows what piece of detritus might be fundamental to building a case.

In the case of the D.C. sniper, many thousands of individual pieces of information were collected by investigators for 27 crime scenes across the country. Information was logged within the various state and local law enforcement systems. Minutia, no matter how small, was available to authorities, especially once federal access to the information made sharing across investigations possible. The following are some of the data points that were collected and led to the conviction of the snipers for the D.C. crimes as well as crimes across the country. (They are all different in nature and had any been discarded out of hand the investigations might have ended differently.)

- Spent rifle rounds from the D.C. shootings;
- A magazine with a fingerprint at a Montgomery, Alabama, liquor store;
- A tip that Malvo had an associate named John Muhammad;
- Muhammad's car and license plate;

- A laptop stolen from one of the victims [September 5, 2002, Paul LaRuffa (survivor), Clinton Maryland].

The lesson learned here, which is identical to the lesson learned within the ABI community, is to collect and keep everything; one never knows if and when it will be important.

18.6 Summary

The horrific events that comprise the D.C. snipers serial killing spree makes an illustrative case study for the application of the ABI pillars. By examining the sequence of events and the analysis that was performed, the following conclusions can be drawn. First, georeferencing all data would have improved understanding of the data and provided context. Unfortunately, the means to do that did not exist at the time. Second, integrating before exploitation might have prevented law enforcement from erroneously tracking and stopping white cargo vans. Again the tools to do this integration do not appear to have existed in 2002.

Interestingly, sequence neutrality and data neutrality were applied to great effect. Once a caller tied two separate crimes together, law enforcement was able to use all the information collected in the past to solve the current crime. Additionally, information that was collected incidentally and not discarded provided the critical piece of information to apprehend Malvo and Muhammad.

The events of October 2002 occurred long before the principles of ABI were codified for intelligence analysis. Today, formal exchange in tradecraft and methods between the intelligence, law enforcement, and homeland security communities normalizes best practices and shares analytic tools—especially spatially indexed databases, geographic information systems, and visual analytic techniques.

18.7 Chapter Author Biography

Mark Phillips is a senior systems engineer with over 30 years of experience in leading and contributing to technical development and acquisition programs for the Department of Defense. Phillips was the principal author of a series of defining strategy papers on ABI issued by the Office of the Under Secretary of Defense for Intelligence. He also served as a subject matter expert on ABI for the Department of Defense. Phillips retired after a 20-year career in the U.S. Navy. He has since supported the U.S. government in various roles including as a senior executive at the Missile Defense Agency, chief engineer on several surveillance programs, and as program manager supporting the intelligence community. Phillips holds a B.S. in physics from the Catholic University of America and an M.S. in electrical engineering from the University of Central Florida.

References

[1] Kovaelski, S. F., and M. E. Ruane, "Before Area Sniper Attacks, Another Deadly Bullet Trail," *The Washington Post,* December 15, 2002, p. A01.

[2] Roberts, J., "Antigua Sniper Connection?," CBS News, November 4, 2002.

[3] Federal Bureau of Investigation, "A Byte out of History, The Beltway Snipers, Pt. 1," FBI, October 22, 2007.

[4] "Beltway sniper attacks," Wikipedia.

[5] Murphy, G. R., and C. Wexler, "Managing a Multijurisdictional Case: Identifying the Lessons Learned from the Sniper Investigation," *Police Executive Research Forum,* U.S. Department of Justice, October 2004.

[6] "Sniper Investigation Timeline," ABC News, January 7, 2006.

[7] "Timeline: Tracking the Sniper's Trail," FoxNews.com, October 29, 2002.

[8] Clines, F. X., "Widening Fears, Few Clues As 6th Death Is Tied to Sniper," *The New York Times,* October 5, 2002.

[9] "Ford E-Series," Wikipedia.

[10] Federal Bureau of Investigation, "A Byte out of History: The Beltway Snipers, Pt. 2," FBI, Oct 2007.

19

Analyzing Transactions in a Network

William Raetz

One of the key differences in the shift from target-based intelligence to ABI is that targets of interest become the output of deductive, geospatial, and relational analysis of activities and transactions. As RAND's Gregory Treverton noted in 2011, imagery analysts "used to look for things and know what we were looking for. If we saw a Soviet T-72 tank, we knew we'd find a number of its brethren nearby. Now…we're looking for activities or transactions. And we don't know what we're looking for" [1, p. ix]. This chapter demonstrates deductive and relational analysis using simulated activities and transactions, providing a real-world application for entity resolution and the discovery of unknowns.

19.1 Analyzing Transactions with Graph Analytics

Graph analytics—derived from the discrete mathematical discipline of graph theory—is a technique for examining the relationship between data using pairwise relationships. Numerous algorithms and visualization tools for graph analytics have proliferated over the past 15 years. This example demonstrates how simple geospatial and relational analysis tools can be used to understand complex patterns of movement—the activities and transactions conducted by entities—over a city-sized area. This scenario involves an ABI analyst looking for a small "red network" of terrorists hiding among a civilian population.

This example uses a synthetic data set created by the Institute for Defense Analysis (IDA). The IDA dataset covers 5,445 individual locations, 4,623 entities, and 116,720 vehicle tracks, over the course of three days [2]. The location covered is a portion of Baghdad (see Figure 19.1). The synthetic tracks are

Figure 19.1 Sample of synthetic track data over a map of Baghdad. (Map data source: OpenStreetMap. Track data courtesy of IDA [2].)

considered "ground truth"—that is devoid of sensor artifacts and noisy data typical of real-world situations; however, the metadata and typical values simulate a realistic scenario. Hidden within the normal patterns of the 4,623 entities is a malicious network. The purpose of this exercise is to analyze the data using ABI principles to unravel this network: to discover the signal hidden in the noise of everyday life.

The concepts of "signal" and "noise," which have their origin in signal processing and electrical engineering, are central to the analysis of nefarious actors that operate in the open but blend into the background. Signal is the information relevant to an analyst contained in the data; noise is everything else. For instance, a "red," or target, network's signal might consist of activity unique to achieving their aims; unusual purchases, a break in routine, or gatherings at unusual times of day are all possible examples of signal.

Criminal and terrorist networks have become adept at masking their signal—the "abnormal" activity necessary to achieve their aims—in the noise of the general population's activity. To increase the signal-to-noise ratio (SNR), an analyst must determine inductively or deductively what types of activities constitute the signal. In a dynamic, noisy, densely populated environment, this is difficult unless the analyst can narrow the search space by choosing a relevant area of interest, choosing a time period when enemy activity is likely to be greater, or beginning with known watch listed entities as the seeds for geochaining or geospatial network analysis.

19.2 Discerning the Anomalous

Separating out the signal from the background noise is as much art as science. As an analyst becomes more familiar with a population or area, "normal," or background, behavior becomes inherent through tacit model building and hypothesis testing. For example, commutes from the suburbs into the city during the week may be normal activity in Atlanta, while commutes that begin and end within the city may be normal in New York. In ABI analysis understanding normal behavior is important; identifying the background will help draw attention to the abnormal.

The goals and structure of the target group define abnormal activity. For example, the activity required to build and deploy an improvised explosive device (IED) present in the example data set will be very different from money laundering. A network whose aim is to build and deploy an IED may consist of bomb makers, procurers, security, and leadership within a small geographic area. Knowing the general goals and structure of the target group will help identify the types of activities that constitute signal.

In this sample scenario, the terrorist network is well established and has carried out IED attacks in the recent past. The analyst hypothesizes that they require dedicated safe house locations to warehouse materials and build devices. He also assumes that these discrete locations are not used for any other purpose. In order to maximize the SNR, the focus will be on locations instead of entities—georeferencing the activities around these locations to discover specific entities and their patterns. Nondiscrete locations where many people meet will have a more significant activity signature. The analyst will also have to consider how entities move between these locations and discrete locations that have a weaker signal but contribute to a greater probability of resolving a unique entity of interest. An abnormal pattern of activity around these discrete locations is the initial signal the analyst is looking for.

At this point, the analyst has a hypothesis, a general plan based on his knowledge of the key types of locations a terrorist network requires. He will search for locations that look like safe house and warehouse locations based on events and transactions. When the field has been narrowed to a reasonable set of possible discrete locations, he will initiate forensic backtracking of transactions to identify additional locations and compile a rough list of *red network* members from the participating entities. This is an implementation of the "where-who" concept from Chapter 5.

19.3 Becoming Familiar with the Data Set

After receiving the data and the intelligence goal, the analyst's first step is to familiarize himself with the data. This will help inform what processing and analytic tasks are possible; a sparse data set might require more sophistication, while a very large one may require additional processing power. In this case, because the available data is synthetic, the data is presented in three clean comma-separated value (.csv) files (Table 19.1). Analysts typically receive multiple files that may come from different sources or may be collected/created at different times. Sometimes, extensive data conditioning (Chapter 12) is required before the files can even be opened in the analyst's exploitation environment.

In this dataset, the Area of Interest (AOI) is the center of Baghdad, shown in Figure 19.1. Because this is a synthetic dataset, each building has been given a unique ID that is used to mark the beginning and end of vehicle tracks in the tracks file.[1] This means that the analyst can easily observe transits between locations. Otherwise, he would have to write an algorithm to infer the beginning and end locations based off of foundation data and the coordinates of the vehicle track.

Because ABI analysis is inherently spatial and temporal, one of the first steps in data familiarization is to get the data onto a map. To get a feel for how these vehicle tracks look, the analyst pulls a sample of track point data from the tracks file and plots these coordinates over a street map of Baghdad using Python and Basemap as shown in Figure 19.1. Python is an easy-to-use open-source programming language increasingly used for simple scripting and large-scale analytics [3]. Basemap is a toolkit for plotting two-dimensional map data in Python and integrating those maps with processing scripts. The open-source

Table 19.1
Data Tables and Contents for the Synthetic Data Set

Table	Properties
Buildings	Building ID, Location ID
Tracks	Track ID, Location IDs
Entities	Entity ID, Location ID

1. The synthetic data set used for this example is idealized, because it represents complete, unambiguous tracks between defined locations. In practice, data sets are typically more sparse and noisy.

framework developed by Jeffrey Whitaker provides similar functionality to the MATLAB mapping toolbox [4].

19.4 Analyzing Activity Patterns

In order to better understand the activity that the analyst will use to find suspicious locations, the next step is to take a look at the activity patterns for some locations to get a broad feel for the patterns of life in this AOI. The analyst plots the magnitude of track activity for a given location over time; in other words, a plot of how busy a location is at each time of day and when each geospatial node is likely to function as a discrete location. Figure 19.2 shows a plot of activity for an example apartment building over the course of a day. Because Python scripts are distributable and flexible, it is easy to produce plots similar to Figure 19.2 with a few keystrokes, allowing rapid triage and characterization of the locations in the data.

To perform the calculation, the analyst writes a script to determine the number of tracks tangent to each location and sums them over time. This script breaks the tracks up by the half-hour, so there are 48 "bins" for activity magnitude. Clearly visible here is a sharp spike at around 8 a.m. and another sustained rise around 8 p.m. This seems to indicate that the residents of this apartment go to work at more or less the same time, and some stay out longer than others. Also visible here is a small bump around lunchtime; some people come home for lunch.

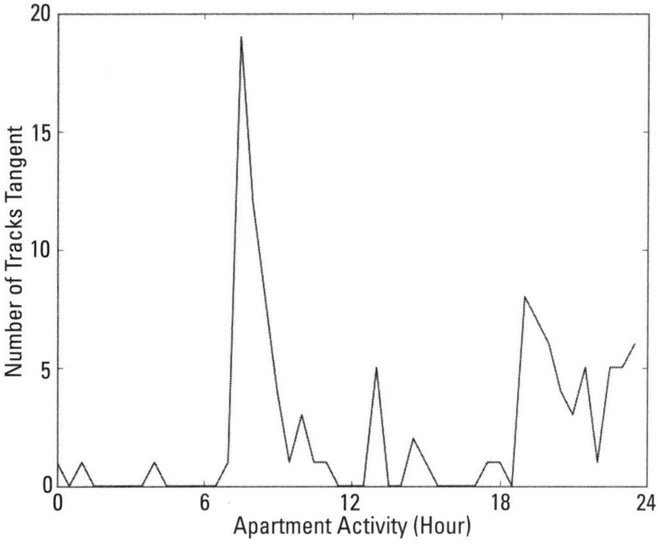

Figure 19.2 Example 24-hr activity pattern: apartment.

To get an idea of how a workplace and home location differ, the analyst creates the same plot for a busy work location, shown in Figure 19.3. Immediately apparent here is the same double peak in activity magnitude, first between 6 a.m. and 9 a.m. when people arrive, and around 6 p.m. when people leave. A small bump is visible here when people to out to get lunch as well. The analyst can assume that this is an active workplace, not somewhere that a terrorist network would choose to use as a warehouse or safe house.

A restaurant (Figure 19.4) exhibits an activity pattern that differs from residences and workplaces. With high levels of activity only at certain times of the day, a pattern of visitors at nontraditional mealtimes might be a clue that this restaurant has an additional purpose. The restaurant in Figure 19.4 displays a clear peak in the morning, at lunch, and a much smaller peak around dinner. This is a very clear example of how activity patterns can change from location type to location type, even for the same class of location.

It is important to note that the activity patterns for a location represent a pattern-of-life element for the entities in that location and for participating entities. The pattern-of-life element provides some hint to the norms in the city. It may allow the analyst to classify a building based on the times and types of activities and transactions (Section 19.4.1) and to identify locations that deviate from these cultural norms. Deducing why locations deviate from the norm—and whether these deviations are significant—is part of the analytic art of separating signal from background noise.

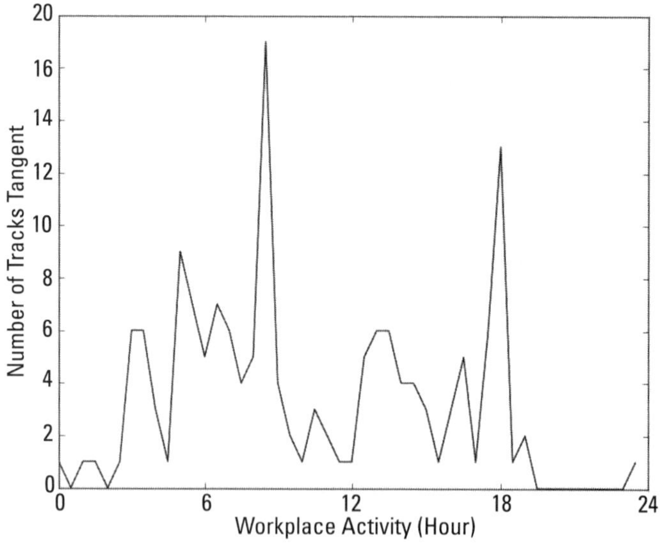

Figure 19.3 Example 24-hr activity pattern: workplace.

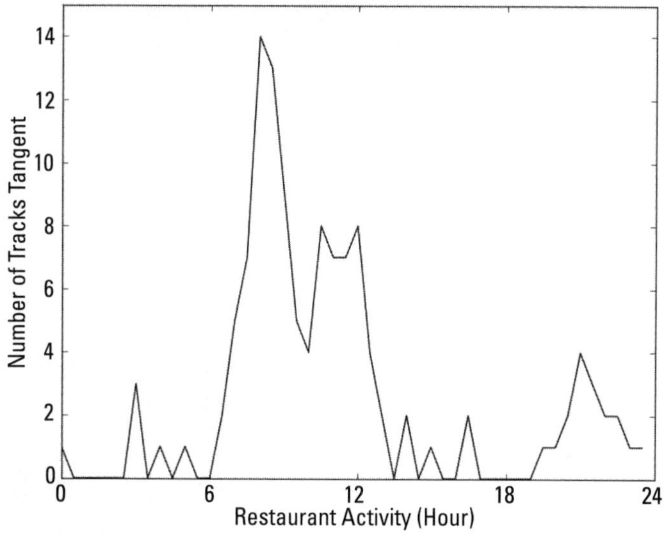

Figure 19.4 Example 24-hr activity pattern: restaurant.

19.4.1 Method: Location Classification

One of the most technically complex methods of finding suspicious locations is to interpret these activity patterns through a series of rules to determine which are "typical" of a certain location type. For instance, if a location exhibits a very typical workplace pattern, as evidenced by its distinctive double peak, it can be eliminated from consideration, based on the assumption that the terrorist network prefers to avoid conducting activities at the busiest times and locations.

In order to interpret noisy activity patterns, the analyst adapts an algorithm originally developed for seismic analysis, from UNC Chapel Hill [5]. First, the signal, here, the *activity pattern,* is smoothed at several different strengths by passing it through a one-dimensional Gaussian kernel at several scales. Figure 19.5 illustrates the original pattern, the dotted line, and its smoothed output, the dashed line. Smoothing the pattern in this way provides more significant peaks and troughs—note that the series of small peaks between 1 p.m. and 6 p.m. are smoothed, while the important morning peak remains recognizable. The local minima and maxima are found for the coarsest smoothed waveform, and then those locations are refined by sequentially reducing the strength of the smoothing function. The end result is an excellent approximation of the waveform peaks.

The locations of these waveform peaks make an excellent tool for comparing many locations quickly and easily to discover temporal pattern anomalies. For instance, the analyst can assume that a typical work location will have two peaks, about eight hours apart. Similarly, a restaurant similar to the one shown

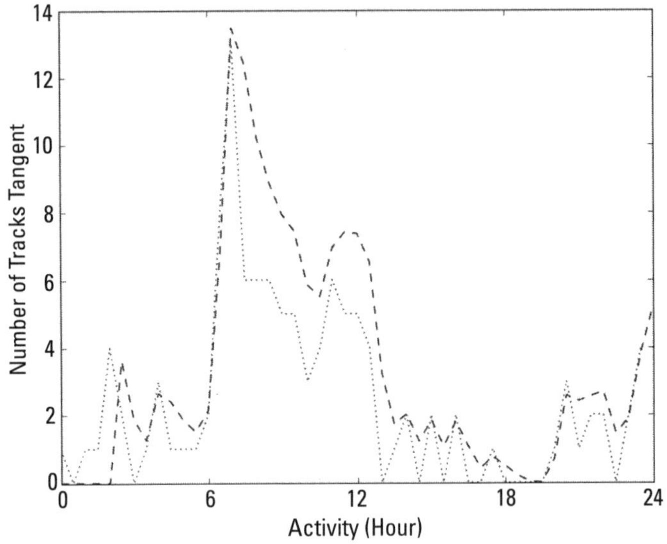

Figure 19.5 Smoothed 24-hr activity pattern.

previously might only have one peak, in the morning. In fact, these peak locations can be used as a rough filter to remove all of the "typical" locations from the list of possibilities. After all, the odds of a terrorist warehouse getting a rush of people at 9 a.m. and 6 p.m. are pretty slim.

The analyst implements a filter to remove locations that have two peaks between five and nine hours apart. He also runs this filter on only locations that have enough activity to be analyzed correctly (essentially nondiscrete locations). Since some home locations have three or four tracks that give false positives, locations with less than 50 activities are discarded. This is a safe assumption to make, as a review of the data reveals most typical work locations have at least 100 activities. The results of this filter are displayed in Table 19.2.

This filter seems to be very accurate, with few false positives for such a simple filter. However, it does not filter out that many possible locations.

In this example, the analyst only has access to location data (buildings) and tracks between them (GEOINT). The filter could be significantly improved by combining this analysis with other forms of activity like communications.

Table 19.2
Filter Results

Location Type	Number
Locations, total	5,445
Locations, filtered	105
False positives	5

19.4.2 Method: Average Time Distance

The method outlined in Section 19.4.1 is an accurate but cautious way of using activity patterns to classify location types. In order to get a different perspective on these locations, instead of looking at the peaks of activity patterns, the analyst will next look at the average time between activities.

In order to develop this filter, the analyst makes a few assumptions; work locations will be used often and will have a small average time between activities. They are almost always busy during the workday. Single-family homes will have a larger average, as they are used only by a few people. He also assumes that the suspicious locations are not being visited regularly, like a single family home, and that there is not a lot of activity, unlike a workplace, café or soccer field. Using these basic assumptions, nondiscrete and discrete locations can be separated based on their pattern of activity. Figure 19.6 shows a histogram of the average time between activities for a selection of 50 single-family home locations.

A majority of locations in the sample had average activity spacing between five and 10 hours. No locations had less than three hours between activities. Nondiscrete locations like apartment buildings, soccer fields, and other community locations will have far more activity, and correspondingly far lower average time between activities. The group average for a small sample of apartments is below one hour.

Because there is a distinctive and statistically significant difference between discrete and nondiscrete locations using the average time distance

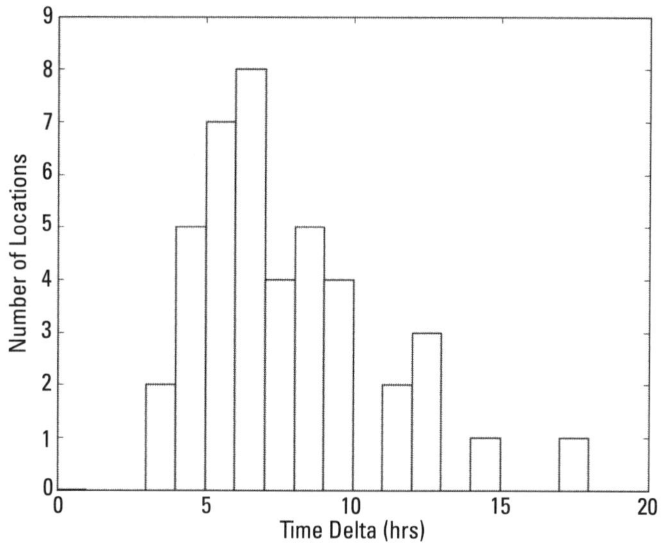

Figure 19.6 Average time delta for activities from a sample of single family homes.

technique, the analyst can use the average time between activities to identify probable home locations. He calculates the average time between activities for every available uncategorized location and treats all the locations with an average greater than three as single-family home locations.

Of the 5,445 locations in the sample, 4,800 are categorized as single-family homes with 45 false positives. The analyst then considers additional analysis questions that further refine the filter: Single-family homes are by far the largest single location type. What other data can make this type of filter more precise? Is there a way to break up single-family homes into subcategories? How would the addition of cell records change this filter? What could explain the second peak at 13 hours in Figure 19.6?

Recall that because the analyst initially identified safe houses and possible warehouses as the signal of interest, he filters the 4,800 projected single-family homes from this step of the analysis—although they can form the end of back-tracked transactions in future analysis.

19.4.3 Method: Activity Volume

The first steps of the analysis process filtered out busy workplaces (nondiscrete locations) and single-family homes (discrete locations), leaving the analyst with a subset of locations that represent unconventional workplaces and other locations that may function as safe houses or warehouses. While it may be possible to craft a series of filters of the type outlined in Section 19.4.1 to account for all of the various unconventional workplaces, the analyst's final filter capitalizes on a simpler assumption. Any warehouses or safe houses will have far lower levels of activity than busy barbershops or religious locations. The analyst uses an activity volume filter to remove all of the remaining locations that have many more activities than expected. He also removes all locations with no activities, assuming the red network used a location shortly before its attack.

A volume filter with a threshold of 20 activities per day, applied to the remaining locations, removes all but 109 locations. These locations are a mixture of single-family homes and the few apartments, markets, barbers, and workplaces that are both atypical and have abnormally low activity. Each of these locations exhibit suspicious activity patterns as defined by the analyst's initial assumptions. He considers additional questions: Is it possible to narrow this list further? What additional collection would be most useful? How can the addition of foundation data increase the accuracy of this filter?

19.4.4 Activity Tracing

The analyst's next step is to choose a few of the best candidates for additional collection. If 109 is too many locations to examine in the time required by the customer, he can create a rough prioritization by making a final assumption about the behavior of the red network by assuming they have traveled directly between at least two of their locations. If both of those locations are in the list of 109 interesting locations the analyst has compiled, there may be a transit (or several) between two locations on the list. This is a risky, but easy way to create a rough prioritization for further investigation by focusing first on the discrete locations of greatest likelihood of involvement with the red network.

By filtering the location list to only those remaining discrete locations connected by at least one track, the subset is further narrowed to 42 locations. These may be prioritized for additional analysis and collection.

19.5 Analyzing High-Priority Locations with a Graph

Three locations (identified by their Building IDs -93844, -93838, and -10532) from the subset of 42 stand out based on the number of transactions between them over the three-day period. To get a better understanding of how these locations are related, and who may be involved, the analyst creates a network graph of locations, using tracks to infer relationships between locations. The network graph for these locations is presented in Figure 19.7.

It is clear from this graph that the three locations further collection flagged as suspicious are closely related. It is also apparent that two other locations, -93832 and -93792, may be related as well. A consultation with foundation data reveals that both of these locations are home locations. The "entities" table shows one person associated with each -93832 and -93792, giving two people associated with a suspicious location.

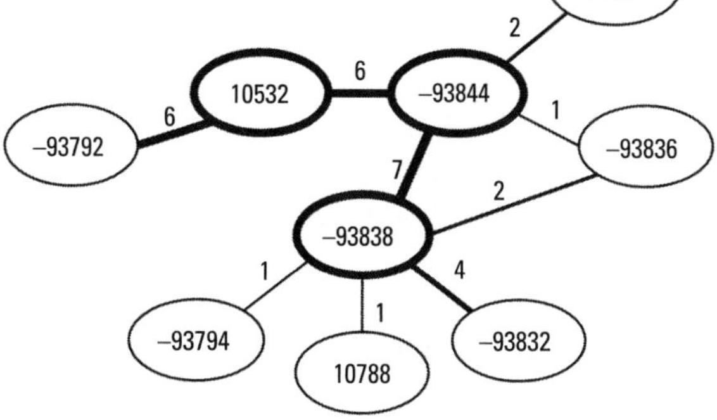

Figure 19.7 Network graph of transits between suspicious locations.

19.6 Validation

At this point, the analyst has taken a blank slate and turned a hypothesis into a short list of names and locations. A net reduction in volume from 5,445 to 42 represents a 99.2% decrease, without losing the important information. Following well-founded hypotheses and making smart assumptions allowed the analyst to tune out the noise and focus on the relevant behavior. These identities and locations provide an excellent jumping-off point for further collection, turning a difficult question into real answers.

Although the real world doesn't have truth data for testing hypotheses, the "ground truth," distributed with this synthetic data set, confirms that these are indeed members of the red network. Locations -10532, -93844, and -93838 are used by the red network, and the filtered list of 42 locations included three out of four total red network locations. While this type of analysis did not display only the important locations, it is important to remember that often analysis serves as a means for tasking more collection; as an analyst, presenting a collection manager with a list of 42 interesting locations is far more efficient than collecting over the entire neighborhood.

19.7 Summary

This example demonstrates deductive methods for activity and transaction analysis that reduce the number of possible locations to a much smaller subset using scripting, hypotheses, analyst-derived rules, and graph analysis. To get started, the analyst had to wrestle with the data set to become acquainted with the data and the patterns of life for the area of interest. He formed a series of assumptions about the behavior of the population and tested these by analyzing graphs of activity sliced different ways. Then the analyst implemented a series of filters to reduce the pool of possible locations. Focusing on locations and then resolving entities that participated in activities and transactions—georeferencing to discover—was the only way to triage a very large data set with millions of track points. Because locations have a larger activity signature than individuals in the data set, it is easier to develop and test hypotheses on the activities and transactions around a location and then use this information as a tip for entity-focused graph analytics.

Through a combination of these filters the analyst removed 5,403 out of 5,445 locations. This allowed for highly targeted analysis (and in the real world, subsequent collection). In the finale of the example, two interesting entities were identified based on their relationship to the suspicious locations. In addition to surveilling these locations, these entities and their proxies could be targeted for collection and analysis.

19.8 Chapter Author Biography

William Raetz is currently a systems engineer at the MITRE corporation. He has been a principal investigator for a research and development team studying ABI and analytic tools at Northrop Grumman, and he was a member of the team developing a prototype ABI analytic tool in 2011 and 2012. He used that experience to write a paper on ABI analysis methods that was later published in IEEE Explore [6]. He holds a B.A. in general engineering from Johns Hopkins University.

References

[1] Moore, D. T., *Sensemaking: a Structure for an Intelligence Revolution,* Washington, D.C.: National Defense Intelligence College Press, 2011.

[2] "Baghdad Synthetic Activity-Based Intelligence Data Set," Institute for Defense Analyses, 2010.

[3] "About PythonTM," Python.org.

[4] "Introduction — Basemap Matplotlib Toolkit 1.0.8 documentation," web. Available: http://matplotlib.org/basemap/users/intro.html.

[5] "Edge Detector 1D Tutorial," CISMM, web. Available: http://cismm.cs.unc.edu/resources/tutorials/edge-detector-1d-tutorial/.

[6] Raetz, W., "A New Approach to Graph Analysis for Activity Based Intelligence." Presented at the *2012 IEEE Applied Imagery Pattern Recognition Workshop (AIPR),* Washington, D.C., October 9–11, 2012.

20

ABI and the Search for Malaysian Airlines Flight 370

Alex Shernoff

Malaysia Airlines Flight 370 (MH370) was scheduled to land in Beijing, China, at 22:30 UTC on March 7, 2014. When the wheels of the Boeing 777-200ER aircraft carrying 239 crew and passengers never touched the waiting runway, a multinational investigation into its whereabouts was ignited. The plausibility of such a large, seemingly well-tracked object vanishing into thin air seemed to be nearly impossible given today's technology-fueled world where even our smartphones can find themselves. Several ABI principles were demonstrated in the search for the missing airliner. This chapter examines the incident chronologically as the events unfolded to illustrate how an ABI analyst would implement the four pillars of ABI in a time-dominant analysis scenario.

20.1 Introduction

MH370 was cruising at 35,000 ft and 542 miles per hour, heading northeast from Kuala Lumpur toward Ho Chi Minh City. At 17:19 UTC, MH370's last voice message was broadcast to air traffic control: "Good night Malaysian Three Seven Zero." Three minutes later, the transponder and the automatic dependent surveillance-broadcast (ADS-B) systems stopped transmitting [1]. The first sign of trouble came when MH370 failed to check into air traffic control in Ho Chi Minh City as it crossed into Vietnamese airspace. Vietnamese air traffic controllers requested the captain of another aircraft to attempt to contact MH370 using the international distress frequency. That captain reported success in the contact but just received mumbling and static in response [2]. The next attempt to contact the cockpit was in the form of a message instructing the pilots to contact Vietnam air traffic control which went unanswered [3]. There were at least two additional attempts to contact the cockpit of MH370 using the aircraft's satellite link: one at 18:25 UTC and one at 23:13 UTC [4]. After that, it is unclear what air traffic controllers and officials exactly knew the night of the flight, but there were no additional reported attempts to contact the plane the night of the disappearance. The key events recorded on the night of the disappearance are summarized in Table 20.1.

20.2 Data Sparsity, Suppositions, and Misdirections

MH370 officially was declared missing by Malaysia Airlines at 23:24 UTC on March 7, nearly an hour after the plane missed its scheduled arrival in Beijing [4]. The next days and weeks exposed how the lack of data and information available to lead investigators caused unintended delay and confusion. Investigators also had to separate the "who and why" from the "where and when." There may also have been distraction from the primary mission of locating the plane as the media, conspiracy theorists, and international governments made up their own version of "who and why" amidst the lack of information to the contrary.

At this point in the search, there was limited information available to officials. Stress, confusion, and panic mounted as the minutes and hours slowly ticked by with little additional information. The public was surprised at the lack of relevant information available immediately following the disappearance. Contrary to popular belief, commercial and military aircraft are not persistently

Table 20.1
Event Table Early in the MH370 Investigation

Event	Event Type	Location	Time (UTC)	Confidence
Takeoff from Kuala Lumpur	Physical	KUL	16:41	Confirmed
Crew Check-In	Comms-ATC	Enroute to PEK	17:01	Confirmed
Last ATC Transmission	Comms-ATC	Enroute to PEK	17:19	Confirmed
Last Transponder Contact	Radio	6°55′15″N 103°34′43″ E	17:21	High
Transponder and ADS-B turned off	Radio	?	17:22	High
Attempt to Contact	Radio	?	17:30	High
Message to Aircraft	Comms	?	18:03	High
Satellite Phone Attempt 1	Comms	?	18:25	High
Missed PEK Arrival Time	Physical	PEK	22:30	Confirmed
Satellite Phone Attempt 2	Comms	?	23:13	High

tracked throughout their flight regime. There are long periods, especially during overseas flights when the aircraft have limited contact with ground control and monitoring stations. With cell phones on "airplane mode," thus disabling most or all location tracking services, there are few electronic signals that can be tracked outside of aircraft's transponder and air-traffic control systems like the ADS-B.

Soon after the disappearance, conspiracy theories abounded. A widely circulated story noted that the transponder was "turned off" by the flight crew, but the only data available said it stopped transmitting (an example of humans developing a narrative to match data). Some news outlets postulated the plane was hijacked and taken to a secret runway in Afghanistan, a plot line more appropriate for a Tom Clancy thriller novel or James Bond movie [5]. In a rapidly evolving near real-time situation, there is little "historical" data to give context to the current situation. The absence of data causes hypotheses with little evidence to appear plausible, making them harder to debunk in the future and potentially derailing valid analysis of real data.

As events unfolded, the types of data considered during this mystery shifted from obvious to obscure sources of data, underscoring the importance of data neutrality. GPS tracks, radar returns, or voice reports of the plane's status or location—raw data that would have been valuable—was not available. Analysts turned to processed and derived data from obscure sources as the case progressed. Methods to process the sparse, available data in unique ways would take weeks or months to develop. If the passengers and crew were alive, their time was running out.

20.3 The Next Days: Fixating on the Wrong Entity

The main focus of the in the hours and days following the disappearance of MH370 was on resolving whether the pilots or passengers on board were involved in the incident. Given previous terrorist events surrounding major airliners, inductive reasoning led many to focus on either the pilots or passengers as the cause of the missing—and at this point presumed "hijacked" flight. Investigators focused on determining all they could about pilots and passengers, determining the details of "who" was involved as the trail on "what" and "where" got cold.

The pilot, Capt. Zaharie Shah had a homemade flight simulator, seized by Malaysian authorities and whisked away to an FBI crime lab in Quantico, Virginia, and Shah was named as the "chief suspect" in the disappearance [6]. 9/11 taught the world that hijackers practice in commercial flight simulators. A three-week search resulted in "nothing suspicious whatsoever" [7]: a failure of inductive reasoning.

Two Iranian men on MH370 were flying on stolen Italian and Austrian passports with tickets purchased from a Thai travel agency by an Iranian middleman [8, 9]. Induction: Middle-eastern men traveling on stolen passports are up to no good. Efforts to resolve their real identities found that the men were likely immigrating to Europe. Transiting through Malaysia was a frequent practice for illegal migration. After an investigation, Malaysian police chief Khalid Abu Bakar announced that there was no link to terrorist organizations [9]: another failure of inductive reasoning.

The hijacking/terrorism plotline postulated early on by the media, focusing the investigation on resolving the entity and intent was actually a detour that cost investigators valuable time, energy, and effort. In this case, the most important entity in the investigation was not the people involved. It was the plane.

In an investigation trying to resolve the plane's unknown location or to associate spurious multisource data with the missing airliner, an analyst would grab as much data as possible and georeference that data to discover any possible relationships. The type of data the analysts may be looking for wouldn't be available for several days or even weeks after the event occurred. As the searches began and the international media started to be more involved, the analyst will have to parse through data to determine the validity of each type and source of the data. This is a relatively daunting challenge for the analyst, and there may be several starts and stops as promising leads turn into dead ones.

A caution: ABI analysts must avoid entity fixation. Preconceived notions of what has happened in the past cause humans to extrapolate patterns into different situations. The ABI methodology guides the analyst to begin with the data, using deductive reasoning to eliminate what the answer is not. A more judicious way to locate the entity (the plane) would have been to map known information and eliminate where the aircraft could not be, focusing the search by correlating orthogonal data sources.

This brings us to the largest ABI mistake of the MH370 investigation. The night of the disappearance and for several days afterward, there was little or no data available to officials to help geolocate the aircraft. Thai military radars that could have tracked the plane produced ambiguous data and may have even been turned off. Other countries bordering the tense South China Sea refused to acknowledge whether their radars tracked or were capable of tracking the aircraft. There was a failure to integrate all of the available data that was streaming to and from MH370 in a centralized manner. Data was sparse, concealed, missing, dirty, and as we would later learn, hidden inside a signal never meant to be exploited for geolocation.

20.4 Wide Area Search and Commercial Satellite Imagery

Starting the night of the disappearance and for several days after, an international search began looking for MH370. The initial search area was combed from March 7 to 9 in the Gulf of Thailand. On March 9, the search area was expanded to include the Strait of Malacca. This search area was tipped off by a claim that a military radar indicated that MH370 may have turned west from its initial flight plan and original flight path. The next day, March 10, the Chinese tasked satellites to take imagery of the South China Sea. This satellite imagery would start a trend of breaking news reports that ultimately ended in false alarms [10].

At first blush, exploiting imagery for MH370 seems like a relatively easy task. After all, how hard can it be to find a 209-ft, 656,000-lb metallic object, a modern airliner equipped with dozens of avionics, radios, and positioning systems? Despite its massive size, contrast, material, and other signatures, investigators quickly discovered how even a jumbo jet becomes a weak signature against the backdrop of the hundreds of thousands of square miles of the Indian Ocean where oil slicks, boats, and debris are common. According to CIA imagery analyst Stephen Wood, whitecaps from wave swells look like aircraft debris in medium-resolution panchromatic imagery [11].

20.4.1 A Tradecraft Breakthrough: Crowdsourced Imagery Exploitation

With a search area rapidly approaching a quarter of the Earth's surface, investigators were overwhelmed by the prospect of exploiting the volumes of imagery available to them. They simply did not have the bandwidth or manpower to accomplish such an enormous wide area search task.

To address this challenge, a commercial company called Tomnod assisted investigators by implementing a large-scale "crowdsourcing" platform to solicit contributions from a large online community. Tomnod was born out of a University of San Diego research project and was bought by commercial satellite maker DigitalGlobe in 2013. They use DigitalGlobe's network of four commercial electro-optical satellites to provide support to officials during crisis situations [12].

Tomnod's technology focuses the attention of a small cadre of highly trained imagery analysts by relying on several million average people to weed out the images that lack discernable signatures. Figure 20.1 shows the process Tomnod uses to deliver meaningful images to experts.

The first step delivers a single DigitalGlobe image within the search area of interest to a user in the crowd. That user then has the opportunity to mark the image with a geopositioned marker identifying four differing objects: wreckage, rafts, oil slicks, or other. If a user marks an image, that mark is recorded in

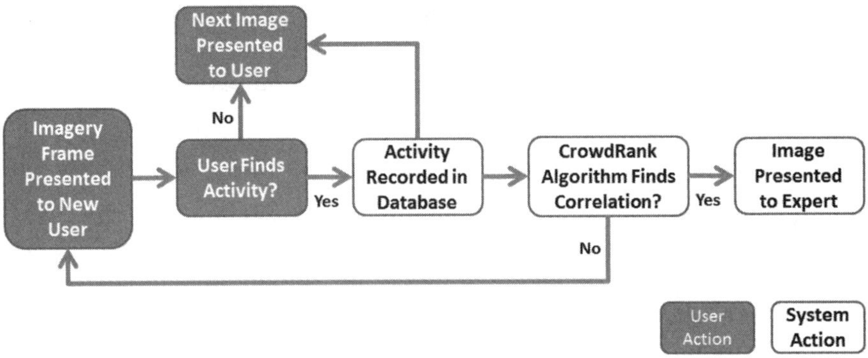

Figure 20.1 Tomnod's process for crowdsourced imagery exploitation. (Adapted from [13, 14].)

a database, and the user is presented with a new image to exploit. Figure 20.2 shows an example of a crowdsourced imagery tagging platform similar to the Tomnod application. A key attribute of the approach is to keep the tags simple. In the MH370 search, allowable tags were limited to oil slick, wreckage, raft, and other.

Once a user marks a frame of imagery, that image is presented several other times to different users. Tomnod runs an algorithm called CrowdRank to determine the locations of maximum agreement within all the marks in their database. The CrowdRank algorithm takes into account two pieces of information when determining whether or not correlation has been reached. The first

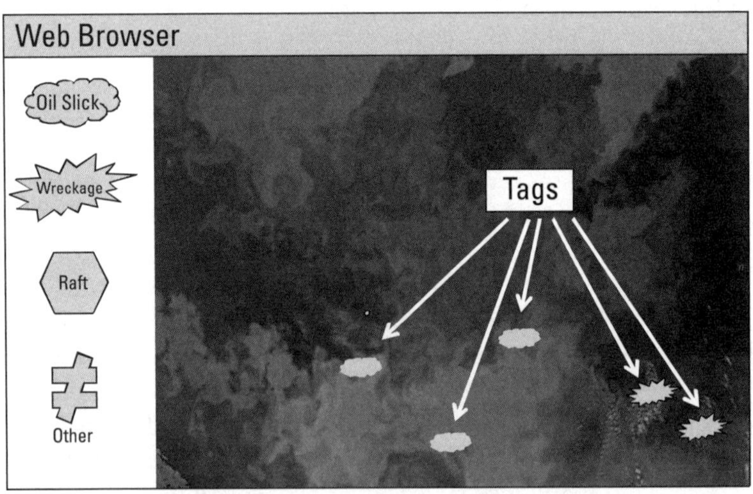

Figure 20.2 Example of a crowdsourced imagery tagging platform for ocean search. (Adapted from [14]. Imagery Source, NASA/NOAA [15].)

piece of information is how right the user usually is. Users get "nods" within Tomnod anytime any other user agrees with marks that they place. The higher the "nods," the more credible a particular user's marks are within the CrowdRank algorithm.

The CrowdRank algorithm also notes the image tiles marked most often. These areas are marked as a high-priority area to be evaluated by a professional imagery analyst. According to Luke Barrington, CTO and cofounder of Tomnod, "...what we've learned from the satellite image companies is that only about 2 percent of this data that they collect every day ever gets looked at by humans. Our ambition at Tomnod is discovering how we can tap into those other pixels, and find value in there and find interesting things" [12].

Images from highly tagged areas can be queued for a trained imagery analyst to exploit using software packages like ArcGIS, RemoteView, or SOCET GXP. Using the untrained crowd for screening and a highly trained imagery analyst for validation and exploitation is similar to the concepts of automated data conditioning in Chapter 14. In this case, the "automation" is a low-fidelity ranking by the crowd. They narrow the search area so professionals can focus their attention on the areas—similar to ABI's discrete locations—where the entity is most likely to be located.

20.4.2 Lessons Learned in Crowdsourced Imagery Search

Tomnod's MH370 search started on March 10 where the Gulf of Thailand meets the South China Sea. In just five days, there had been more the 190 million map views and every pixel available in the original search area had been seen by a human eye at least 30 times. Despite the fact that Tomnod provided several viable leads, the effort ended on May 5, after the decision was made to end the surface-based ocean search. The crowdsourced project had over 8 million users and searched 1,007,750 km^2 of high-resolution imagery—an area one-eighth the size of the continental United States. Additionally, amateur analysts placed almost 13 million marks placed on commercial imagery, as shown in Table 20.2.

Although the Tomnod platform was used for imagery search for several years, the MH370 search was the widest deployment of the technology to date and exposed the world to this innovative tradecraft. It also demonstrated the ABI pillar of data neutrality. Commercial satellite imagery is a "standard" source, but the veracity of tags from mostly untrained "photo interpreters" is derided in the GEOINT community as unreliable and not useful. Certainly, using untrained imagery analysts for confirmation of conclusions derived from imagery is ill-advised, but treating this data as a source of georeferenced information for further analysis is an example of data neutrality.

Table 20.2
Tag Count from the Tomnod MH370 Search Campaign [16]

Tag Name	Tag Count
Raft	520,052
Wreckage	2,853,507
Oil Slick	876,268
Other	855,061
TOTAL TAGS	12,804,888

This example also shows the power of deductive reasoning with large volumes of georeferenced data. Image tiles with no tags could likely be discarded from further analysis (i.e., if 30 pairs of eyes found nothing, put that tile at the bottom of the queue). Data triage: eliminating where the entity is not is an example of deductive analysis in action.

Despite the participation of more than 8 million amateur analysts in the Tomnod imagery search, there was little progress in conclusively locating MH370. At this point in the MH370 investigation, the imagery-based searches in the Bay of Bengal, Gulf of Thailand, and the South China Sea turned up empty [10].

There is a critical moment in almost every ABI investigation where a roadblock occurs, which usually results from approaching the problem from a single-INT focus. The process continues when the analyst again implements the data neutrality principle and considers integrating sources of data whose value was initially ignored.

20.5 A Breakthrough: Sequence and Data Neutral Analysis of Incidentally Collected Data

Several weeks into the search, dozens of ships had scoured thousands of square kilometers of open ocean looking for any signs of the missing airliner, but this was a tiny fraction of the hundreds of thousands of square kilometers that remained. Given all of the satellite phones and radio communication devices on the plane, there had to be some metadata somewhere that could lead investigators to better refine the massive search areas they were currently focusing on.

Inmarsat, a British satellite telecommunications company, provides telephone and data services via portable terminals that communicate to ground stations through a network of geostationary telecommunications satellites. Fifteen days after the airliner first went missing, engineers from Inmarsat approached the United Kingdom's Air Accidents Investigation Branch (AAIB) with an innovative proposal. They had developed a new processing technique that derived additional metadata from routine communications reports from an aircraft's onboard systems.

The avionics communication data-link on the MH370 Boeing 777 aircraft was provided by Inmarsat's Classic Aero service. Although the onboard aircraft communications addressing and reporting system (ACARS) was disabled, the Inmarsat relay from the onboard satellite data unit (SDU) to the Inmarsat-3 F1 satellite was still operational [10]. Could this tiny amount incidentally collected data be repurposed and reanalyzed to unravel the mystery of the missing jet?

The Inmarsat processing technique—based on reprocessing differential Doppler signals between the aircraft and the satellite—was novel. Like in ABI, analysts don't know what kind of data reprocessing will help them until they have a better understanding of what the problem is and what kind of information they need to solve that problem. Had the pings and the associated metadata been left on the cutting room floor, the ability to utilize specialized processing would likely have been lost—a serious proof of the power of indexing incidentally collected data for future forensic analysis. This enabled revolutionary data and sequence neutral analysis on data that was never considered for tracking and locating commercial aircraft.

The Inmarsat data was able to provide some critical information in the absence of traditional means. Investigators knew that the last radar contact of MH370 was from a military radar northwest of Malaysia somewhere in the Andaman Sea [17]. Velocity and direction allowed investigators to project the aircraft's future position using dead reckoning. They knew the maximum distance the aircraft could have traveled based on the fuel load, wind conditions, and possible altitudes, but how long did it really fly?

Inmarsat's data told investigators that the plane flew for several hours after the last radar contact because the satellite recorded several "handshakes" between the Inmarsat SDU and the ground station about an hour apart. These handshakes contained metadata about each call and the time that they took place, essentially documenting events. In sum, they represented a transaction [18].

The metadata contained within each of the handshakes was the key to helping investigators narrow down the search area and determine probable travel routes. It contained the time the handshake left the ground station and the time the aircraft received the information. Based on this time difference and the known location of the Inmarsat-3 F1 satellite, engineers were able to plot out two probable paths in wide arcs known as the "northern" and "southern" corridors, because the triangulation problem had two possible solutions. The southern corridor was ruled most probable by investigators and became the primary focus of the search as shown in Figure 20.3.

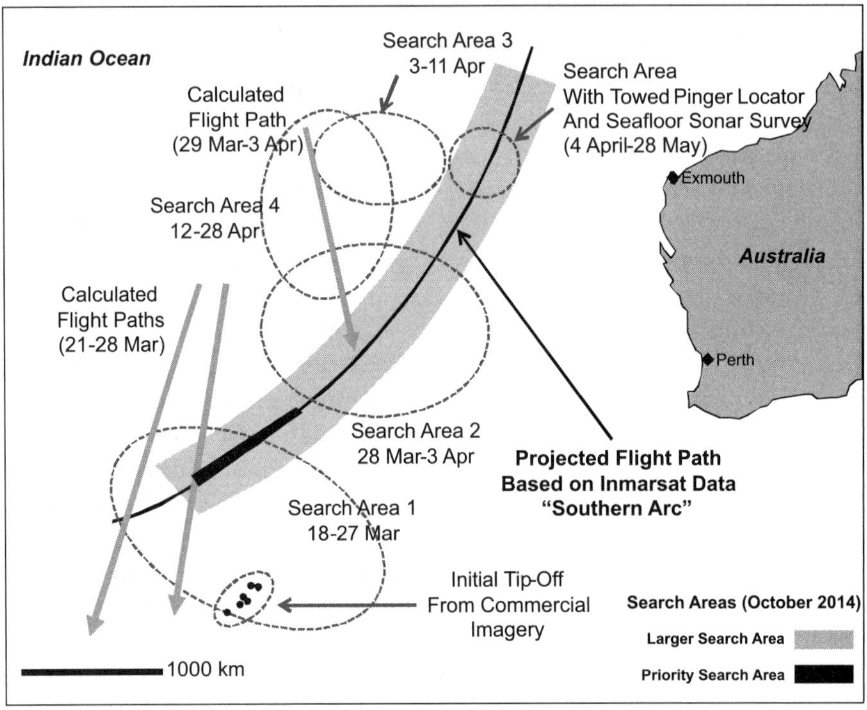

Figure 20.3 Projected flight paths and historical/current search areas. (Adapted from [10, 19–21].)

This "multiple answer" scenario occurs often in ABI investigations. When trying to condition ambiguous data to resolve an entity, often times the result(s) will also be ambiguous. ABI analysts will have to momentarily consider multiple possible solutions until they can use deductive reasoning to eliminate improbable solutions.

The velocity and direction, metadata for the last known reliable position of MH370 prior to disappearing from radar, indicated it was headed in a southerly direction. To determine the likely location of the flight's termination, engineers extrapolated the velocity and likely position of the final Inmarsat handshake to narrow the search area within 100 square kilometers in the Indian Ocean. With the search area narrowed, investigators could focus on the unique signature of the aircraft's "black boxes" or use advanced underwater imaging techniques to look for the aircraft's wreckage.

20.6 Summary: The Search Continues

The story of MH370 is not a problem in intelligence analysis, but like many intelligence problems, it is an investigation dominated by sparse data, unresolvable entities, misleading hypotheses, wide area search, and leadership pressure to produce results as quickly as possible. This example highlights the ABI principles of georeference to discover, data neutrality, sequence neutrality, and integration before exploitation. It also shows how inductive reasoning on old hypotheses steers analysts awry. Deductive analytics focuses a large, complex search problem to a manageable one. Incidental collection, data neutrality, and sequence neutrality played a major role in the primary breakthrough in the MH370 case, reducing the search area by more than 90%.

The search for MH370 demonstrated how multidisciplinary teamwork and unconventional analytic techniques are necessary to resolve unknown information. In the case of MH370, Inmarsat engineering and math experts had to reprocess data with a unique method to get the information investigators needed to accomplish their search and rescue jobs. Likewise for hard ABI-related problem sets, intelligence analysts, engineers, and program managers must work side by side to understand the art of the possible when it comes to combining technology and tradecraft.

The facts behind the disappearance of MH370 may remain a mystery forever, but the search developed new techniques. Civil aviation authorities considered new ways to track aircraft and index the data from their flight paths. Airlines and aircraft designers considered new hypotheses on what might cause an aircraft to go missing. Even when analysis hits a dead end, it causes introspection: Our lessons learned may prevent future catastrophes.

As of this writing, a wing component from a Boeing 777—confirmed to be a part of the missing MH370—was found on Reunion Island in the Indian Ocean. The Australian government announced that the search may end in 2016 if no credible leads are found.

20.7 Chapter Author Biography

Alex Shernoff is a mission engineer for BAE Systems, specializing in information technology systems enabling ABI tradecraft. He holds an MBA and an M.S. in information technology from Marymount University and a B.S. in electrical engineering from Ohio State University.

References

[1] "Malaysia MH370 Preliminary Report." Office of the Chief Inspector of Air Accidents Ministry of Transport, March 2014.

[2] "MISSING MH370: Pilot: I Established Contact With Plane," *New Straits Times*, March 9, 2014.

[3] Watson, I., "MH370: Plane Audio Recording Played in Public for First Time to Chinese Families," CNN, April 29, 2014.

[4] "Signalling Unit Log for (9M-MRO) Flight MH-370," Inmarsat/Malaysia Department of Civil Aviation.

[5] Payne, S., "Malaysia Airlines Plane MH370: Russia FSB Claim Jet Hijacked and Flown to Afghanistan," *International Business Times,* April 14, 2014.

[6] Nelson, D., "MH370 captain plotted route to southern Indian Ocean on home simulator," *The Telegraph,* June 22, 2014. http://www.telegraph.co.uk/news/worldnews/asia/malasia/10917868/MH370-captain-plotted-route-to-southern-Indian-Ocean-on-home-simulator.html.

[7] Thomas, P., and J. Margolin, "FBI Finishes Probe into Malaysia Airlines Captain's Flight Simulator," ABC News, April 2, 2014.

[8] "INTERPOL Confirms at Least Two Stolen Passports Used by Passengers on Missing Malaysian Airlines Flight 370 Were Registered in its Databases," Interpol, March 9, 2014.

[9] "Malaysia Airlines MH370: Stolen passports 'No Terror Link,'" BBC News, March 11, 2014.

[10] "MH370—Definition of Underwater Search Areas," Australian Transport Safety Bureau, June 26, 2014.

[11] Shoichet, C. E., M. Pearson, and J. Mullen, "Flight 370 Search Area Shifts After 'Credible Lead,'" CNN, March 28, 2014.

[12] "Crowdrank Algorithm Used for Searching," web. Available: http://ainibot.com/osgoh/crowdrank-algorithm-used-for-search-flight-mh370.

[13] Barrington, L., N. Ricklin, and S. Har-Noy, Crowdsourced Search and Locate Platform, U.S. patent application, US20140233863, 2014.

[14] Barrington, L., N. Ricklin, and S. Har-Noy, Crowdsourced Image Analysis Platform, U.S. patent application US20140237386, 2014.

[15] Imagery of the Barents Sea by the MODIS Instrument, NASA/NOAA, 2010.

[16] Merelli, A., "Using Crowdsourcing to Search for Flight MH370 Has Both Pluses and Minuses," web. Available: http://qz.com/188270/using-crowdsourcing-to-search-for-flight-mh-370-has-both-pluses-and-minuses/.

[17] Forsythe, M., and M. Schmidt, "Radar Suggests Jet Shifted Path More Than Once," *The New York Times,* March 14, 2014.

[18] "MH370 Flight Path Analysis Update," Australian Transport Safety Bureau, 2014.

[19] "MH370 Search." Australian Maritime Safety Authority (AMSA), 2014.

[20] "Differential Doppler Study from Inmarsat Concerning MH370," Inmarsat, March 23, 2014.

[21] "Search Continues for Malaysian Flight MH370," web. Available: http://www.jacc.gov.au/media/releases/2014/april/mr040.aspx.

21

Visual Analytics for Pattern-of-Life Analysis

This chapter integrates concepts for visual analytics with the basic principles of georeference to discover to analyze the pattern-of-life of entities based on check-in records from a social network. It presents several examples of complex visualizations used to graphically understand entity motion and relationships across named locations in Washington, D.C., and the surrounding metro area. The purpose of the exercise is to discover entities with similar patterns of life and cotraveling motion patterns—possibly related entities. The chapter also examines scripting to identify intersecting entities using the R statistical language.

21.1 Applying Visual Analytics to Pattern-of-Life Analysis

Visual analytic techniques provide a mechanism for correlating data and discovering patterns. They are especially poignant for spatiotemporal analytics when using georeferenced data. Sections 21.1 and 21.2 describe some of the steps undertaken to understand and extract value from a large dataset, and provide insight into the "art" of ABI analysis and the often circuitous, exploratory path followed by an ABI analyst.

21.1.1 Overview of the Data Set

Consider a data set of 6,442,890 worldwide location-based "check-ins" from social network Gowalla from February 2009 to October 2010 archived by Leskovec [1, 2] Records in this data set take the form:

UserID	Latitude	Longitude	Date/Time	CheckInLoc

The data scientist begins with the question: "How many users (unique UserID) checked in at the same CheckInLoc around the same Date/Time?" Answering this question through scripting (coding) is easier but less literal and interactive. Visualization requires multiple steps to filter down to information of interest. Because it is difficult to comprehend worldwide data at the entity level, the analyst filters to the United States (3,675,742 points) and then further filters to the Washington, D.C., metro area (102,114 points). Visual analytics provides information about a user's pattern of life when the tools are used to drill down to a particular user and his/her behavior.

21.1.2 Exploring the Activities and Transactions of Two Randomly Selected Users

Two users were selected at random. User 3243 checks in most often in the Waldorf, Maryland, area. User 39005 checks in most often in the Woodbridge, Virginia, area. Figure 21.1 illustrates their pattern of check-ins using the JMP bubble plot where the size of the bubble indicates the total number of check-ins for all users at that location. User 3243 checks in at a very popular Washington, D.C., location and many less popular locations near Waldorf, Maryland. User

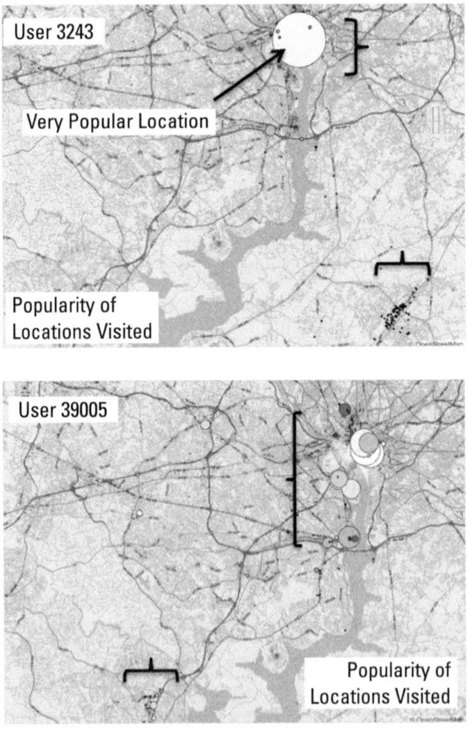

Figure 21.1 Pattern of life of two social network users. (Map data source: OpenStreetMap.)

39005 checks in at many places in the Woodbridge area (less popular) and at many popular locations around Crystal City and Alexandria, Virginia, and Washington, D.C.

A bivariate scatterplot of the CheckInLoc versus DateTime allows examination of intersections at the same location and time as shown in Figure 21.2. The horizontal stripes across Figure 21.2 show where each user tends to check in most of the time (they do not line up). It is very difficult to discern overlaps from the chart.

Script processing counts the total number of check-ins at each location for each user. The product of the total, if nonzero, identifies a CheckInLoc frequented by both users as shown in Table 21.1. There are only two such locations (17173, 347463).

A mosaic plot visualizes the overlap—or in this case lack thereof—of simultaneous check-ins as shown in Figure 21.3. The top half of the plot shows location 17173, which user 39005 visited twice and user 3243 visited once (not simultaneously). A similar behavior is evident in location 347463.

In this example, the data scientist found that although the two randomly selected users visited the same location, they never checked in there at the same day or time. This data set demonstrates the principle of sparse data: It is limited to locations registered with the social network and smartphone-savvy users who opt to manually check in at those locations.

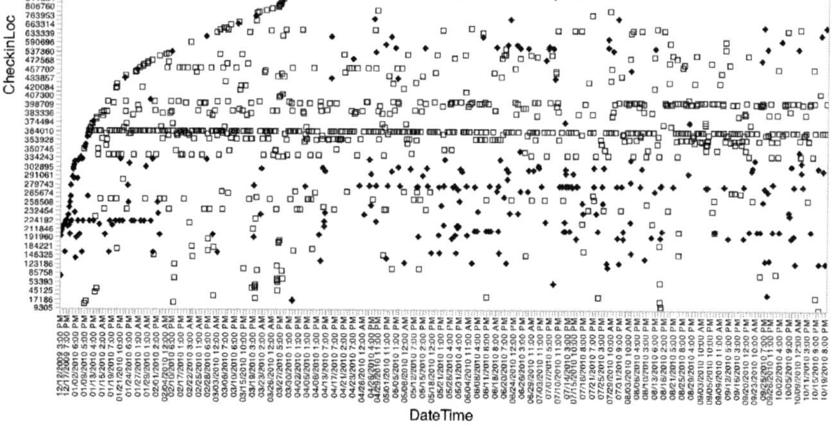

Figure 21.2 Bivariate plot of two randomly selected users pattern of activity.

Table 21.1
Subset of Tabulated Results for the Intersection of Pattern of Check-ins Between Two Social Network Users

CheckinLoc	3243	39005	Product
9305	0	1	0
16206	0	1	0
17173	1	2	2
17186	1	0	0
19180	0	1	0
21622	0	3	0
347463	2	4	8

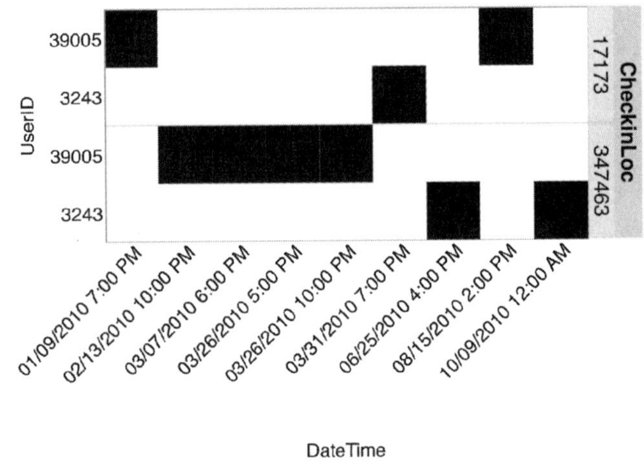

Figure 21.3 Mosaic plot of location-based, temporal check-ins of two social network users.

21.1.3 Identification of Cotravelers/Pairs in Social Network Data

Visual analytics can be used to identify cotravelers, albeit with great difficulty. First, the 102,114 D.C.-area data points are filtered to the most popular locations (those with more than 200 check-ins). This results in 3,876 check-ins at the top 14 D.C.-metro area locations by total number of check-ins. A mosaic plot like the one in Figure 21.3 is created for each of the 14 locations. The full plot is too difficult to interpret. Figure 21.4 shows the mosaic plot for a popular check-in location, where they Y-axis illustrates the percentage of check-ins due to a single user ID. When the plot is divided in half, two user IDs were present at the same time. When divided into thirds, three user IDs were present at the same time.

Further drill down (Figure 21.5) reveals 11 simultaneous check-ins, including one three-party simultaneous check-in at a single popular location.

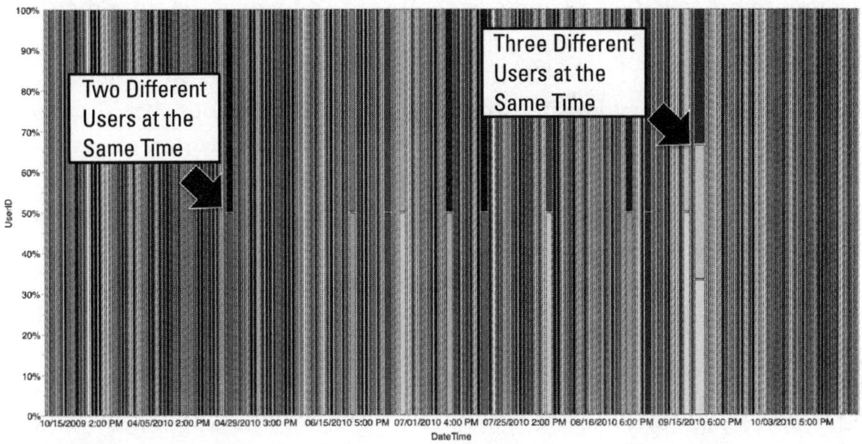

Figure 21.4 Mosaic plot for a popular check-in location.

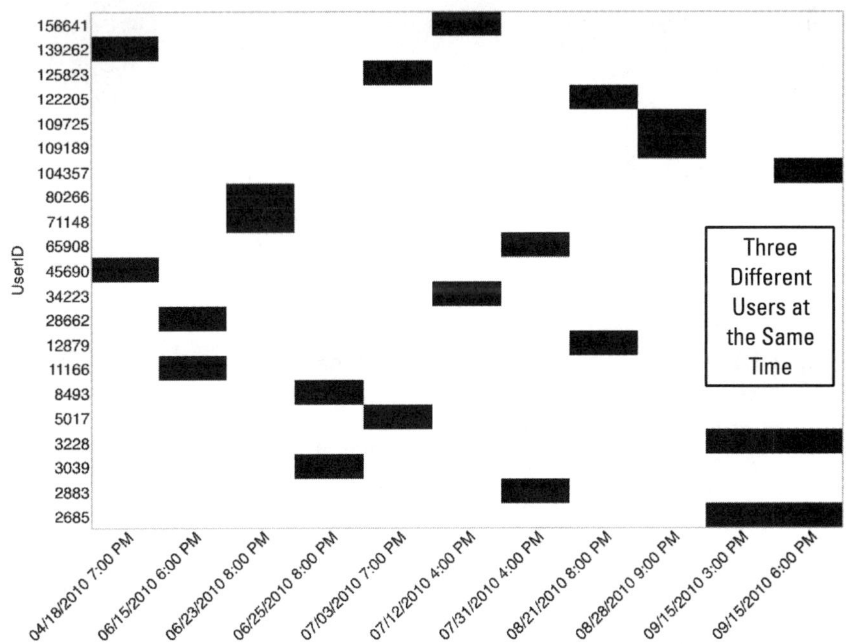

Figure 21.5 Drill down of the mosaic plot highlighting 11 multiuser simultaneous check-ins.

The next logical question—an application of the where-who-where concept discussed in Chapter 5—is "do these three individuals regularly interact?" With the three user IDs in hand, the analyst zooms back out from the 11 case data set to the full 102,114 and selects all cases containing the three user IDs, resulting in 359 records. These are shown in the bubble plot in Figure 21.6.

On September 15, 2010, at 6:00 p.m., all three users were present at a single location in downtown Washington, D.C. Throughout that afternoon, user 2685 and 3228 traveled together along the black line from east to west. After midnight they arrived at the cluster of points in the upper left of the map. The pair cotraveled during the next day, checking in at a number of locations along Connecticut Avenue.

The third traveler never reintersects the pair. Perhaps this person is a long-lost friend meeting up for a drink? A fan of a local sports team partaking in a social activity on a Wednesday night? Or perhaps just a random traveler that frequents popular locales? While we can generally conclude that users 2685 and 3228 are related (they spend evenings and nights together), we cannot make a judgment about user 104357, but a durable relationship is unlikely based on a single geotemporal intersection. This is especially true when only a single source is used.

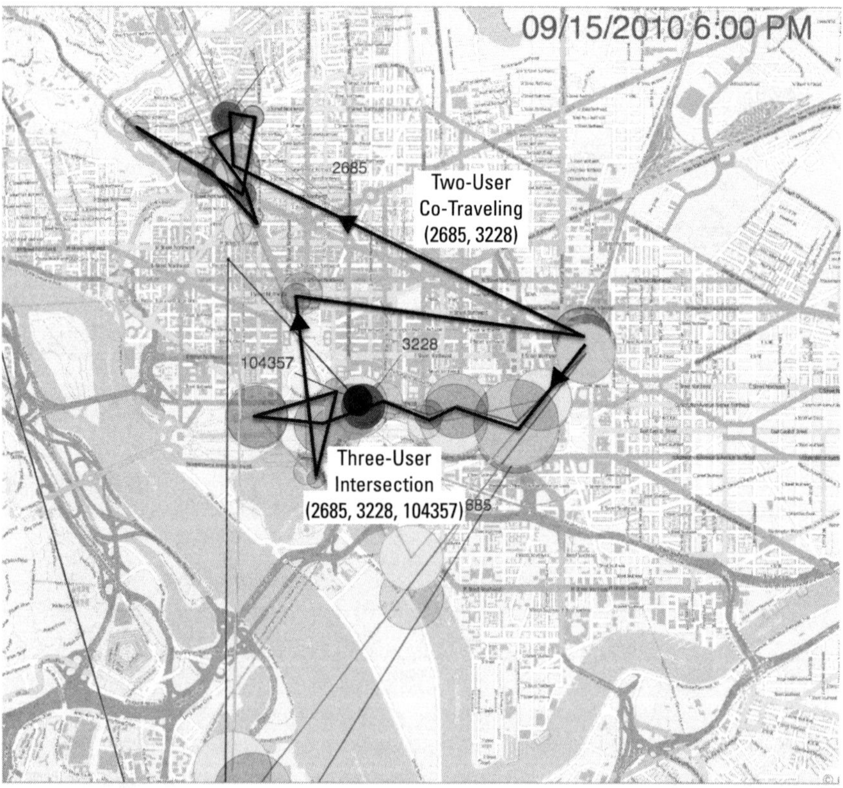

Figure 21.6 Bubble plot of multiuser intersection and two-user cotraveling.

21.2 Discovering Paired Entities in a Large Data Set

Visual analytics is a powerful, but often laborious and serendipitous approach to exploring data sets. An alternative approach is to write code that seeks mathematical relations in the data. Often, the best approach is to combine the techniques.

It is very difficult to analyze data with statistical programming languages if the analysts/data scientists do not know what they are looking for. Visual analytic exploration of the data is a good first step to establish hypotheses, rules, and relations that can then be coded and processed in bulk for the full dataset.

Scripting against a 100,000 case subset of check-ins in the Washington, D.C., area found 696 instances where two users checked in within two minutes of one another. (This illustrates the difficulty of triaging data from a large dataset: The subject of the single query returned less than 1% of the total cases in the dataset).

Figure 21.7 shows the top 10 users and their coincident check-ins with each other. Presumably, some of the 10 entities in this matrix are related to one another, as they are all users of Gowalla, voluntarily check in at nondiscrete locations, and were simultaneously located with another user on more than one occasion. User 124862 simultaneously checks in with six unique users (total links). Thirty-six of his/her 43 total check-ins are with user 124863 (note, the user IDs are sequential).

The second most well connected user is 39005—coincidentally the same user that we randomly selected in Section 21.1.2 . He/she checks in at a variety of shopping centers in Stafford, Virginia, and nondiscrete locations in Washington, D.C., with 16 other users as shown in Figure 21.8.

The user visited church at the same time as user 137427 on one occasion. He/she also went to a movie at the same time as user 1893. He/she cotraveled with 37398 and 129395 on multiple occasions. Figure 21.9 shows the paired behavior of 39005 and 129395, who traveled to multiple locations together on March 27, 2010.

UserID	Total Checkins	Total Links	124863	129395	37398	39005	124862	3228	36259	576	2685	129399
124862	43	6	36	0	0	0	3	0	0	0	0	0
39005	34	16	0	9	10	0	0	1	1	0	0	2
129399	30	9	0	16	1	7	0	0	0	0	0	1
2685	21	8	0	0	1	0	0	14	0	1	0	0
124863	21	5	5	0	0	0	12	0	0	0	0	0
576	17	5	0	0	0	0	0	0	13	1	0	0
36259	16	10	0	0	0	0	0	0	2	6	0	0
37398	16	6	0	0	4	8	0	0	0	0	0	0
3228	14	7	0	0	0	0	0	0	0	0	8	0
129395	10	6	0	0	0	1	0	0	0	0	0	5

Figure 21.7 Top 10 users with coincident check-ins.

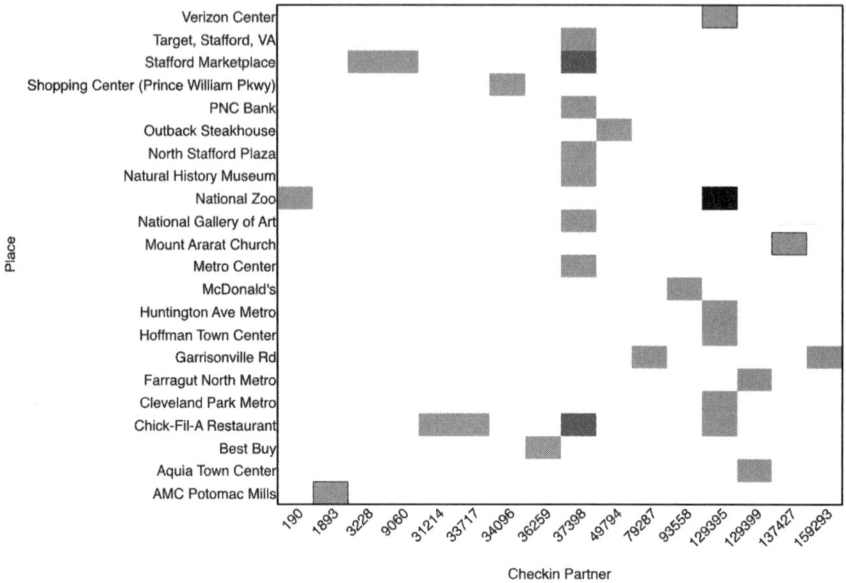

Figure 21.8 Nondiscrete locations user 39005 visited with other users.

Integrating open-source data, the geolocations can be correlated with named locations like the National Zoo and the Verizon Center. Open-source data also tells us that the event at the Verizon Center was a basketball game between the Utah Jazz and Washington Wizards. The pair cotravel only for a single day over the entire data set. We might conclude that this entity is an out-of-town guest. That hypothesis can be tested by returning to the 6.4-million-point worldwide dataset.

User 129395 checked in 122 times and only in Stafford and Alexandria, Virginia, and the District of Columbia. During the day, his or her check-ins are in Alexandria, near Duke St. and Telegraph Rd. (work). In the evenings, he or she can be found in Stafford (home). This is an example of identifying geospatial locations based on the time of day and the pattern-of-life elements present in this self-reported data set.

Note that another user, 190, also checks in at the National Zoo at the same time as the cotraveling pair. We do not know if this entity was cotraveling the entire time and decided to check in at only a single location or if this is an unrelated entity that happened to check in near the cotraveling pair while all three of them were standing next to the lions and tigers exhibit. The full data set finds user 190 all over the world, but his or her pattern of life places him or her most frequently in Denver, Colorado.

And what about the other frequent cotraveler, 37398? The pair coincidentally checked in 10 times over a four-month period, between the hours

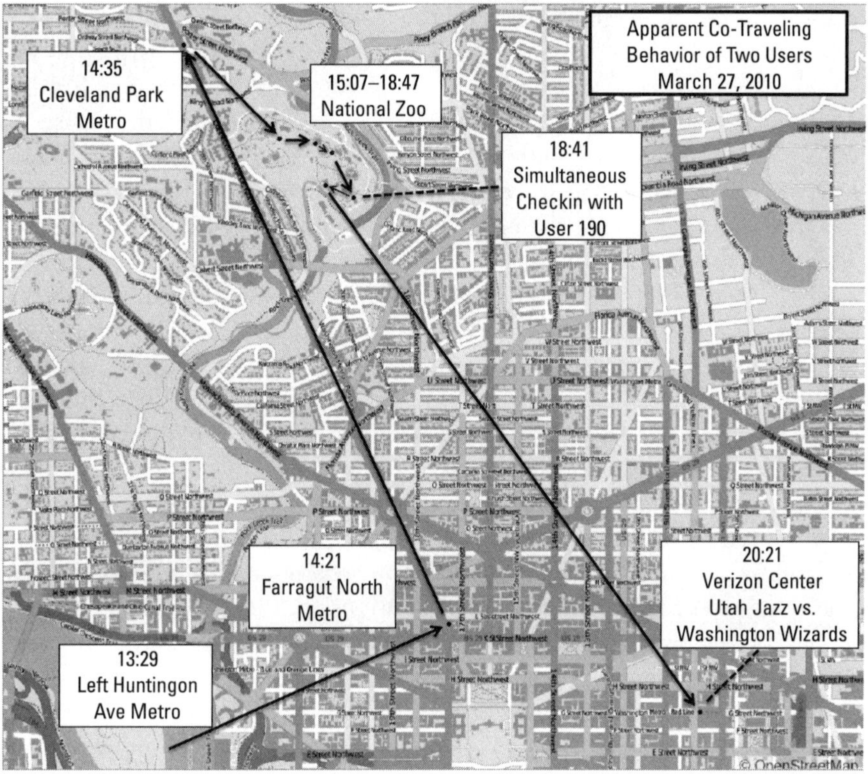

Figure 21.9 Apparent cotraveling behavior of two users. (Map data source: OpenStreetMap.)

of 14:00 and 18:00 and 21:00 and 23:59 at the Natural History Museum, Metro Center, the National Gallery of Art, and various shopping centers and restaurants around Stafford, Virginia. We might conclude that this is a family member, child, friend, or significant other.

21.3 Summary

This example demonstrates how a combination of spatial analysis, visual analytics, statistical filtering, and scripting can be combined to understand patterns of life in real "big data" sets; however, conditioning, ingesting, and filtering this data to create a single example took more than 12 hours. Most software tools have difficulty loading and manipulating the full 6.4-million record data set. Visualizations of the unfiltered data set are impossible for humans to comprehend.

A combination of techniques was required in the analysis-synthesis cycle. First, we filtered data spatially to develop a hypothesis: Related entities are likely to appear often at the same places and times. The more frequently they are co-located, the more likely they are to be related. Then, a script was written to statistically process the data to find intersections. After drilling down on a handful of entities, visual analytics was brought in again to understand the dataset in a spatiotemporal context (Figure 21.9). Finally, to test the "out of town guest" hypothesis, the analyst had to return to the full 6.4-million point data set to map the locations of user 190 (a Denver resident) and user 129395 (a Stafford, Virginia, resident).

The key to drilling down to an entity of interest was to begin with the assumption that intersections are likely to take place at nondiscrete locations with a high number of check-ins. The subsequent filtering and analysis steps were essentially hypotheses that were rapidly tested using visual analytics techniques. When analyzing large data sets, the analyst must simply explore: Start somewhere, try something, and see where it takes you.

The data set also illustrates a major challenge for geotemporally referenced data. The check-ins are extremely precise in time (to the second), but the latitude and longitude of the check-in may correspond to multiple nearby locations—especially when nondiscrete locations like shopping centers are considered.[1] The most popular locations in the data set are those with the most total check-ins—typically nondiscrete locations and major tourist sites like the American Art Museum, the U.S. Capitol, or the National Zoo. Translating the numbered (but unnamed) location ID to an identified location is easy in these cases. Some of the check-in locations have coordinates that map to many potential nearby locations.

Because this data requires voluntary check-ins at registered locations, it is an example of the sparse data typical of intelligence problems. If the data consisted of beaconed location data from GPS-enabled smartphones, it would be possible to identify multiple overlapping locations. On the other hand, the 6.4-million points in this data set were spaced by hours or days. Beaconed GPS data reported per second creates data sets of enormous scale. Imagine the computing power required to calculate the interuser distance between all movers in real time![2] Businesses that implement such large-scale analytics to integrate data to understand patterns of life are poised to outperform their competition in a "big data" world.

1. The locations have unique identifiers, but they are not recorded in the dataset. The data can be geocoded to obtain addresses, and those addresses can be correlated with known establishments, but not always unambigously.
2. Social networking sites like Facebook, Foursquare, Gowalla, and Twitter have an innovative trick to narrow this problem to a more tractable one. They generally limit facial recognition, recommendations, and location-based information to the much smaller network of users connected within an isolated graph (for example, a friends list). Partitioning information by some known relationship and processing on the basis of that relationship significantly improves the utility of big data spatial analytics.

21.4 Acknowledgements

Benjamin West, a senior intelligence analyst, and systems engineer, developed the R script for the data processing in Section 21.2 .

References

[1] Leskovec, J., and A. Krevl, "SNAP Datasets: Stanford Large Network Dataset Collection," 2014.

[2] Cho, E., S. A. Myers, and J. Leskovec, "Friendship and Mobility: User Movement in Location-Based Social Networks," presented at the *ACM SIGKDD International Conference on Knowledge Discovery and Data Mining (KDD),* 2011.

22

Multi-INT Spatiotemporal Analysis

A 2010 study by OUSD(I) identified "an information domain to combine persistent surveillance data with other INTs with a ubiquitous layer of GEOINT" as one of 16 technology gaps for ABI and human domain analytics [1]. This chapter describes a generic multi-INT spatial, temporal, and relational analysis framework widely adopted by commercial tool vendors to provide interactive, dynamic data integration and analysis to support ABI techniques.

22.1 Overview

ABI analysis tools are increasingly instantiated using web-based, thin client interfaces. Open-source web mapping and advanced analytic code libraries have proliferated. Software like Adobe Flash and the advancement of HTML5 allow enhanced visualization and even complex analytics within modern browsers like Firefox and Chrome. Thin client tools have outstripped "thick client" desktop tools because licensing and updating desktop tools has become complex in government environments. Analysts lack the administrative privileges required to make changes to their desktop systems. Also, as the military and intelligence communities move to lower-cost information technology resources, low-powered terminals are quickly replacing expensive and cumbersome desktop machines. Web-based tools also allow all users to share the same databases. Software patches and updates can be rolled out to all users on the same server.

22.2 Human Interface Basics

A key feature for spatiotemporal-relational analysis tools is the interlinking of multiple views, allowing analysts to quickly understand how data elements are located in time and space, and in relation to other data.

22.2.1 Map View

An "information domain for combining persistent surveillance data on a ubiquitous foundation of GEOINT" drives the central feature of the analysis environment to the map (Figure 22.1) [1]. Spatial searches are performed using a bounding box (1). Events are represented as geolocated dots or symbols (2). Short text descriptions annotate events. Tracks—a type of transaction—are represented as lines with a green dot for starts and a red dot or X for stops (3). Depending on the nature of the key intelligence question (KIQ) or request for information (RFI), the analyst can choose to discover and display full tracks or only starts and stops. Clicking on any event or track point in the map brings up metadata describing the data element. Information like speed and heading

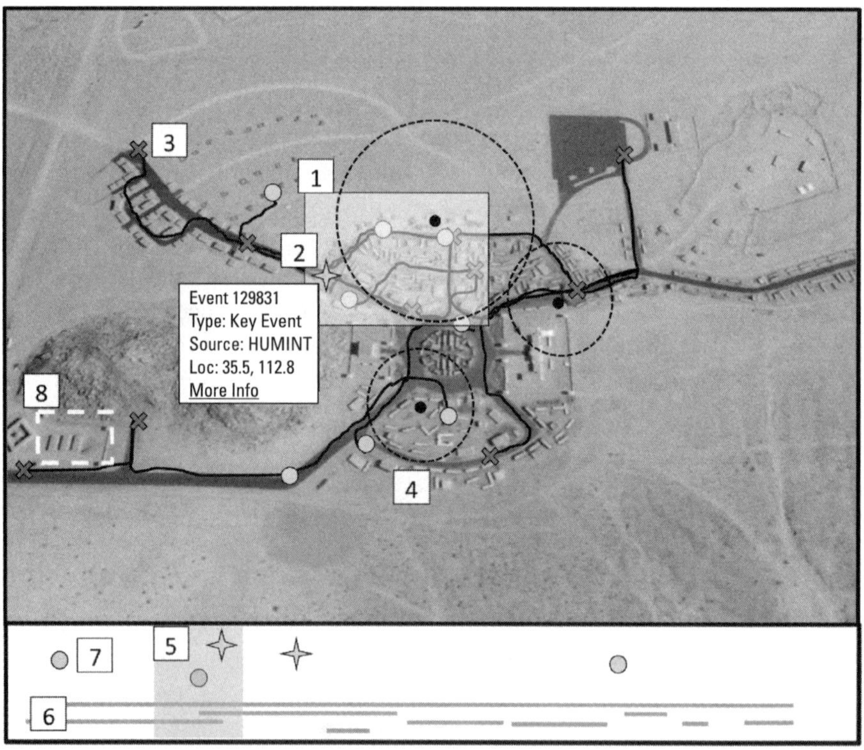

Figure 22.1 Map and timeline view. (Image source: USDA Farm Service Agency.)

accompany track points. Other metadata related to the collecting sensor may be appended to other events and transactions collected from unique sensors. Uncertainty around event position may be represented by a 95% confidence ellipse at the time of collection (4).

Map layers use the OpenLayers standard—a high-performance Javascript library for web-based map applications. Layers may consist of rasters/tiles or vectors, either base maps or reference imagery. Raster data typically served out using the Open Geospatial Consortium (OGC) Web Map Service (WMS) format. This is useful for background imagery, contextual data, foundation GEOINT, and reference data that does not change significantly at small time slices. OpenStreetMaps and Google Maps are two examples of WMS data that is widely used for foundational GEOINT. The OGC web feature service (WFS) is used to serve out individual events and transactions. Instead of producing an entire static raster layer, WFS data represents each activity as a point, line, or polygon. Vector data uses the GeoJSON, TopoJSON, KML, GML, and a variety of other formats.

22.2.2 Timeline View

Temporal analysis requires a timeline that depicts spatial events and transactions as they occur in time. Many geospatial tools—originally designed to make a static map that integrates layered data at a point in time—have added timelines to allow animation of data or the layering of temporal data upon foundational GEOINT. Most tools instantiate the timeline below the spatial view (Google Earth uses timeline slider in the upper left corner of the window).

The timeline filter (5)—the shaded area in Figure 22.1—allows the analyst to select a slice in time to animate the dataset. Filtering on the timeline removes all data from the map that does not cross through the time window. Tracks and events with duration appear as lines on the timeline (6). Point events like geolocated signals or event markers with a single time value appear as points (7). A track that crosses through the window remains on the map in its entirety, but a small white circle is used to represent the location of the object along the track at the mean value of the filtered slice.

22.2.3 Relational View

Relational views are popular in counterfraud and social network analysis tools like Detica NetReveal and Palantir. By integrating a relational view or a graph with the spatiotemporal analysis environment, it is possible to link different spatial locations, events, and transactions by relational properties.

The grouping of multisource events and transactions is an activity set (Figure 22.2). The activity set acts as a "shoebox" for sequence neutral analysis. In the course of examining data in time and space, an analyst identifies data that appears to be related, but does not know the nature of the relationship. Drawing a box around the data elements, he or she can group them and create an activity set to save them for later analysis, sharing, or linking with other activity sets.

By linking activity sets, the analyst can describe a filtered set of spatial and temporal events as a series of related activities. Typically, linked activity sets form the canvas for information sharing across multiple analysts working the same problem set. The relational view leverages graphs and may also instantiate semantic technologies like the RDF to provide context to relationships.

22.3 Analytic Concepts of Operations

This section describes some of the basic analysis principles widely used in spatiotemporal analysis tools. Analysts typically use a variation of Figure 22.1 on their primary display and the focus of their attention. Secondary and tertiary monitors are used for detailed analysis of individual events, source reports, or relational views like the one shown in Figure 22.2.

22.3.1 Discovery and Filtering

In the traditional, target-based intelligence cycle, analysts would enter a target identifier to pull back all information about the target, exploit that informa-

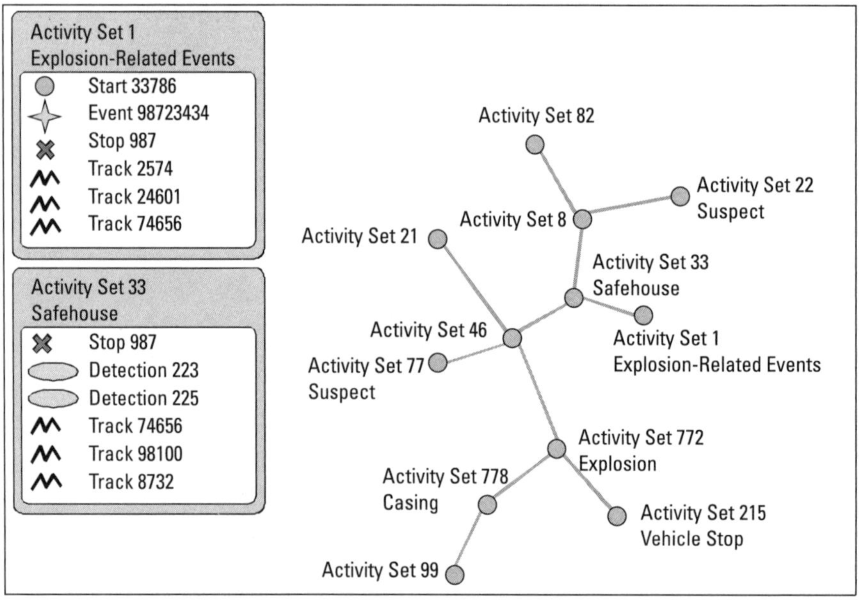

Figure 22.2 Activity set and network relational view.

tion, report on the target, and then go on to the next target. In ABI analysis, the targets are unknown at the onset of analysis and must be discovered through deductive analytics, reasoning, pattern analysis, and information correlation.

Analysts begin the discovery process by highlighting an area on the map, selecting a slice in time, and identifying the data sources of interest using a panel or drop-down menu. Searching for data may result in querying many distributed databases. Results are presented to the user as map/timeline renderings. Analysts typically select a smaller time slice and animate through the data to exploit transactions or attempt to recognize patterns. This process is called data triage. Instead of requesting information through a precisely phrased query, ABI analytics prefers to bring all available data to analysts' desktop so they can determine if the information has value. This process implements the ABI principles of data neutrality and integration before exploitation simultaneously. It also places a large burden on query and visualization systems—most of the data returned by the query will be discarded as irrelevant. However, filtering out data a priori risks losing valuable correlatable information in the area of interest.

One way analysis tools have adapted to this concept of operations is to prerender vector/feature information as raster layers at high zoom levels. If an analyst requests all the events related to unrest in Syria, thousands or millions of points may be returned [2]. Rendering each one as an individual vector is not helpful when the analyst views data at the country level. When analysis is focused to an area of interest, the framework switches from WMS images (tiles) to WFS features to allow precise selection of individual activities for correlation and metadata analysis.

As described in Chapter 4, ABI analysts practice a dynamic method of analysis and synthesis in very tight time cycles. They query for a large volumes of data, filter and triage it to a few key events of interest, tag those events in activity sets, and then repeat the process for a different related area or slice of time. The synthesis process links activity sets and reconstructs a spatial story that may be animated to understand and report on patterns of life.

22.3.2 Forensic Backtracking

Analysts use the framework for forensic backtracking, an embodiment of the sequence neutral paradigm of ABI. PV Labs describes a system that "indexes data in real time, permitting the data to be used in various exploitation solutions... for backtracking and identifying nodes of other multi-INT sources" [3]. Exelis also offers a solution for "activity-based intelligence with forensic capabilities establishing trends and interconnected patterns of life including social interactions, origins of travel and destinations" [4].

Key events act as tips to analysts or the start point for forward or forensic analysis of related data. From a key event located at t=0 on the timeline, the analyst can swipe left to animate events and transactions leading up to the event, reversing time [5]. Moving objects track backward from their destination to the origin. These locations can be queried, researched, or tagged with a watchbox. Using this technique, analysts reconstruct events and locate spatial locations that were not known before the key event took place. See Figure 22.3 for an example of forensic backtracking from a key event.

22.3.3 Watchboxes and Alerts

A geofence is a virtual perimeter used to trigger actions based on geospatial events [6]. Metzger describes how this concept was used by GMTI analysts to provide real-time indication and warning of vehicle motion [7]. The technique can also be used in the ABI method.[1] Upon identifying areas of interest through deductive analysis of events and transactions in the spatiotemporal framework, the analyst may set up watchboxes around smaller regions (item 8 in Figure

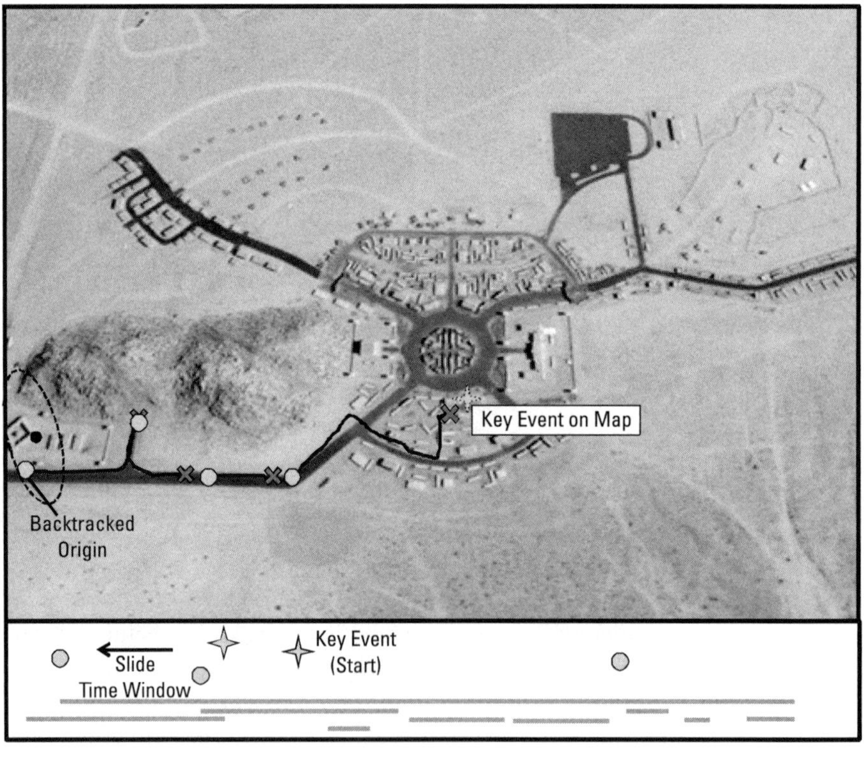

Figure 22.3 Example of forensic backtracking from a key event. Image (Source: USDA Farm Service Agency.)

1. Watchboxes can be a useful tool, but the student is cautioned that very large watchboxes trigger many false alarms. Beginning with watchboxes forces the analyst back into a target-based model. They should be used *after* deductive analysis has identified regions of interest.

22.1). Watchboxes identify discrete locations and function as standing queries for additional data that results from subsequent collection, processing, or analysis. Essentially, if information appears in the watchbox that matches user-defined filters, the analyst is alerted via email, RSS, or other means that new data has arrived related to a location of interest [8].

It is important to note that watchboxes are generally focused on known areas of interest—either defined initially through target-based analysis or discovered through deductive analytics on events and transactions. Watchboxes tend to support inductive reasoning based on expected outcomes and inhibit discovery of the unknown (you have to set the threshold of the watchbox to something, so it will only alert on what is expected). Top analysts continually practice discovery and deductive filtering to update watchboxes with new hypotheses, triggers, and thresholds.

Alerts may result in subsequent analysis or collection. For example, alerts may be sent to collection management authorities with instructions to collect on the area of interest with particular capabilities when events and transactions matching certain filters are detected. When alerts go to collection systems, they are typically referred to as "tips" or "cues."

22.3.4 Track Linking

As described in Chapter 12, automated track extraction algorithms seldom produce complete tracks from an object's origin to its destination. Various confounding factors like shadows and obstructions cause track breaks. A common feature in analytic environments is the ability to manually link tracklets based on metadata. WAMI-derived tracks are often accompanied by an image chip of the detected object at each slice in time [9, 10]. Chips are extracted from the full frame image. Sometimes, the background pixels may also be removed from the chip to further reduce file size.

Clicking the last track point at the end of a track break loads a chip of the detected object. By clicking start points on nearby tracklets, the analyst can attempt to match the broken tracklets through visual characteristics and associated metadata like speed and heading. Different colors or symbols differentiate linked tracks to maintain provenance of the data so analysts can determine which tracklets were algorithmically generated and which were validated by a human. Figure 22.4 shows an example of track linking in the environment.

22.4 Advanced Analytics

Another key feature of many ABI analysis tools is the implementation of "advanced analytics"—automated algorithmic processes that automate routine functions or synthesize large data sets into enriched visualizations.

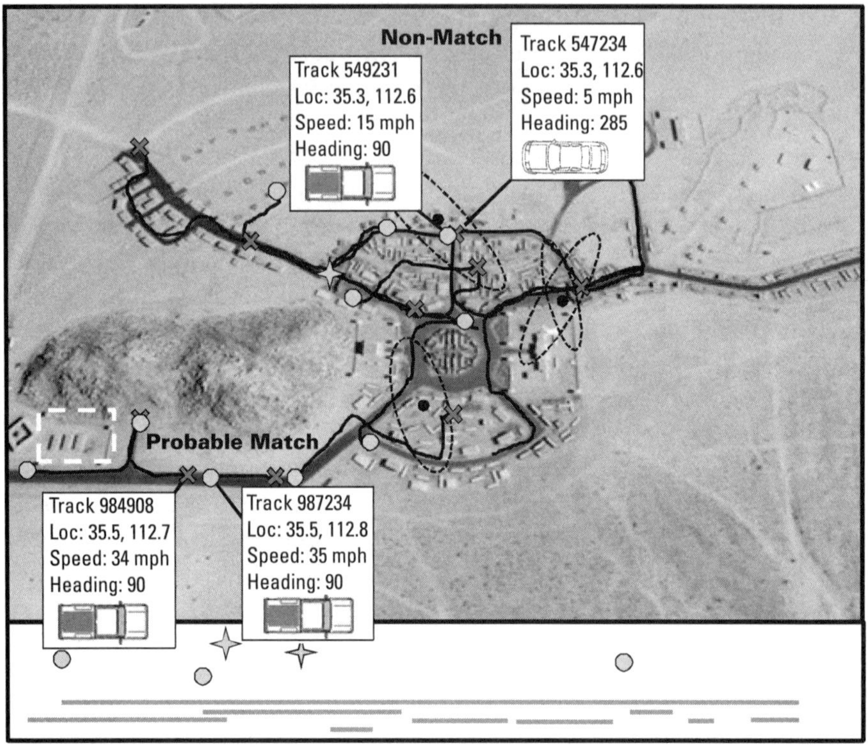

Figure 22.4 Track linking with WAMI data in a spatiotemporal analytic environment. (Image source: USDA Farm Service Agency.)

Density maps allow pattern analysis across large spatial areas. Also called "heat maps," these visualizations sum event and transaction data to create a raster layer with hot spots in areas with large numbers of activities. Data aggregation is defined over a certain time interval. For example, by setting a weekly time threshold and creating multiple density maps, analysts can quickly understand how patterns of activity change from week to week.

Density maps allow analysts to quickly get a sense for where (and when) activities tend to occur. This information is used in different ways depending on the analysis needs. If events are very rare like missile launches or explosions, density maps focus the analyst's attention to these key events.

In the case of vehicle movement (tracks), density maps identify where most traffic tends to occur. This essentially identifies nondiscrete locations and may serve as a contraindicator for interesting nodes at which to exploit patterns of life. For example, in an urban environment, density maps highlight major shopping centers and crowded intersections. In a remote environment, density maps of movement data may tip analysts to interesting locations.

Other algorithms process track data to find intersections and overlaps. For example, movers with similar speed and heading in close proximity appear as cotravelers. When they are in a line, they may be considered a convoy. When two movers come within a certain proximity for a certain time, this can be characterized as a "meeting." Mathematical relations with different time and space thresholds identify particular behaviors or compound events. Large data sets can be quickly parsed and triaged with these analytics to identify new key events and further focus the analyst's attention for follow-on exploitation.

Advanced user interfaces have also been implemented. In 2015, Ringtail Design and PIXIA Corporation demonstrated a unique touch-screen interface that uses the HiperStare® WAMI Viewer with the Ringtail "Replay" real-time analysis and visualization tool to rapidly and easily exploit multiple data sets on a single pane of glass [11, 12].

22.5 Information Sharing and Data Export

Many frameworks feature geoannotations to enhance spatial storytelling. These geospatially and temporally referenced "callout boxes" highlight key events and contain analyst-entered metadata describing a complex series of events and transactions.

Not all analysts operate within an ABI analysis tool but could benefit from the output of ABI analysis. Tracks, image chips, event markers, annotations, and other data in activity sets can be exported in KML, the standard format for Google Earth and many spatial visualization tools. KML files with temporal metadata enable the time slider within Google Earth, allowing animation and playback of the spatial story.

22.6 Summary

Over the past 10 years, several tools have emerged that use the same common core features to aid analysts in understanding large amounts of spatial and temporal data. At the 2014 USGIF GEOINT Symposium, tool vendors including BAE Systems [13, 14], Northrop Grumman [15], General Dynamics [16], Analytical Graphics [17, 18], DigitalGlobe, and Leidos [19] showcased advanced analytics tools similar to the above [20]. Georeferenced events and transactions, temporally explored and correlated with other INT sources allow analysts to exploit pattern-of-life elements to uncover new locations and relationships. These tools continue to develop as analysts find new uses for data sources and develop tradecraft for combining data in unforeseen ways.

References

[1] Arbetter, R., "Understanding Activity-Based Intelligence and the Human Dimension," presented at the *2010 GEOINT Symposium,* New Orleans, LA, November 1, 2010.

[2] "The GDELT Project," web. Available: http://gdeltproject.org/.

[3] "PV Labs," corporate overview.

[4] "CorvusEye 1500: Persistent Real-Time Intelligence Over a Wide Area (Spec Sheet)," Exelis Corporation, October 2014.

[5] Ratches, J. A., R. Chait, and J. W. Lyons, "Some Recent Sensor-Related Army Critical Technology Events," National Defense University, Defense & Technology 100 Paper, Center for Technology and National Security Policy, Feb. 2013.

[6] Ijeh, A. C., et al., "Geofencing in a Security Strategy Model," in *Global Security, Safety, and Sustainability,* Berlin ; New York: Springer, 2009.

[7] Metzger, P. J., "Automated Indications and Warnings from Live Surveillance Data," Lincoln Laboratory Journal, Vol. 18, No. 1, 2009.

[8] Rise of the Drones (NOVA), Public Broadcasting System, 2013.

[9] "LVSD Motion Imagery Streaming, MISB RP 1011.1," IC/DoD/NGA Motion Imagery Standards Board, February 27, 2014.

[10] "MISB Engineering Guideline 0810.2, Profile 2: KLV for LVSD Applications." 11 Jun 2010.

[11] Ringtail Design and PIXIA Corporation, "Wide Area Motion Imagery Analysis," web. Available: https://www.youtube.com/watch?v=qn_50VPHg6A.

[12] Ringtail Design, "Replay for Real-Time Situational Awareness," web. Available: https://www.youtube.com/watch?v=_SgbJJYJbW0.

[13] Ratzer, C., "BAE Systems Honored for Outstanding Achievement in Aerospace & Defense," web. Available: http://www.baesystems.com/article/BAES_167280/bae-systems-honored-for-outstanding-achievement-in-aerospace--defense. [Accessed: 10-Mar-2015].

[14] BAE Systems, "Activity-Based Intelligence: Get Answers to Your Most Difficult Questions (SOCET GXP 4.0 Brochure)," 2012.

[15] Mitchell, L. T., "From Messy to Intelligible Data: A Look at Northrop Grumman's ABI R&D," *Trajectory Magazine,* web. Available: http://trajectorymagazine.com/got-geoint/item/1786-from-messy-to-intelligible-data.html. [Accessed: March 10, 2015] .

[16] "Multi-INT Analysis and Archive SystemTM (MAAS®)—Operationally Deployed Full Motion Video (FMV) and Imagery Processing, Exploitation and Dissemination (PED) Information Sheet," web. Available: http://www.gd-ais.com/Products/ISR-Imagery-Analysis/MAAS.

[17] Analytical Graphics, Inc., "A Case Study: Patterns-of-Life and Activity Based Intelligence Analysis. Suritec Integrates STK Engine to Create Multi-INT Fusion Framework," March 13, 2013, web. Available: http://www.agi.com/downloads/support/productSupport/literature/pdfs/CaseStudies/031313_CaseStudy_Suritec.pdf.

[18] Claypoole, S., "STK and Activity-Based Intelligence," April 25, 2013.

[19] Leidos Corporation, "Advanced Analytics Suite."

[20] Gerber, C., "Video Program Expands Imagery," Geospatial Intelligence Forum, Vol. 10, No. 7, October 2012.

23

Pattern Analysis of Ubiquitous Sensors

The "Internet of Things" is an emergent paradigm where sensor-enabled digital devices record and stream increasing volumes of information about the patterns of life of their wearer, operator, holder—the so-called user. We, as the users, leave a tremendous amount of "digital detritus" behind in our everyday activities. Data mining reveals patterns of life, georeferences activities, and resolves entities based on their activities and transactions. This chapter demonstrates how the principles of ABI apply to the analysis of humans, their activities, and their networks...and how these practices are employed by commercial companies against ordinary citizens for marketing and business purposes every day.

23.1 Entity Resolution Through Activity Patterns

Progressive Insurance's Snapshot program seeks to reward good drivers and punish bad ones: "Snapshot personalizes your insurance rate based on your actual driving (called usage-based insurance). The better you drive, the more you can save" [1].

The company mails each subscribing driver a transponder that plugs into the onboard diagnostic (OBD2) port, a standard piece of equipment in most vehicles produced since the 1980s. According to Progressive, the device records information like distance, time of the day, hard acceleration/deceleration, and the vehicle identification number (VIN) (a unique identifier for the vehicle and a proxy for the driver). The device is equipped with a cellular data connection that beams logged data back to Progressive in real time. It personalizes your rate based on "your number of hard brakes, the number of miles you drive, and the amount of time you spend driving between midnight and 4 a.m." [1]. Users of the device are typically monitored for 75 days, at which time they return the device to Progressive and a modified "usage-based" rate is calculated based on the driver's pattern of activities.

Progressive claims the device does not contain a GPS sensor or other tracking technology and that it does not collect "speed" information. The device does collect distance and duration (for which an average speed can be collected). We call this "quasi-spatial" data because the distance values reveal details about the origin and destination: You can (generally) assume that the first trip on the day originates from the target's bed-down location and that the last trip terminates at the same location. For office workers, the first and last trip are of equal distance (excluding midday trips and intermediate stops). The times associated with each trip are also a type of start/stop data.

The author and his spouse signed up for persistent monitoring by Progressive Insurance and hosted a friendly competition: "who is the better driver?" After 75 days of monitoring, each user received the same discount. Forensic analysis of the data showed no discernable difference in the mean number of "hard brakes" between drivers. The two vehicles took approximately the same number of trips (149 versus 159) and traveled approximately the same distance (1,030 miles versus 1,216 miles) during the sampling period. But there is more to the story.

The Snapshot program is based on the assumption that the VIN is a proxy for the driver and his/her behavior. In this case, the author has a single-car garage where both vehicles are parked in tandem. Instead of a vehicle being selected based on the "ownership" proxy, the rear vehicle is always selected by the first driver to leave the house. Therefore, although each driver prefers a vehicle ("my car"), both entities swap proxies based on convenience, averaging out the pattern of activity across both entities, confounding Progressive's analysis. Another question arises: Can the driver of the vehicle be uniquely identified by the behavior of the vehicle? No spatial information is recorded, so the entity cannot be identified by georeferencing with known spatial nodes.

Figure 23.1 depicts the pattern of activities for each vehicle over time. The boxed letters highlight grouped patterns and the letters with subscripts highlight an individual day in a five-day work week. Weekends are removed for clarity. Color shades on the distance bars show the time of day of the trip. Lighter is earlier, and darker is later. The shaded boxes highlight the results of the forensic entity identification—the entity can be disambiguated based on a pattern of life for all but two days in the sample set.

The relationship between patterns A and B are the first clue. At B, one of the drivers regularly leaves in the early morning, departs at night, and drives between 20 and 40 miles per day—we will hypothesize this as the pattern for driver 1. This pattern appears again at E, F, I, J/K, N, P/Q, $R_{1\text{-}3}$, $S_{4,5}$, and U.

Another pattern is the light use of the other vehicle (also dominated by short, midday trips) as seen at A. We assign this occasional use pattern to driver

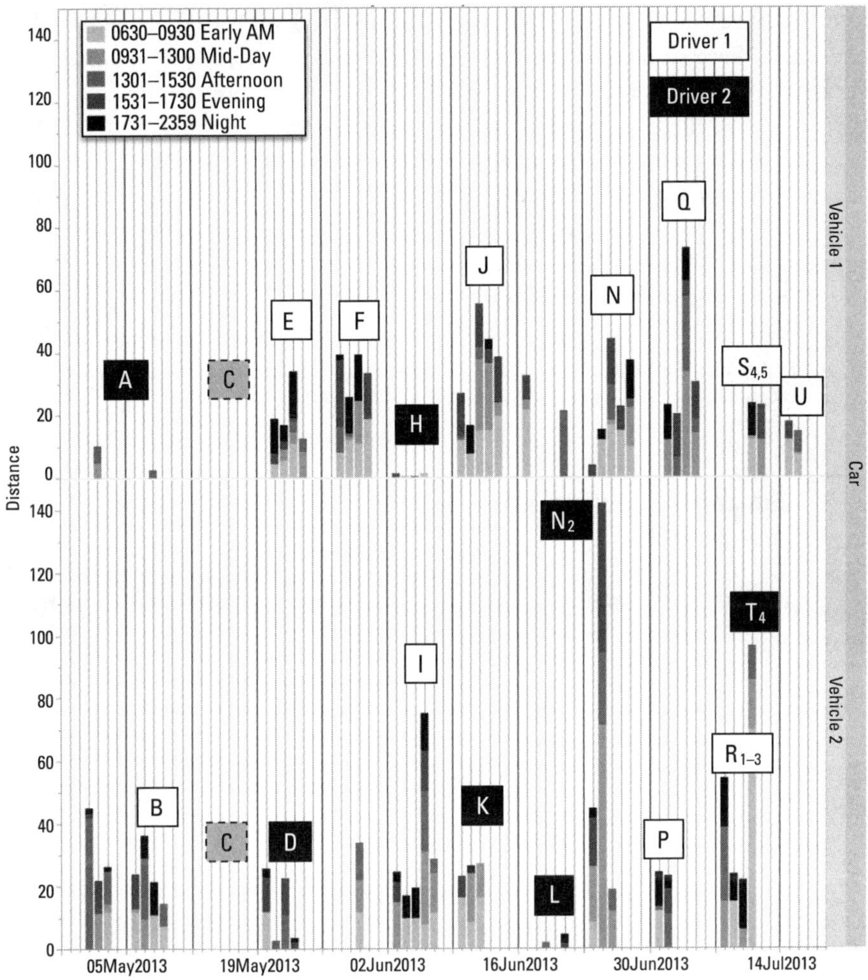

Figure 23.1 Progressive snapshot travel data for two drivers.

2 and also see it at D, H, K, and L. Deductive reasoning can be used to make vehicle/driver pairings in the case where one pairing is known (D). In some instances where the pattern is not obvious. For example, J/K and P/Q look similar.

Another case of deductive reasoning happens at R and S, which follow driver 1's pattern of life precisely. Therefore, T4 must be driver 2. If T4, a 100-mile round trip, is an activity of driver 2, does that reveal anything about N2, the mysterious event on June 25? This is an example of deductive analysis revealing new questions.

Upon initial examination of the data, approximately 50–70% of the patterns were uniquely resolvable to a single entity based on a simple, generalizable rule. Another 20–25% of the data could be resolved with some deduction and hypothesis generation but usually required some knowledge about the individual entities (the tendency to swap vehicles; driver 2 works from home). The last 5–10% of the data required MHT, the integration of contextual data like federal holidays, or correlation with another data set like flight records. The last bits of data required the most effort to resolve and some were ultimately unresolvable. These ratios are typical for ABI analysis and reflect the realities of assessing noisy, sparse, difficult data sets.

The addition of a single additional data source—for example geolocation—would have dramatically simplified the task by allowing the association of each trip with a home, work, or intermediate location. This is the value of georeference to discover and integration before exploitation.

23.2 Temporal Pattern of Life

Another source of highly temporal, weakly spatial activity data comes from health monitoring systems. Pedometers, once obscure devices to measure a user's steps per day, are becoming commonplace as individuals are increasingly conscious of their health. Figure 23.2 shows a histogram of step data from an office worker and reveals much about pattern of life and individual activities and transactions. The data is very accurate temporally and is weak spatially, as the number of steps measures the distance of the transaction and tells something about the distance between two events.

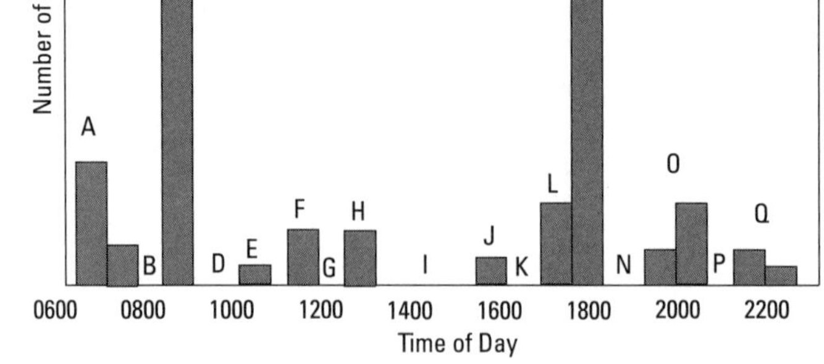

Figure 23.2 Histogram of pedometer data.

In ABI, the lack of collection does not imply lack of activity. This concept reminds the analyst that the absence of data does not mean "nothing happened." It might just mean that the activities and transactions were not observable by the sensor. The inverse of this principle is illustrated in Figure 23.3: Lack of activity does not mean lack of collection. Rather, because the pedometer persistently measures multiple direction changes as steps, lack of data means lack of movement, itself an activity.

Consider for example the activity at A, B, C, and D as a related sequence of activities. Collection begins around 6:00 a.m. as the user moves around (A). Then, the user does not move for approximately 30 minutes. Then, the user moves approximately 800 steps and then ceases moving for an hour. Activity B is the drive to work, arriving around 8:00 a.m. C is the walk from the parking lot to the user's desk, and D is an hour of work at a desk without getting up. When these four activities are considered as a sequence, the user's pattern of life is evident. The principle of "fact of one means fact of two" is applied to consider the trip to and from work as a complete transaction. If this data represents the pattern of life of an office worker, and C is the walk from the parking lot, then M is an activity paired with C: the walk back to the car. This makes N the drive home (note that N is longer than B). Activities F and H could be paired. When G is considered in context with the time of day, these three activities could

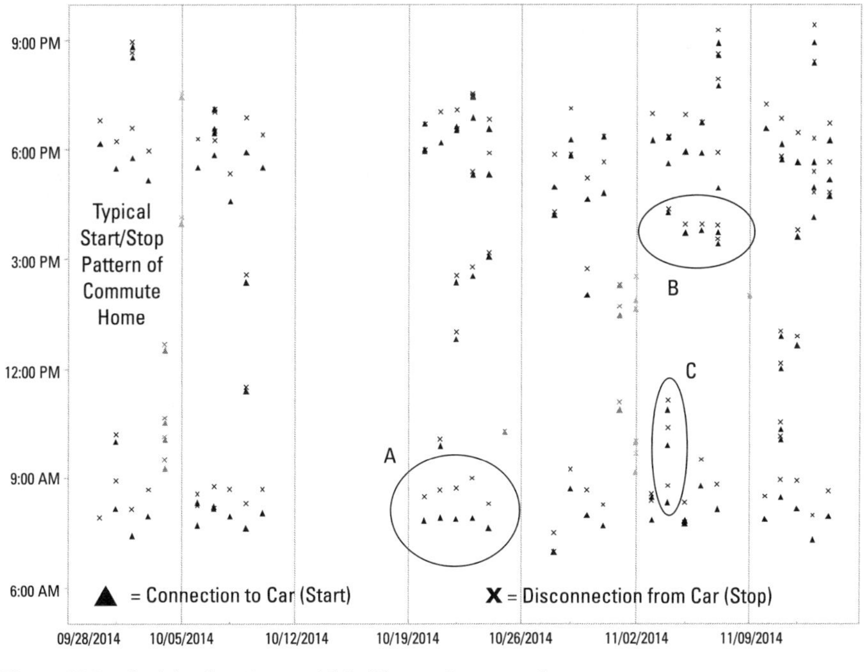

Figure 23.3 Activity data from vehicle Bluetooth connections.

represent walking to lunch, eating, and returning to the desk for multiple hours of sedentary work at I.

Digital devices record the user's pattern of life whether he or she realizes it or not. Consider an export of the activity log of the "Trigger" Android app on a user's mobile device. Trigger launches activities when events are triggered, for example, connecting or disconnecting from a Bluetooth device like the entertainment system in a modern automobile. Because the vehicle is a proxy for the individual, and because the mobile device logs an event for every connection and disconnection, these events act as proxies for starts and stops—the connection is activated and deactivated with the ignition of the vehicle. This data is shown in Figure 23.3.

In Figure 23.3, triangles represent starts and X's represent stops. Weekend days are lightly shaded. Area A shows a typical pattern of arrival and departure from home to work around 8:00 a.m. (Thursday took a little longer than usual and Friday the entity left home a few minutes early). The pattern at B during the week of November 2 introduces a new behavior. The entity appears to leave work between 3 and 4 p.m., drives a short distance (less than 10 minutes), disconnects for an hour or so, and then commutes home. The behavior at C indicates some kind of midday trip, perhaps from the office to another location and back. This entity also seldom connects the mobile device to the vehicle on weekends and there is a large gap with no usage in mid October. This type of data represents another pattern-of-life element—it allows the analyst to estimate where the entity is located most of the time, and because he or she tends to follow predictable patterns, it presents an opportunity for collection en route or at one of the suspected endpoints of the transaction.

23.3 Integrating Multiple Data Sources from Ubiquitous Sensors

Most of the diverse sensor data collected by increasingly proliferated commercial sensors is never "exploited." It is gathered and indexed "just in case" or "because it's interesting." When these data are combined, it illustrates the ABI principles of integration before exploitation and shows how a lot of understanding can be extracted from several data sets registered in time and space, or simply related to one another. The pedometer data in Figure 23.2 was integrated with Facebook history, Gmail-sent mail logs, a Microsoft Outlook schedule and e-mail records, the vehicle/mobile device synchronization in Figure 23.3, and text message and phone logs for a single day to illustrate activity resolution for a known entity (the author) in Figure 23.4.

The entity wakes around 6 a.m., putters around the house commenting on Facebook posts and sending a half dozen e-mails before departing for work around 7:15. After arriving at work, he sends a few text messages and emails,

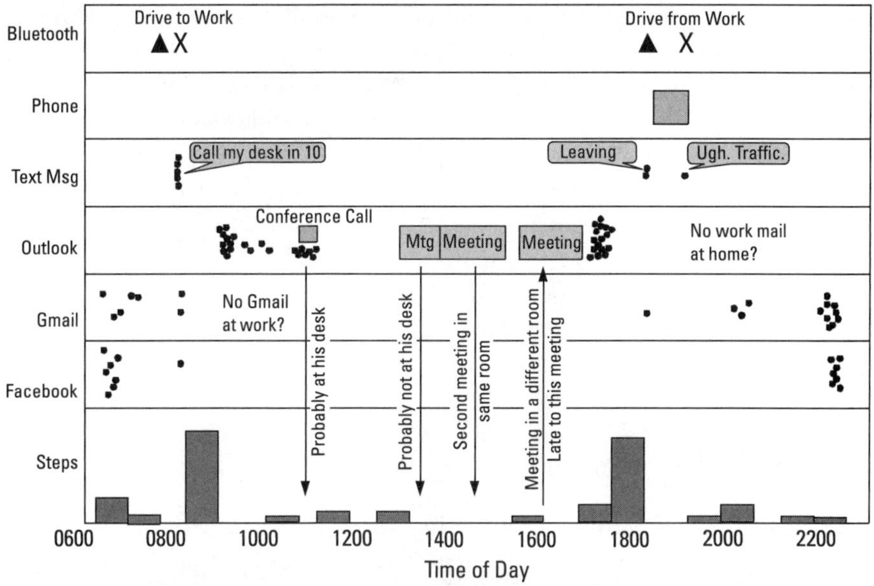

Figure 23.4 Pattern of activities for a single entity in a single day.

then walks to his desk. Upon logging in, he answers a string of e-mails, then gets up to get a cup of coffee or use the restroom. Then, he is probably at his desk when he participates in a conference call (he multitasks by sending seven e-mails). Then he walks somewhere—possibly for a 25-minute lunch—and then walks somewhere else. This location is probably not his desk, because the Outlook calendar says there is a meeting. Because there are no steps between adjacent meetings, the second meeting is probably in the same room. Then, he walks to another location (possibly his desk) for another meeting but arrives late. At the end of the day, he sends a dozen e-mails, walks to the car, texts "leaving" and makes a phone call while driving. Upon arriving home he putters around the house, sends a few emails, remains still for about 45 minutes (possibly watching television or eating), and then sends a flurry of emails and online messages before going to bed.

The semantic understanding of the day's activities can be deduced from integrating multiple sources of data to paint a complete picture. Certain key sensors like the pedometer identify when the entity must be in the same place because he has not moved. Other sensors like the transaction records from the two e-mail systems help identify locations. The entity is able to send Gmail from home or from a mobile device but cannot access Gmail at work. Therefore, when he sends Gmail, he cannot be located at work. Similarly, a temporal record of Outlook e-mail messages means that he must be at his desk and logged into the corporate e-mail system.

Analysis of related entities could uncover similar patterns. For example, for each e-mail there must be at least one recipient. Some of the recipients might be related to the meeting attendees. Another entity was informed to "call my desk in 10." The same or a different entity was notified when the entity left work and that he was stuck in traffic, explaining his late arrival home.

Emerging research in semantic trajectories describes a pattern of life as a sequence of semantic movements (e.g., "he went to the store") as a natural language representation of large volumes of spatial data [2]. Some research seeks to cluster similar individuals based on their semantic trajectories rather than trying to correlate individual data points mathematically using correlation coefficients and spatial proximities [3].

23.4 Summary

ABI data from digital devices, including self-reported activities and transactions, are increasingly becoming a part of analysis for homeland security, law enforcement, and intelligence activities. The proliferation of such digital data will only continue. Methods and techniques to integrate large volumes of this data in real time and analyze it quickly and cogently enough to make decisions are needed to realize the benefit these data provide. This chapter illustrated visual analytic techniques for discovering patterns in this data, but emergent techniques in "big data analytics" are being used by commercial companies to automatically mine and analyze this ubiquitous sensor data at network speed and massive scale.

References

[1] "FAQs: Snapshot Discount, Pay As You Drive®, Usage-Based Insurance," Progressive Insurance.

[2] Sabarish, B. A., R. Karthi, and T. Gireeshkumar, "A Survey of Location Prediction Using Trajectory Mining," in *Artificial Intelligence and Evolutionary Algorithms in Engineering Systems* (eds. Suresh, L. P., S. S. Dash, and B. K. Panigrahi), Vol. 324, Springer India, 2015, pp. 119–127.

[3] Ying, J. J.-C., et al., "Semantic Trajectory Mining for Location Prediction," *Proceedings of the 19th ACM SIGSPATIAL International Conference on Advances in Geographic Information Systems,* 2011, p. 34.

24

ABI Now and Into the Future

Patrick Biltgen and David Gauthier

The creation of ABI was the proverbial "canary in the coal mine" for the intelligence community. Data is coming; and it will suffocate your analysts. Compounding the problem, newer asymmetric threats can afford to operate with little to no discernable signature and traditional nation-based threats can afford to hide their signatures from our intelligence capabilities by employing expensive countermeasures. Since its introduction in the mid 2000s, ABI has grown from its roots as a method for geospatial multi-INT fusion for counterterrorism into a catch-all term for automation, advanced analytics, anticipatory analysis, pattern analysis, correlation, and intelligence integration. Each of the major intelligence agencies has adapted a spin on the technique and is pursuing tradecraft and technology programs to implement the principles of ABI. This chapter describes some of those nascent efforts. The increasing diversity of threats facing national security and the pressure to do more as budgets and contractor staffs contract requires a change in business model. ABI—a set of methods or series of principles for large-scale deductive analysis and correlation of multi-INT data—provides a promise to address urgent challenges across multiple intelligence problems. More importantly, the core tenets of ABI become increasingly important in the integrated cyber/geospace and consequent threats emerging in the not too distance future.

24.1 An Era of Increasing Change

At the 2014 IATA AVSEC World conference, DNI Clapper said, "Every year, I've told Congress that we're facing the most diverse array of threats I've seen in all my years in the intelligence business" [1]. In 2012, the National Intelligence Council (NIC) released Global Trends: 2030, the fifth installment of the intelligence community's critical assessment of future trends and drivers. The report was "intended to stimulate thinking about the rapid and vast geopolitical changes characterizing the world today and possible global trajectories during the next 15–20 years" [2]. They postulated several megatrends and game changers, including individual empowerment, increasing global instability, a diffusion of international power, and the revolutionary impact of new technologies. Although technologies such as information technology provide increased productivity and quality of life, the NIC proposed that the "fear of the growth of an Orwellian surveillance state may lead citizens particularly in the developed world to restrict or dismantle big data systems" [2]. The report also clearly stated that the rate of rapid change and diversity of threats was expected to continue into the future.

On September 17, 2014, Clapper unveiled the 2014 National Intelligence Strategy (NIS)—for the first time, unclassified in its entirety—as the blueprint for IC priorities over the next four years. The NIS describes three overarching mission areas, strategic intelligence, current operations, and anticipatory intelligence, as well as four mission focus areas, cyberintelligence, counterterrorism, counterproliferation, and counterintelligence [3, p. 6]. For the first time, the cyberintelligence mission is recognized as co-equal to the traditional intelligence missions of counterproliferation and counterintelligence (as shown in Figure 24.1). The proliferation of state and non-state cyber actors and the exploitation of information technology is a dominant threat also recognized by the NIC in *Global Trends 2030* [2].

Incoming NGA director Robert Cardillo said, "The nature of the adversary today is agile. It adapts. It moves and communicates in a way it didn't

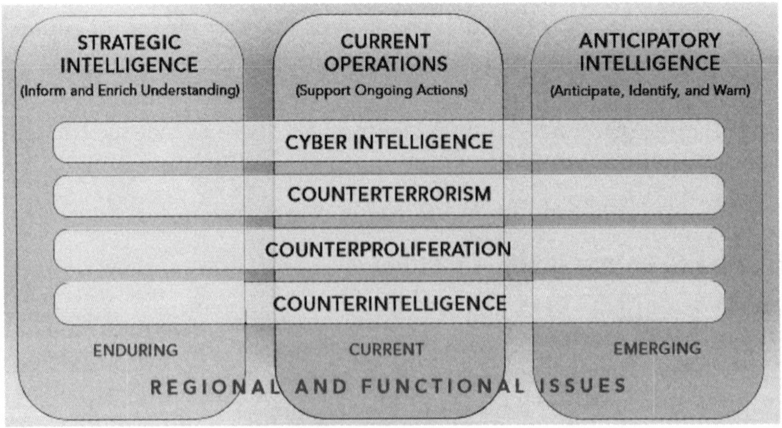

Figure 24.1 Mission objectives identified in the 2014 NIS [3, p. 6].

before. So we must change the way we do business" [4]. ABI represents such a change. It is a fundamental shift in tradecraft and technology for intelligence integration and decision advantage that can be evolved from its counterterrorism roots to address a wider range of threats.

24.2 ABI and a Revolution in Geospatial Intelligence

The ABI revolution at NGA began with grassroots efforts in the early 2000s and evolved as increasing numbers of analysts moved from literal exploitation of images and video to nonliteral, deductive analysis of georeferenced metadata. In April 2011, former NGA director Letitia Long introduced activity-based GEOINT to the public, saying:

> We are being called upon to think differently, work differently and collaborate differently… We are being challenged to think in terms of activity-based GEOINT rather than target-based GEOINT and to explain not only where something is happening, but also why [5].

Long describes ABI as a way to manage the flood of "big data" and transform it into "big value," saying, "We cannot deliver 'big value' with our traditional target-centric, product-focused approach to analysis. Rather, we must implement an activity-based intelligence approach so we can collaborate more widely, understand more deeply, and discover more vital information." She describes ABI as a fundamental shift in analytic strategy because "the targets and threats we face are no longer stable, predictable, or so slow-moving that we can gradually build insight with traditional intelligence techniques" [6].

The so-called equation chart in Figure 24.2 [7] shows how georeferenced content and persistent GEOINT content form the basis of big data, which much be ingested, managed, and accessed using the methods described in Chapter 10. New capabilities for activity analytics, object discovery, and modeling/network analytics leverage analytic techniques, automation, and fusion methods from Chapters 12–14. Finally, knowledge management techniques described in Chapter 15 create a core foundation of living knowledge that evolves through subsequent collection and analysis to support object-based production and GEOINT production.

Long, in her remarks at the 2014 GEOINT Symposium, introduced NGA's portfolio initiatives [8]. The analytic capabilities portfolio, depicted in Figure 24.3, includes models and strategies, data collection, data analytics, and documentation of structured data. NGA's structured data nomenclature includes traditional GEOINT terms like facility, unit, and equipment—but also includes activities and people (entities) central to ABI tradecraft [9]. R&D for ABI capabilities is included in NGA's analytic capabilities portfolio initiative.

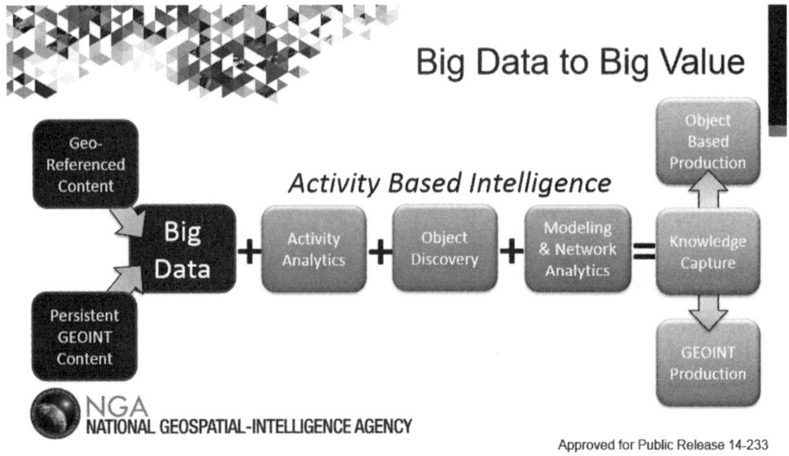

Figure 24.2 Shift from big data to big value. Presented at the USGIF 2013* GEOINT Symposium. (Approved for Public Release [7].)

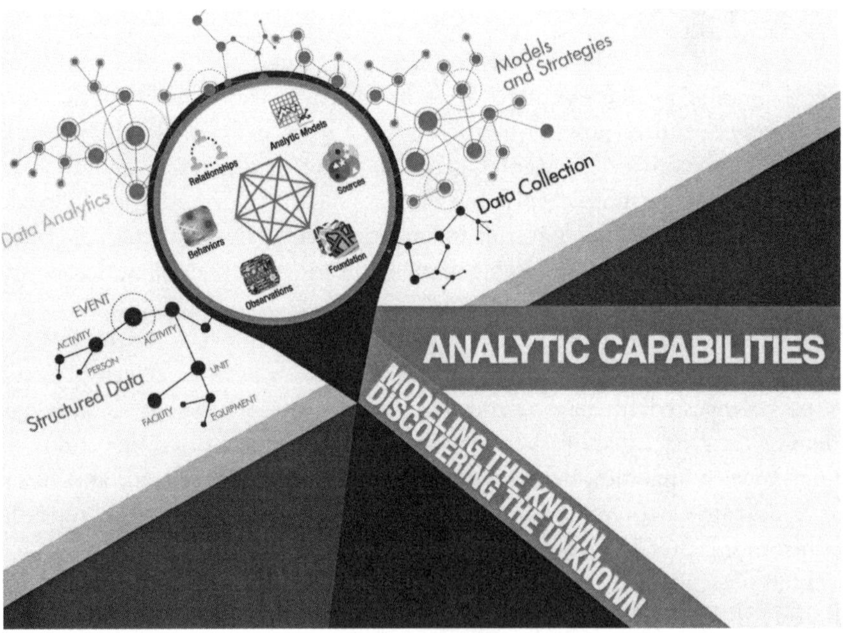

Figure 24.3 NGA analytic capabilities portfolio. (Source: NGA [9].)

The importance of GEOINT to the fourth age of intelligence was underscored by the NGA's next director, Robert Cardillo, who said "Every modern, local, regional and global challenge—climate change, future energy landscape and more—has geography at its heart" [10]. Cardillo also promoted the increasing used of unclassified information in GEOINT analysis, announcing a completely unclassified government-sponsored analysis environment, the GEOINT Pathfinder, on March 16, 2015, which launched in the summer of 2015 [11]. Cardillo highlighted the increasing importance of the "synergy of our new unclassified sources and our sophisticated classified sources to enable more exquisite insights and understanding on the new, higher open playing field" [11].

NGA also released a vision for its analytic environment of 2020, noting that analysts in the future will need to "spend less time exploiting GEOINT primary sources and more time analyzing and understanding the activities, relationships, and patterns discovered from these sources"—implementation of the ABI tradecraft on worldwide intelligence issues [12]. The key technology needs as identified by the agency, including ABI tradecraft enablers, are highlighted in Table 24.1.

Table 24.1
NGA Key Analysis Technology Needs for 2020

Focus Area	Key Technology Needs
Research and Discover	Centralized cross-domain, data-agnostic federated search capability Natural language/semantic query capability Automated metadata tagging to specify spatiotemporal and intelligence/domain content "Deep Learning" algorithms for automated data discovery IC desktop dashboard with real-time, contextually-relevant content streams
Access and Visualize	Aggregation of data access points into a unified/common distribution framework Unification of content visualization into a single, integrated analytic environment Open, standards-based data models for interoperability Service enablement of all NGA data sources Immersive visualization at the analyst's desk Mobile exploitation and analysis capabilities
Exploit and Analyze	Automated algorithms for object (entity and activity) detection, vectorizing, and attribution Automated algorithms for assigning and validating graph-based relationships between objects Enterprise-level database(s) for structured observation capture, sharing, and analysis Decoupling of technical capabilities/algorithms from stove-piped analytical platforms Touch-, gesture- and voice-based interaction with the integrated analytic environment Enterprise-level graph databases Analytic modeling framework for centralized GEOINT model management, sharing, and testing
Publish and Report	Data-centric publishing platform that exposes GEOINT observations and judgments to customers in near-real time Dissemination and messaging services that allow for data publishing and alerting without requiring reformatting, exporting, or porting Interoperable structured data visualization capabilities Automated report generation capability from structured data

Source: [12]. Approved for Public Release, NGA 14-472.

24.3 ABI and Object-Based Production

While NGA has firmly taken the reigns in the definition of ABI tradecraft and its implementation as a GEOINT technique, the NSA and DIA partnered on OBP to organize known information to increase the usefulness of intelligence. ABI and OBP are related as shown Figure 24.4. While ABI is focused on discovery of unknowns through deductive reasoning on activities and transactions, OBP structures and organizes known information to improve reporting with a focus on objects and entities—and their behaviors. Techniques like geospatial visualization, data filtering, correlation and fusion, and network analysis are endemic to both methods. The combination of both sets of methods improves intelligence integration and saves analysts' time. When known objects and entities are identified and organized, it is easier for analysts to identify knowledge gaps and apply ABI methods to close those gaps through analysis and subsequent collection.

Figure 24.4 shows the principle of data neutrality in the form of "normalized data services" and highlights the role for "normalcy baselines, activity models, and pattern-of-life analysis" as described in Chapters 14 and 15. OBP, as shown in the center of Figure 24.4, depicts a hierarchical model, perhaps using the graph analytic concepts of Chapter 15 and a nonlinear analytic process that captures knowledge to form judgments and answer intelligence questions. Chapter 16's concept of models is shown in Figure 24.4 as "normalcy baselines,

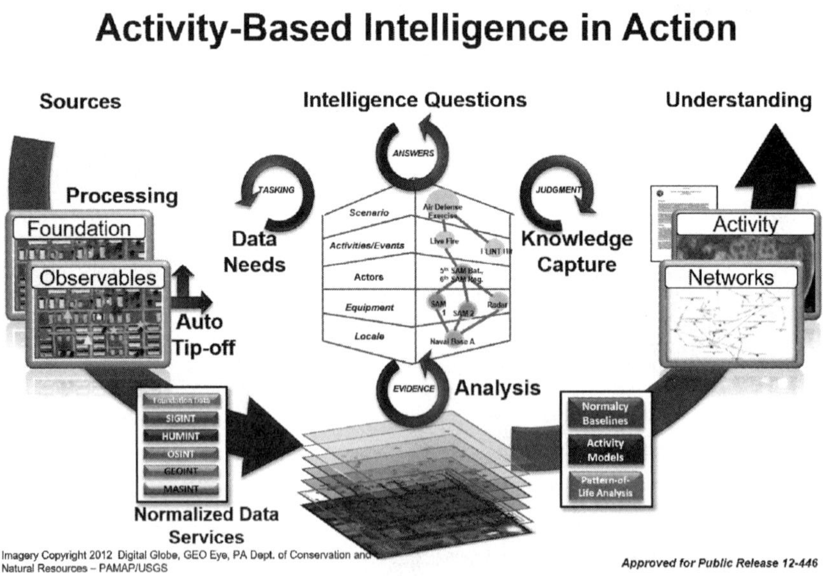

Figure 24.4 Relationship between ABI and object-based production [13].

activity models, and pattern-of-life analysis." As opposed to the traditional intelligence process that focuses on the delivery of serialized products, the output of the combined ABI/OBP process is an improved understanding of activities and networks.

24.4 ABI Applied to Overhead Reconnaissance

Central to the original USD(I) definition of ABI is the notion of "analysis and subsequent collection" or the idea that the linear TCPED process for collecting intelligence can be instantiated to address knowledge gaps as the result of deductive analytics on ABI data. Overhead collection systems developed, acquired, launched, and operated by the NRO have historically been aligned with traditional intelligence problems. Recently, NRO has applied increasing focus to adapting those collectors, their tasking, and their related ground processing systems to enable ABI. NRO director Betty Sapp believes in overhead and ground innovation to enable revolutionary capabilities like ABI, saying, "We intend to provide the U.S. the sensor diversity and on-demand persistence that we have found so valuable in Iraq and Afghanistan [aboard] airborne platforms. We just want to do it from space, with the unique advantages space provides—near-instantaneous global access and access to denied areas" [14, 15]. In 2010, NRO launched the Sentient program (Figure 24.5) to address problem-driven, orchestrated, multi-INT collection [16]. Touting Sentient in a rare public appearance at the GEOINT Symposium in 2014, Sapp said, "We've demonstrated that we cannot only be responsive but predictive in where we aim our space

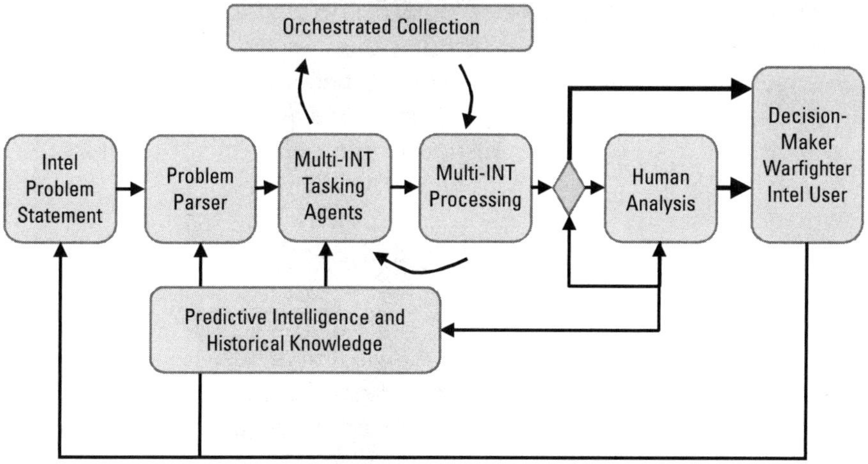

Figure 24.5 Sentient Grand Vision. Adapted from [14]. (Presented by D/NRO Betty Sapp at the 2014 GEOINT Symposium.)

assets." As shown in Figure 24.5, the concepts of "predictive intelligence and historical knowledge" (described in Chapter 16) are key to linking human analysis with multi-INT tasking, orchestrated collection, and multi-INT processing. This critical feedback loop is central to enabling the ABI vision for "analysis and subsequent collection" using automated or semi-automated means.

Sapp also described the operational success story of the Fusion Analysis & Development Effort (FADE) and the Multi-Intelligence Spatial Temporal Toolsuite (MIST), which became operational in 2007 when NRO users recognized that they got more information out of spatiotemporal data when it was animated. NRO designed a "set of tools that help analysts find patterns in large quantities of data" [15]. MIST allows users to temporally and geospatially render millions of data elements, animate them, correlate multiple sources, and share linkages between data using web-based tools. "FADE is used by the Intelligence Community, Department of Defense, and the Department of Homeland Security as an integral part of intelligence cells" [17]. An integrated ABI/multi-INT framework is a core component of the NRO's future ground architecture [18].

24.5 The Future of ABI in the Intelligence Community

In 1987, the television show *Star Trek: The Next Generation,* set in the 24th century, introduced the concept of the "communicator badge," a multifunctional device worn on the right breast of the uniform. The badge represented an organizational identifier, geolocator, health monitoring system, environment sensor, tracker, universal translator, and communications device combined into a 4 cm by 5 cm package.

In 2014, all of these capabilities have been incorporated into the modern smartphone. In South Korea, for example, the penetration rate of smartphone technology exceeded 70% for the first time this year, meaning seven out of 10 subscriptions are for smartphones [19]. In Seoul, a city of 9.8 million, over three million residents commute by private car and five million by public transportation with an average commute time in excess of 40 minutes per day [20]. There, a mobile phone also functions as a mobile payment device, a subway metro card, and a streaming video player—many Koreans watch "T.V." on the train during their commute. The mobile device serves as the ultimate identity proxy.

The latest generation of BMW's ConnectedDrive system features integrated cellular connectivity and GPS hard-wired to the vehicle providing location services and remotely accessible telematics that relay vehicle health and status to authorized service centers [21]. Smart meters, intelligent thermostats like Google's Nest, wi-fi enabled refrigerators, and Amazon's always-on voice-recognizing digital assistant Echo increasingly proliferate as part of our daily lives.

By 2030, living "off the grid" will be impossible. The necessities of life will require a digital connection to your friends, family, finances, and workplace. This introduces a new paradigm for data collection, analysis, and manipulation—for commercial, educational, or intelligence purposes—foreign and domestic. Bloomberg forecasts that by 2030, the world will host 40 megacities, cities with at least 10 million people [22]. Manila has a population density of 43,000 people per square kilometer (twice the density of Paris) [23]. In megacities, tens of thousands of entities may occupy a single building, and thousands may move in and out of a single city block in a single day. The flow of objects and information in and out of the control volume of these buildings may be the only way to collect meaningful intelligence on humans and their networks because traditional remote sensing modalities will have insufficient resolution to disambiguate entities and their activities. Entity resolution will require thorough analysis of multiple proxies and their interaction with other entity proxies, especially in cases where significant operational security is employed. Absence of a signature of any kind in the digital storm will itself highlight entities of interest. Everything happens somewhere, but if nothing happens somewhere that is a tip to a discrete location of interest.

The methods described in this textbook will become increasingly core to the art of analysis. The customer service industry is already adopting these techniques to provide extreme personalization based upon personal identity and location. Connected data from everyday items networked via the Internet will enable hyperefficient flow of physical materials such as food, energy, and people inside complex geographic distribution systems. Business systems that are created to enable this hyperefficiency, often described as "smart grids" and the "Internet of Things", will generate massive quantities of transaction data. This data, considered nontraditional by the intelligence community, will become a resource for analytic methods such as ABI to disambiguate serious threats from benign activities.

Use of nontraditional sources inside the intelligence community will not happen overnight. As with any workforce possessing a strong cultural identity, the organizational change necessary to drive the adoption of outside sources of information coupled with entirely new analytic methods will take time. However, the intelligence community has already recognized that it will no longer hold a monopoly on sophisticated collection sources and therefore must adopt the use of geospatial information from all sources as a new foundation for the knowledge of place and time. A continuous baseline of new information updated from social media and open sources will subsequently be amplified by the use of traditional intelligence systems. This widespread shift to a model based upon data neutrality, a core tenet of ABI, fundamentally changes the nature of intelligence in the future.

Christensen described a method for using S-curves to measure the impact of disruptive technological change within a firm [24]. Today this same curve is often used to describe the process of organizational change in businesses, and NGA has adapted this curve to anticipate the challenge of a wide-scale transition to data-driven analytic methods such as OBP and ABI (Figure 24.6). The first hurdle is both architectural and organizational—providing a standardized data access capability across the diaspora of potential sources of information no matter the source. Standardized OBP and ABI data services are providing initial solutions across the IC.

The second hurdle of change will be the widespread utility of that information inside mission-focused analytics—the pairing of human analysts with machine analytics that will drive automated collection decisions in real-time. NGA is currently using the term "persistent GEOINT" to define this epoch of new capabilities and in May 2014 created an eponymous office to lead the transition. NRO is spearheading research and development of these capabilities with the sentient program (Figure 24.5). The concept of human-machine-teaming, that is, the integration and optimization of man and machine to collaboratively perform tasks for which each is ideally suited is gradually being implemented in commerce, medicine, and military science.

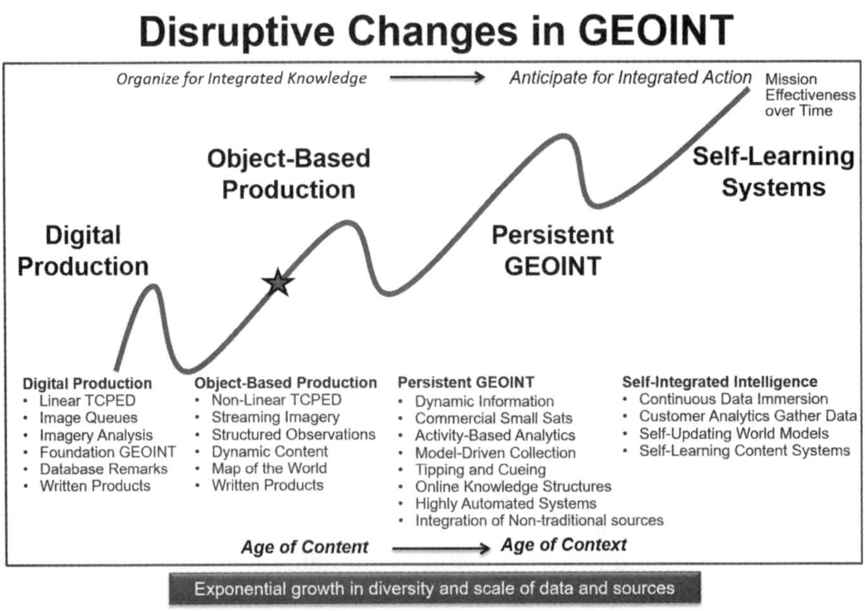

Figure 24.6 Disrupting changes in GEOINT. (Approved for public release. NGA 15-296 [25].)

24.6 Conclusion

The Intelligence Community of 2030 will be entirely comprised of digital natives born after 9/11 who seamlessly and comfortably navigate a complex data landscape that blurs the distinctions between geospace and cyberspace. The topics in this book will be taught in elementary school.

Our adversaries will have attended the same schools, and counter-ABI methods will be needed to deter, deny, and deceive adversaries who will use our digital dependence against us. Devices—those Internet-enabled self-aware transportation and communications technologies—will increasingly behave like humans. Even your washing machine will betray your pattern of life. LAUNDRY-INT will reveal your activities and transactions…where you've been and what you've done and when you've done it because each molecule of dirt is a proxy for someone or somewhere. Your clothes will know what they are doing, and they'll even know when they are about to be put on.

In the not too distant future, the boundaries between CYBERINT, SIGINT, and HUMINT will blur, but the rich spatiotemporal canvas of GEOINT will still form the ubiquitous foundation upon which all sources of data are integrated.

The core principles of this book, georeference to discover, sequence neutrality, data neutrality, and integration before exploitation, will still hold, but the application of the techniques will dramatically expand to include new sources and methods. We will develop new techniques for analysis and anticipation that provide decision advantage, avoid strategic surprise, and improve our understanding across a range of issues in our increasingly dangerous, diverse, and dynamically changing world.

24.7 Chapter Author Biography

David Gauthier is the director of advanced concepts for the office of persistent GEOINT at NGA. Previously, he was the agency's lead for ABI and the chief of integration in the office of special programs. Gauthier has held several assignments supporting community campaign initiatives. He is the recipient of numerous awards including a 2015 National Intelligence Professionals Award and a 2012 Meritorious Unit Citation. Gauthier holds an M.S. in aerospace engineering and an M.S. in telecommunications science from the University of Colorado at Boulder and a B.S. in electrical engineering from Rensselaer Polytechnic Institute. The views presented are those of the author and do not necssarily represent the views of the Department of Defense or its components.

References

[1] Clapper, J., "Remarks as delivered by The Honorable James R. Clapper Director of National Intelligence," presented at the IATA—AVSEC World, Grand Hyatt Hotel, Washington, D.C., October 27, 2014.

[2] "Global Trends 2030: Alternative Worlds," National Intelligence Council, December 2012.

[3] "National Intelligence Strategy of the United States of America (2014)," Office of the Director of National Intelligence, September 2014.

[4] Roop, L., "5 questions with America's New National Geospatial-Intelligence Agency Director," AL.com, November 2014.

[5] Long, L., "On My Mind: Leveraging Technology," *Pathfinder,* Vol. 9, No. 2, April 2011.

[6] Long, L., "Building a New Superhighway for Intelligence Integration, Remarks at the 2013 INSA Leadership Dinner," *INSA Leadership Dinner,* April 30, 2013.

[7] Gauthier, D., "Activity Based Intelligence: Finding Things That Don't Want to be Found." Presented at the *2013* GEOINT Symposium,* Tampa, FL, April 16, 2014. Approved for Public Release. NGA Case #14-233.

[8] Long, L., "Remarks at the 2013* USGIF GEOINT Symposium," Tampa, FL, April 2014.

[9] "Analytic Capabilities Portfolio Overview," National Geospatial-Intelligence Agency, Handout at the *2013* GEOINT Symposium,* approved for public release," April 2014.

[10] Cardillo, R., "Remarks at the Geography 2050 Conference," New York, November 19, 2014.

[11] Cardillo, R., "Remarks as Prepared for Robert Cardillo, Director, National Geospatial-Intelligence Agency for AFCEA/NGA Industry Day 2015," approved for public release, NGA Case #15-281.

[12] "2020 Analysis Technology Plan," National Geospatial-Intelligence Agency, approved for public release. NGA Case #14-472, November 12, 2014.

[13] Gauthier, D., "Activity-Based Intelligence," Presented at the *Unmanned Aircraft Systems Conference,* September 12, 2012, approved for public release, NGA Case #12-446.

[14] Sapp, B., "Keynote Presentation," *Proceedings of the 2013* GEOINT Symposium,* Tampa, FL, April 2014.

[15] Alderton, M., "From Airborne to Spaceborne, NRO Director Shares Recipe for the Next Generation in Space Innovation," *Trajectory Magazine.*

[16] National Reconnaissance Office, "Sentient Enterprise Request for Information," October 20, 2010, web. Available: https://www.fbo.gov.

[17] Carlson, B., "Remarks," at the *2011 USGIF GEOINT Symposium.*

[18] Nakamura, M., "Future Ground Architecture," presented at the *2014 Enterprise Innovation Symposium,* May 7, 2014.

[19] "Korea's Smartphone Population Tops Milestone," *Wall Street Journal,* July 28, 2014.

[20] "Economy/News/News/KBS World Radio," web. Available: http://world.kbs.co.kr/english/news/news_Ec_detail.htm?No=84738.

[21] "BMW ConnectedDrive: Broaden of Access and Expansion of Services Globally Will Include Benefits for US Customers," BMW USA, May 6, 2013.

[22] "Global Megacities," Bloomberg, September 9, 2014.

[23] "List of Cities Proper by Population Density." Wikipedia [Accessed: 16-Nov-2014].

[24] Christensen, C. M., "Exploring the Limits of the Technology S-Curve. Part I: Component Technologies," Production and Operations Management. Vol. 1, No. 4, Fall 1992.

[25] Gauthier, D., "Persistent GEOINT Vision," approved for public release, NGA Case #15-296.

25

Conclusion

In many disciplines in the early 21st century, a battle rages between traditionalists and revolutionaries. The latter is often comprised of those artists with an intuitive feel for the business. The latter is comprised of the data scientists and analysts who seek to reduce all of human existence to facts, figures, equations, and algorithms. In Taleb's world of securities trading, they are called the quants. In Lewis's *Moneyball,* they are the statheads. IBM has invested billions to perhaps one day replace traditional doctors with a computer named Watson.

Activity-Based Intelligence: Principles and Applications introduces methods and technologies for an emergent field but also introduces a similar dichotomy between analysts and engineers. The authors, one of each, learned to appreciate that the story of ABI is not one of victory for either side. In *The Signal and the Noise,* statistician and analyst Nate Silver notes that in the case of *Moneyball,* the story of scouts versus statisticians was about learning how to blend two approaches to a difficult problem. Cultural differences between the groups are a great challenge to collaboration and forward progress, but the differing perspectives are also a great strength. In ABI, there is room for both the art and the science; in fact, both are required to solve the hardest problems in a new age of intelligence.

Intelligence analysts in some ways resemble Silver's scouts. "We can't explain how we know, but we know" is a phrase that would easily cross the lips of many an intelligence analyst. At times, analysts even have difficulty articulating post hoc the complete reasoning that led to a particular conclusion. This, undeniably, is a very human part of nature. In an incredibly difficult profession, fraught with deliberate attempts to deceive and confuse, analysts are trained from their first day on the job to trust their judgment. It is judgment that is oftentimes unscientific, despite attempts to apply structured analytic techniques (Heuer) or introduce Bayesian thinking (Silver). Complicating this picture is the fissure in the GEOINT analysis profession itself, between traditionalists often focused purely on overhead satellite imagery and revolutionaries, analysts concerned with all spatially referenced data. In both camps, however, intelligence analysis is about making judgments. Despite all the automated tools and algorithms used to process increasingly grotesque amounts of data, at the end of the day a single question falls to a single analyst: "What is your judgment?"

Silver's statheads—our programmers, data scientists, and engineers—believe that if they could just get the model right, the questions would answer themselves. Though team science believes it is superior, technologists are no less likely to fall in the same mental traps to which analysts succumb. Engineers have their own mental traps, biases, and beliefs about how the world behaves. The best technologies almost always fail to solve "people problems." One of the greatest discoveries of the early ABI practitioners was that intelligence problems are at their heart people problems.

ABI introduced four pillars: georeference to discover, sequence neutrality, data neutrality, and integrate before exploit as illustrated in Figure 25.1. These pillars provide a methodological framework for analyzing complex problems in a world of "big data."

The ABI framework introduces three key principles of the artist frequently criticized by the engineer. First, it seems too simple to look at data in a spatial environment and learn something, but the analysts learned through experience that often the only common metadata is time and location—a great place to start. The second is the preference of correlation over causality. Stories

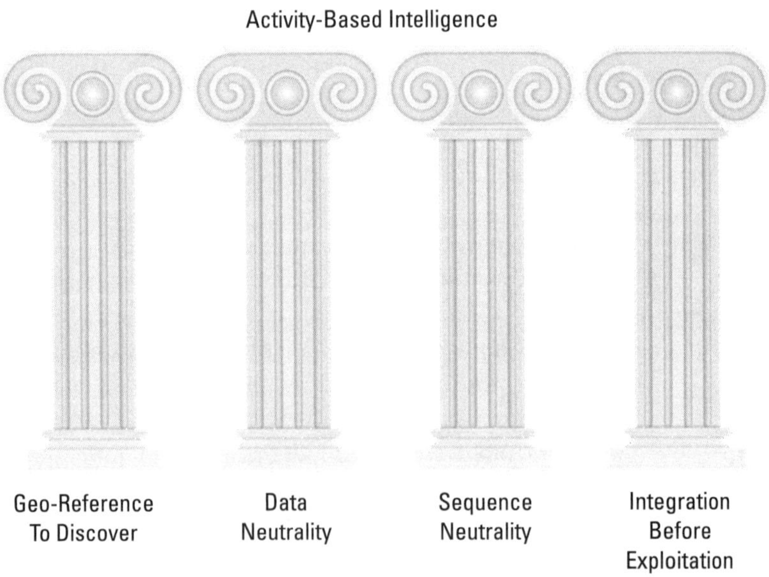

Figure 25.1 The four fundamental tradecraft pillars of ABI.

of intelligence are not complete stories with a defined beginning, middle, and end. A causal chain is not needed if correlation focuses analysis and subsequent collection on a key area of interest or the missing clue of a great mystery. The third oft-debated point is the near-obsessive focus on the entity. Concepts like entity resolution, proxies, and incidental collection focus analysts on "getting to who." This is familiar to leadership analysts, who have for many years focused on high-level personality profiles and psychological analyses. But unlike the focus of leadership analysis—understanding mindset and intent—ABI focuses instead on the most granular level of people problems: people's behavior, whether those people are tank drivers, terrorists, or ordinary citizens. Through a detailed understanding of people's movement in space-time, abductive reasoning unlocked possibilities as to the identity and intent of those same people. Ultimately, getting to who gets to the next step—sometimes "why," sometimes "what's next."

Techniques like automated activity extraction, tracking, and data fusion help analysts wade through large, unwieldy data sets. While these techniques are sometimes synonymized with "ABI" or called "ABI enablers," they are more appropriately termed "ABI enhancers." There are no examples of such technologies solving intelligence problems entirely absent the analyst's touch. Policymakers and military commanders do not ask advice of HAL (the murderous computer from *2001: A Space Odyssey*); they are advised by Jane, Saif, and Miguel. The examples in this book center on the use of advanced analytic techniques to catalyze serendipitous discovery of the unknown. In intelligence analysis, the dots do not connect themselves, and the needles cannot jump out of the haystacks. The tools and techniques in the ABI arsenal focus the analyst's attention so he or she can spend more time analyzing and less time munging ill-conditioned data sets.

The engineer's world is filled with gold-plated automated analytics and masterfully articulated rule sets for tipping and cueing, but it also comes with a caution. In Silver's "sabermetrics," baseball presents possibly the world's richest data set, a wonderfully refined, well-documented, and above all complete set of data from which to draw conclusions. In baseball, the subjects of data collection do not attempt to deliberately hide their actions or prevent data from being collected on them. The world of intelligence, however, is very different. Intelligence services attempt to gather information on near-peer state adversaries, terrorist organizations, hacker collectives, and many others, all of whom make deliberate, concerted attempts to minimize their data footprint. In the world of state-focused intelligence this is referred to as D&D; in entity-focused intelligence this is called OPSEC. The data is dirty, deceptive, and incomplete. Algorithms alone cannot make sense of this data, crippled by unbounded uncertainty; they need human judgment to achieve their full potential.

"Prediction is hard, especially about the future," goes the great maxim. Early ABI analysts loaded spreadsheets into a GIS, analyzing correlations and documenting patterns to better forecast the potential locations of terrorists and insurgents. They made rational decisions about allocating tactical collection assets through a detailed understanding of data collected on the battlefield. The few predictive elements of the approach occurred in the analyst's head; the data alone could not answer the key questions of who and where—identity and location.

These analysts immersed themselves in the data, gathered from the proverbial "cutting room floor" of the modern battlefield with only a loose understanding of key questions to answer. Deductive and abductive reasoning were employed to match the explanatory power of spatiotemporal data: a focus on explanatory hypotheses that account for observed behavior. Implied in this was the multitude of potential explanatory hypotheses waiting to be discovered in the data. Through the process of searching for particular high-value targets, analysts would often discover entities and behaviors based on spatial correlation of individual proxies who were previously unknown. The shift in the focus of analysis to discovery was the primary and enduring breakthrough of the method. We believe this focus is extensible to a wide range of intelligence problems outside the counterterrorism and manhunting arena.

ABI also came to the forefront at a time of major changes in the U.S. intelligence community. Recognizing that processes developed for the Cold War could not provide answers in a nimble and challenging environment dominated by a new type of threat, leaders struggled to redefine old paradigms like TCPED and seized upon ABI as a representation of a "new way of doing business." In so doing, ABI came to be all things in intelligence reform and modernization, and as a consequence became muddied and confused. NGA director Robert Cardillo, speaking at to the Intelligence & National Security Alliance (INSA) in January 2015, stated "TCPED is dead." He went on to state that he was not sure if there would be a single acronym to replace it. "ABI, SOM, and [OBP]—what we call the new way of thinking isn't important. Changing the mindset is," Cardillo stated. This acknowledgement properly placed ABI as one of a handful of new approaches in intelligence, with a specific methodology, specific technological needs, and a specific domain for application. Other methodologies will undoubtedly emerge in the efforts of modern intelligence services to adapt to a continually changing and ever more complicated world, which will complement and perhaps one day supplant ABI.

This book provides a deep exposition of the core methods of ABI and a broad survey of ABI enhancers that extend far beyond ABI methods alone. Understanding these principles will ultimately serve to make intelligence analysts more effective at their single goal: delivering information to aid policymakers and warfighters in making complex decisions in an uncertain world.

About the Authors

Patrick Biltgen is a senior principal systems engineer at Vencore supporting the National Reconnaissance Office (NRO). Previously, he was the senior mission engineer for BAE Systems Intelligence Integration Directorate where he was the subject matter expert on implementation of Activity-Based Intelligence (ABI) capabilities for the National Geospatial-Intelligence Agency (NGA). Dr. Biltgen held a variety of roles within BAE Systems including the development of the initial concepts for automated "pipeline" processing of GEOINT data, multi-INT data discovery and correlation, and object-relationship linking with graph theory. Biltgen also led a forecasting study that examined implications for the Intelligence Community in 2050.

Prior to joining the Intelligence Community, Dr. Biltgen was a research engineer and graduate researcher at the Georgia Institute of Technology. His research work integrated machine learning with aircraft design capabilities to simultaneously optimize tactics and technologies for a long-range bomber. He is an expert in highly-dimensional multidisciplinary design optimization, statistical visual analytic methods, and capability-based trade studies. Biltgen received a B.S., M.S., and Ph.D. in Aerospace Engineering from Georgia Tech. He was recognized with the National Defense Science and Engineering and Graduate Fellowship (NDSEG), the Astronaut Scholarship, a Georgia Tech Presidential Fellowship, and the 2006 American Institute of Aeronautics and Astronautics (AIAA) Orville and Wilbur Wright award. Biltgen was also a fellow in the Sam Nunn Security Program. He and his wife, Janel, live in northern Virginia.

Stephen Ryan is a mission engineering manager in the Intelligence, Surveillance, & Reconnaissance (ISR) Division of Northrop Grumman Corporation's Information Systems sector. He leads and advises on Northrop Grumman's activity-based intelligence (ABI) initiatives and manages a portfolio of R&D programs focused on multi-INT processing and analysis. Prior to joining

Northrop Grumman, he was a decorated intelligence analyst with the National Geospatial-Intelligence Agency (NGA), where he specialized in terrorism analysis focused on Afghanistan and Pakistan, and served as one of the intelligence community's leading experts on ABI. He spent almost a year deployed to Afghanistan during Operation ENDURING FREEDOM in support of Special Operations Forces (SOF). He is the two-time recipient of the Joint Civilian Service Commendation Award and has also received the NATO Medal, the Secretary of Defense Meal for the Global War on Terror, and numerous team and individual awards for excellence in analysis and program support.

Ryan was scenario director and a senior analyst for TRIDENT SPECTRE, an interagency exercise hosted by the SOF community focused on technology and methodology evaluation and development. In 2012, he served on an interagency team sponsored by the National Intelligence Council with five analysts and an outside expert that re-evaluated Al-Qaeda in light of the events of the Arab Spring. In 2013, he helped create the Intelligence Community's first formal training course in activity-based intelligence, dubbed "ABI 101." He received his M.A. in Security Studies, with honors, from the Walsh School of Foreign Service at Georgetown University. He also holds a B.A., *summa cum laude*, in International Affairs from The George Washington University's Elliott School of International Affairs, where he graduated with university and departmental honors, was Chair of the International Affairs Society, and was elected a member of Phi Beta Kappa. He currently lives in Los Angeles, CA.

Biltgen and Ryan received the National Intelligence Meritorious Unit Citation and the National Intelligence Integration Award for their contributions to intelligence operations and activity-based intelligence.

Index

The Artech House Electronic Warfare Library

Activity-Based Intelligence: Principles and Applications, Patrick Biltgen and Stephen Ryan

Advances in Statistical Multisource-Multitarget Information Fusion, Ronald P. S. Mahler

Antenna Systems and Electronic Warfare Applications, Richard A. Poisel

Electronic Intelligence: The Analysis of Radar Signals, Second Edition, Richard G. Wiley

Electronic Warfare for the Digitized Battlefield, Michael R. Frater and Michael Ryan

Electronic Warfare in the Information Age, D. Curtis Schleher

Electronic Warfare Receivers and Receiving Systems, Richard A. Poisel

*Electronic Warfare Target Location Methods,*Richard A. Poisel

EW 101: A First Course in Electronic Warfare, David L. Adamy

EW 103: Tactical Battlefield Communications Electronic Warfare, David L. Adamy

EW 104: EW Against a New Generation of Threats, David L. Adamy

Foundations of Communications Electronic Warfare, Richard A. Poisel

High-Level Data Fusion, Subrata Das

Human-Centered Information Fusion, David L. Hall and John M. Jordan

Information Warfare and Organizational Decision-Making, Alexander Kott, editor

Information Warfare and Electronic Warfare Systems, Richard A. Poisel

Information Warfare Principles and Operations, Edward Waltz

Introduction to Communication Electronic Warfare Systems, Richard A. Poisel

Knowledge Management in the Intelligence Enterprise, Edward Waltz

Mathematical Techniques in Multisensor Data Fusion, Second Edition, David L. Hall and Sonya A. H. McMullen

Modern Communications Jamming Principles and Techniques, Richard A. Poisel

Principles of Data Fusion Automation, Richard T. Antony

Stratagem: Deception and Surprise in War, Barton Whaley

Statistical Multisource-Multitarget Information Fusion, Ronald P. S. Mahler

Tactical Communications for the Digitized Battlefield, Michael Ryan and Michael R. Frater

Target Acquisition in Communication Electronic Warfare Systems, Richard A. Poisel

For further information on these and other Artech House titles, including previously considered out-of-print books now available through our In-Print-Forever® (IPF®) program, contact:

Artech House
685 Canton Street
Norwood, MA 02062
Phone: 781-769-9750
Fax: 781-769-6334
e-mail: artech@artechhouse.com

Artech House
16 Sussex Street
London SW1V 4RW UK
Phone: +44 (0)20-7596-8750
Fax: +44 (0)20-7630-0166
e-mail: artech-uk@artechhouse.com

Find us on the World Wide Web at: www.artechhouse.com